Agroecology: Agroecosystems and Sustainability

Agroecology: Agroecosystems and Sustainability

Editor: Bella Harper

www.callistoreference.com

Callisto Reference,
118-35 Queens Blvd., Suite 400,
Forest Hills, NY 11375, USA

Visit us on the World Wide Web at:
www.callistoreference.com

ISBN: 978-1-64116-515-0 (Hardback)

Cataloging-in-Publication Data

Agroecology : agroecosystems and sustainability / edited by Bella Harper.
p. cm.
Includes bibliographical references and index.
ISBN 978-1-64116-515-0
1. Agricultural ecology. 2. Agriculture--Environmental aspects. 3. Sustainable agriculture. I. Harper, Bella.
S589.7 .A37 2021
577.55--dc23

Table of Contents

Preface

The study of ecological processes that are applied to agricultural production is referred to as agroecology. The basic unit of study in agroecology is an agroecosystem. The study of living and non-living components that are involved in that unit as well as their interactions is also conducted within this domain. A variety of agroecosystems are studied under agroecology. Some of the system properties of agroecosystems that are studied in this field are sustainability, stability, productivity and equitability. It is a multidisciplinary field which includes various sciences such as ecology, agronomy, economics, sociology and many others. Agroecology uses these sciences in order to understand the elements of ecosystem such as plant-insect interactions and soil properties. It also uses social sciences to understand the effects of farming practices on rural communities, economic constraints to developing the production methods and the cultural factors which determine the farming practices. This book contains some path-breaking studies in the field of agroecology. While understanding the long-term perspectives of the topics, it makes an effort in highlighting their impact as a modern tool for the growth of the discipline. This book is a vital tool for all researching and studying this field.

The researches compiled throughout the book are authentic and of high quality, combining several disciplines and from very diverse regions from around the world. Drawing on the contributions of many researchers from diverse countries, the book's objective is to provide the readers with the latest achievements in the area of research. This book will surely be a source of knowledge to all interested and researching the field.

In the end, I would like to express my deep sense of gratitude to all the authors for meeting the set deadlines in completing and submitting their research chapters. I would also like to thank the publisher for the support offered to us throughout the course of the book. Finally, I extend my sincere thanks to my family for being a constant source of inspiration and encouragement.

Editor

Genetic Diversity Analysis of Mangaba (*Hancornia Speciosa* Gomes), an Exotic Brazilian Tropical Species

Ana Veruska Cruz da Silva[1], Allívia Rouse Carregosa Rabbani[2], José Guedes de Sena-Filho[1], Camila Santos Almeida[2], Rosana Barroso Feitosa[3]

[1]*Embrapa Coastal Tablelands*
[3] Universidade Federal do Vale do São Francisco
[2] *Universidade Federal de Sergipe (UFS)*
**Corresponding Author*

SUMMARY

Twenty genotypes from natural populations of mangaba located in Itaporanga D'Ajuda (Sergipe, Brazil) were analyzed using RAPD markers. Polymorphism, genetic diversity and structure parameters were determined to characterize the differences between individual plants. 60 DNA fragments were generated by 10 primers, 85% of which were polymorphic. Results show a quantitative genetic diversity value of 0.35 and a Shannon index of 0.46 in the population. The similarity among the specimens according to Jaccard's coefficient ranged from 0.36 to 0.87. Using cluster analysis, it was possible to indentify five groups. Three individuals also stand out because they presented significant divergence from the groups. The PCoA formed four groups, with three of the groups isolated from the others. This investigation showed that genetic diversity was relatively large among these individuals. In addition, the results demonstrated that RAPD markers are a useful tool for evaluating the genetic diversity and relationships among mangaba.

Key words: Apocynaceae; genetic variability; RAPD.

INTRODUCTION

Hancornia speciosa Gomes (Apocynaceae), commonly known as mangaba, is a native Brazilian tree, distributed mainly in areas of open vegetation such as savannas, dunes, coastal tablelands and coastal lowlands. The fruit provides a rich source of protein (up to 3% on a fresh weight basis), has a highly desirable flavor for fresh consumption and is highly used in the production of juice and ice cream (Caldas *et al.*, 2009). Mangaba is also one of the most promising fruits for programs of sustainable harvesting. Additionally, the bark of this species is used as a traditional medicine to treat hypertension, gastric ulcers and inflammatory diseases (Almeida *et al.*, 1998; Mello Moraes et.al. 2008).

Because of these attractive features, the mangaba plant is under constant anthropogenic pressure in the agroecosystems where they occur, and is a species with the potential need for a conservation strategy. In response to a lack of published information concerning the agronomic characteristics of mangaba or genetic studies, several research projects have been developed to facilitate the collection and genetic characterization of mangaba specimens *in situ* and *ex situ*. The information provided by these studies will be extremely useful for the development of conservation strategies and will help the process of domestication and breeding of the species (Ganga *et al.*, 2009).

Given the social, economic and cultural importance of mangaba, intense harvesting has put pressure on natural populations of this species, and the International Catalog call "Arca do Gosto" states that it is in danger of extinction unless a sustainable management plan is developed (Slow Food, 2011). Mangaba, like other native species in Brazil, are poorly understood and lack genetic studies. Research at the genetic level is essential to the development of effective conservation strategies and future domestication efforts.

RAPD (Random Amplified Polymorphism DNA) has played a key role in identifying genotypes and understanding relationships among close individuals. The technique is also attractive because it does not depend on large quantities of genomic DNA for polymerase chain reaction (PCR) amplifications (Fritsch and Reiseberg 1996). It has been reported (Silveira *et al.*, 2009; Maina *et al.*, 2009) that RAPD analysis has successfully been used to characterize the genetic variability within and among populations of caroa (*Neoglaziovia variegate* Mez) and japanese apricot (*Prunus mume* Sieb. Et Zucc.) (Mayer *et al.*, 2008), as well as blueberry (*Vaccinium* spp.) (Silva *et al.*, 2008). Other examples using RAPD analysis in the population and conservation genetics of plants have also been reported (Rossetto *et al.*, 1995; Cardoso *et al.*, 1998; Lacerda *et al.*, 2001). Additionally, the genetic relationships of *Jatropha curcas* germplasm specimens (Kumar *et al.*, 2009) and mangaba genotypes (Moura et al., 2005) were successfully estimated using this technique.

Genetic diversity analysis of mangaba using RAPD markers is anticipated to increase the knowledge of genetic population structure and will contribute to the development of strategies for seed collection, which may be used to study future progenies and guide the allocation of conservation areas in a sustainable management program. Here, we report the molecular characterization of 20 genotypes of mangaba from a natural population located in the state of Sergipe in Brazil, using RAPD markers.

MATERIAL AND METHODS

Young leaves of 20 genotypes of mangaba were collected in the natural population from the Reserva do Caju located in the city of Itaporanga d'Ajuda (Sergipe, Brazil), in december of 2008 (Table 1). The DNA extractions were performed according to Doyle and Doyle (1991). The PCR reactions were performed according to Williams *et al.*, (1990) and modified to a final volume of 25 μL. To detect molecular polymorphisms among and within accessions, ten primers were tested and used for the amplification of polymorphic loci, which were: IDT 2, IDT 3, IDT 4, IDT 6, IDT 8, IDT 9, IDT 10, IDT 11, IDT 14, and IDT 20 (IDT DNA Technologies, USA). The amplification products (bands) were separated in 1.5% agarose gel, stained with ethidium bromide and visualized under ultraviolet light using a Gel Doc L-Pix image system (Loccus Biotecnologia, Brazil).

Table 1. Location of mangaba (*Hancornia speciosa Gomes*) native from Reserva Ecológica do Caju, Itaporanga d'Ajuda (Sergipe, Brazil).

Genotype	Latitude	Longitude
RC 1	11°06'39.2''	37°11'13.3''
RC 2	11°06'39.3''	37°11'14.4''
RC 3	11°06'37.8''	37°11'15.0''
RC 4	11°06'38.1''	37°11'16.4''
RC 5	11°06'38.3''	37°11'18.4''
RC 6	11°06'39.0''	37°11'18.1''
RC 7	11°06'39.7''	37°11'18.6''
RC 8	11°06'40.6''	37°11'18.8''
RC 9	11°06'41.6''	37°11'18.6''
RC 10	11°06'41.3''	37°11'18.3''
RC 11	11°06'40.8''	37°11'16.9''
RC 12	11°06'40.8''	37°11'16.7''
RC 13	11°06'41.6''	37°11'16.2''
RC 14	11°06'41.7''	37°11'16.3''
RC 15	11°06'41.3''	37°11'15.4''
RC 16	11°06'40.6''	37°11'15.3''
RC 17	11°06'40.9''	37°11'14.1''
RC 18	11°06'41.5''	37°11'13.8''
RC 19	11°06'39''	37°11'13.1''
RC 20	11°06'37.6''	37°11'13.3''

The presence or absence of RAPD band was scored as "1" or "0", respectively. The minimum number of fragments to conduct a study of genetic diversity was based on Oliveira *et al.* (2006), and these were analyzed by the bootstrap method (Manly, 1997) with a resampling of 1000, as recommended by Moura *et al.* (2005), using the software Genes (Cruz, 2006). The number of polymorphic fragments were considered optimal when the assumed stress value was less than 0.05 (Kruskal, 1964).The data matrix of the RAPD scores was generated and similarity coefficients were calculated using Jaccard's arithmetic complement index (1908). The dendrogram was constructed using the Unweighted Pair Group Method with Arithmetic Mean (UPGMA) cluster algorithm. In order to determine the robustness of the dendrogram, the data was bootstrapped with 10,000 replications using FreeTree software (http://web.natur.cuni.cz/flegr/programs/freetree.htm) and for visualization of the cluster we used the TreeView package (http://web.natur.cuni.cz/flegr/programs/freetree/TreeView.exe). Principal Coordinates Analysis (PCoA) was performed using the software XLSTAT (http://www.xlstat.com/) based on similarity matrix of Jaccard. Polymorphic information content (PIC) is a parameter that provides an estimate of the discriminatory power of molecular marker per primer and was calculated according to Ghislain *et al.*, (1999). The marker index (MI) was determined as a product of PIC and the number of polymorphic bands per assay unit as described in Zhao *et al.*, (2007). To measure the population genetic structure, we used Genalex v.6.3 (www.anu.edu.au/BoZo/GenAlEx/) and calculated the Shannon Index (I) (Brown and Weir 1983) as well as genetic diversity (H) as described by Lynch and Milligan (1994) and Maguire *et al.*, (2002) for dominant markers.

RESULTS AND DISCUSSION

The primers used resulted in a banding pattern that was distinct, as seen in Figure 1 which shows the RAPD profiles of the primer IDT 3.

A total of 60 DNA fragments were generated by the 10 primers, 85% of which were polymorphic. Results of the bootstrap analyses of were plotted (Figure 2). We observed that there was a direct relationship between the number of fragments analyzed and the magnitude of correlation values, and was consistency from 48 polymorphic fragments, where the correlation (r) was 0.97 and 0.047 to stress value (SV) (Kruskal, 1964). The genetic diversity estimates from 55 polymorphic fragments among the mangaba genotypes showed excellent precision in our study.

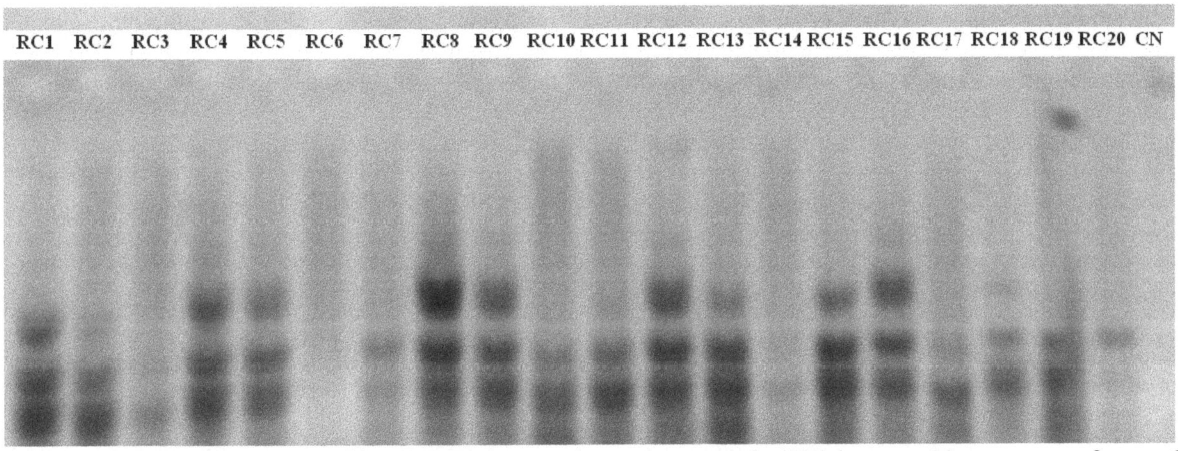

Figure 1. RAPD profiles generated by IDT 3 primer and negative controls (CN) between 20 genotypes of mangaba (*Hancornia speciosa* G.) from Reserva Ecológica do Caju, Itaporanga d'Ajuda (Sergipe – Brazil).

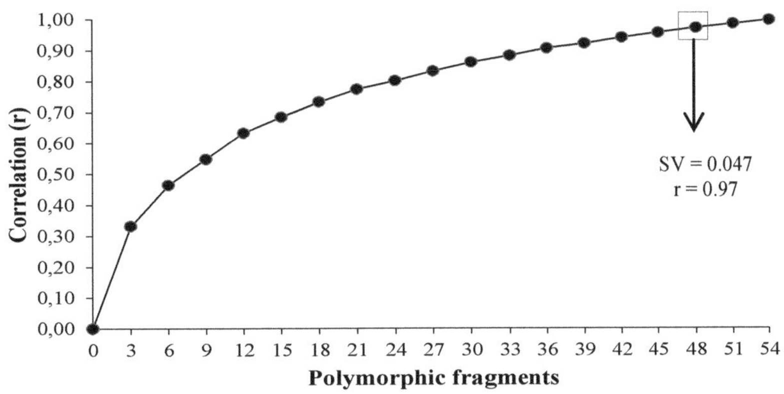

Figure 2. Correlation estimate (r), polymorphic fragments, and the stress value (SV) between the similarity obtained by bootstraps to increasing numbers of polymorphic RAPD fragments to get the minimum fragments number to study of genetic diversity of mangaba genotypes (*Hancornia speciosa* G.) from Reserva Ecológica do Caju, Itaporanga d'Ajuda (Sergipe, Brazil).

IDT20 presented the greatest number of fragments (11), resulting in a polymorphism of 100%, as well as the sequences IDT 6 and IDT 14. The PIC value was 0.321 (0.116 - 0.48), and the global value MI was 1.31 (0.66 - 3.53). For the genetic structure, the Shannon's Index (I) was 0.46 (±0.041) and the genetic diversity (H) was 0.35 (±0.026) (Table 2).

High polymorphism (85%) was identified in mangaba from Sergipe, Brazil, which corroborates previous studies by Silveira *et al.*, (2009), who found similarly high polymorphism (93%) in *Neoglaziovia variegate* in 501 DNA fragments generated by 36 primers. Additionally, a study of *Butia capitata* from Nunes *et al.*, (2008) obtained 136 fragments obtained from 21 primers, where 77 were polymorphic. Mayer *et al.*, (2008), however, evaluated three japanese apricot genotypes and found that only 23.6% were polymorphic using 42 primers generating a total of 432 profiles. According to Nei (1987), the proportion of polymorphic loci is not a good measure of genetic variation, and the analyses more appropriate to measure is by gene diversity (H). In the present study, genetic diversity observed (H - 0.35 ±0.026) in the population indicated that it does not fit the criteria of the Hardy-Weinberg equilibrium, suggesting an excess of homozygotes or heterozygotes among individuals (Kageyama *et al.*, 2003). This is expected, considering that individuals are likely to incorporate or losing alleles by genetic drift in natural conditions.

Shannon's Index was calculated to provide a relative estimate of the degree of variation in the genotypes. The values of Shannon's Index (I) found in mangaba (0.46 ±0.041) were similar to values in previously published studies: 0.31 in *Taxus baccata* using 120 genotypes (Zarek, 2009), and 0.36 in *Jatropha curcas* species using 13 genotypes (Gupta *et al.*, 2008). Others studies reported higher values in other plant species (Jogait *et al.*, 2006), e.g., *Vaccinium oxycoccus* (0.19) using 56 samples (Žukauskien *et al.*, 2009). The Shannon index provides information applicable to other fields in ecology, is relatively insensitive to the effects of distortion caused by the inability to detect heterozygous loci (Dawson *et al.*, 1995), and is a good tool in the study of populations. Additionally, estimates of diversity using the Shannon index do not rely on Hardy-Weinberg equilibrium (Bussell, 1999) and can be compared across species with different breeding systems.

The uses of genetic parameters, such as e.g. genetic diversity and Shannon's index, are convenient in the study of populations, (Zhao *et al.*, 2006). Regarding the ecological factors which determine the genetic structure in plant populations (Loveless and Hamrick, 1984), the population of mangaba presented high values, regarding I (0.43 ±0.041) and H (0.35 ±0.026), which might be related to presence of important ecological characteristics of mangaba due to the conversation of genotypes and allelic movement.

The values from the genetic similarity obtained by the Jaccard coefficient ranged from 0.36 (RC13 - RC19) to 0.87 (RC16 - RC17). The average genetic similarity was equal to 0.59. After UPGMA analyses, it was possible to visualize the formation of five sub-clusters, I (RC12), II (RC1, RC13, RC3, RC2, RC4, RC5, RC11, RC14, and RC15), III (RC10, RC6, RC7, RC8, and RC9), IV (RC20) and V (RC19, RC18, RC16 and RC17). The most significant distance was RC12 (0.57), RC1 (0.62), RC19 (0.62) and RC20 (0.59). The natural population sampled here had a wide genetic variability, and these results may be used to assist breeding programs and propagation studies and/or cloning mangaba species (Figure 3).

Table 2. Primers, totals (TF) and polymorphic fragments (PF%), polymorphic information content (PIC), marker index (MI), Shannon index (I) and genetic diversity (H) between 20 genotypes of mangaba (*Hancornia speciosa* G.) from Reserva Ecológica do Caju, Itaporanga d'Ajuda (Sergipe, Brazil).

Primers	Sequence 5' – 3'	TF	PF%	PIC	MI	I		H	
						Value	Variance	Value	Variance
IDT 2	TGC CGA GCT G	4	50	0,48	0,96	0,42	0,045	0,25	0,031
IDT 3	GTT TCG CTC C	3	67	0,43	0,86	0,65	0,048	0,46	0,029
IDT 4	TGA TCC CTG G	5	60	0,25	0,74	0,40	0,020	0,25	0,012
IDT 6	GTG AGG CGT C	6	100	0,35	2,08	0,53	0,000	0,35	0,000
IDT 8	GGA CCC AAC C	5	80	0,16	0,66	0,40	0,091	0,25	0,048
IDT 9	CCC AAG GTC C	4	75	0,38	1,14	0,58	0,047	0,39	0,028
IDT 10	GGT GCG GGA A	7	86	0,25	1,48	0,27	0,073	0,14	0,034
IDT 11	ACG GAT CCT G	7	86	0,30	1,85	0,67	0,000	0,48	0,000
IDT 14	GGC ACT GAG G	8	100	0,29	2,30	0,62	0,000	0,43	0,000
IDT 20	GGA GGA GAG G	11	100	0,32	3,53	0,53	0,036	0,35	0,023
Total		60	85%	0,32	1,31	0,46	0,041	0,35	0,026

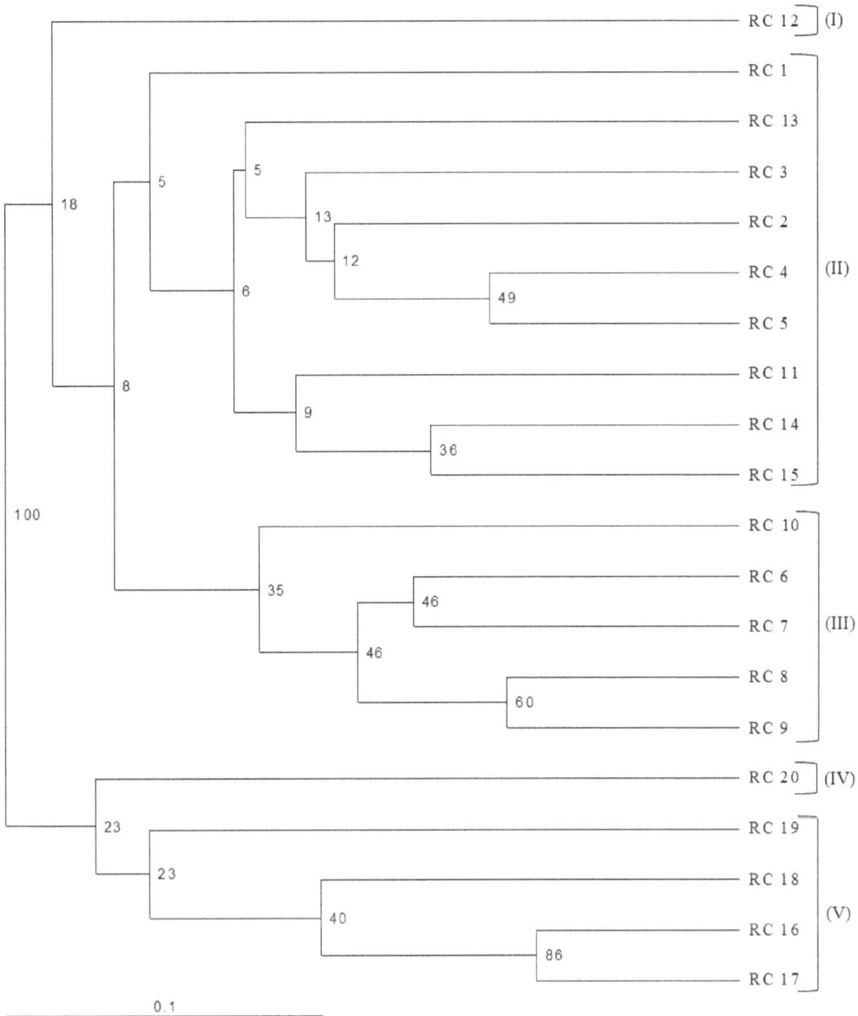

Figure 3. Cluster generated from the Jaccard similarity coefficient and UPGMA clustering using 10 RAPD primers between 20 genotypes of mangaba (*Hancornia speciosa* G.) from Reserva Ecológica do Caju, Itaporanga d'Ajuda (Sergipe, Brazil).

The wide variation in genetic distance among the different mangaba species reflected a high level of polymorphism at the DNA level. The similarity coefficients (0.59) found in our study were to previous studies using RAPD markers. Santana *et al.*, (2008) (studying *Enterolobium contortisiliquum*) found a similarity of 0.67 and Caixeta *et al.*, (2003) (evaluating *Eucalyptus* genotype) found values greater or equal to 0.70.

Principal coordinate analysis (PCoA) was performed in order to determine the genetic relationships among the accessions with minimum distortion. The average distance among samples was plotted in a two-dimensional plot, and explains 40% of the total variation (Figure 4).

The PCoA resulted in a clear separation of groups, corroborating the distribution obtained from the cluster analysis. The individuals were separated into four distinct groups, according to genetic similarity: I (RC1, RC2), II (RC14, RC4, RC13, RC5, RC15, RC13), III (RC7, RC6, RC8, RC9, RC16) and IV (RC17, RC 18, RC19). Three individuals were not grouped (RC12, RC10, RC11), and corresponded to the most genetically divergent specimens.

Grouping of individuals within a natural population may be expected due to many factors, for example, anthropic factors and selection inside of the population. According to UPGMA, some individuals were close (e.g., RC16 - RC17), and also in the PCoA (RC4 - RC13, as well as RC15 - RC3).

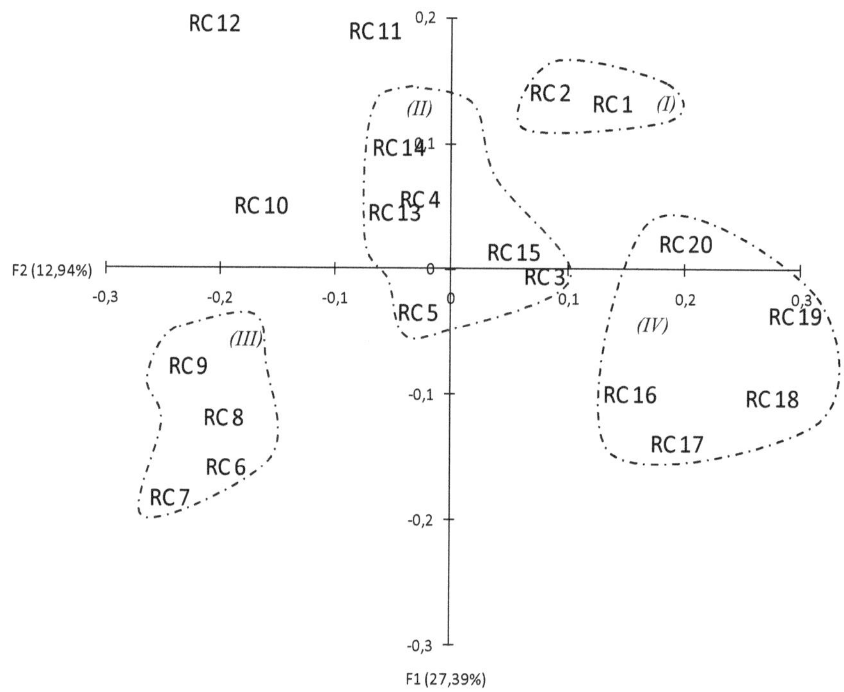

Figure 4. Principal Coordinates Analysis (PCoA) obtained from the Jaccard similarity coefficient using 10 RAPD primers between 20 genotypes of mangaba (*Hancornia speciosa* G.) from Reserva Ecológica do Caju, Itaporanga d'Ajuda (Sergipe, Brazil).

The high genetic variability in this population may be a consequence of adaptation to changes of habitat. Theoretically, variations caused by selection pressures can maintain genetic polymorphism (Sheng *et al.*, 2004). An alternative hypothesis is that, although has many factors that influence the population, the genotypes of mangaba are very active genetically, which provides the development of the variation observed. According to Darrault and Schlindwein (2005) the pollination mechanism in mangaba increases pollination efficiency because it reduces pollen loss and prevents autogamy, thereby favoring cross pollination and contributing to the genetic dynamics. The other hypothesis is that various anthropogenic selection pressures provided this alteration. In the future, more research with the species is needed to support our hypothesis, and to further explore this variation and how it influences the mangaba population.

In the plant breeding field, a key step to obtain genetic variability among individuals of a specific species or among groups is interspecies crosses. The genetic structure of plant populations reflects the interactions of many different processes, such as the long-term evolutionary history of the species (e.g., shifts in distribution, habitat fragmentation, and/or population isolation), mutation, geographical, ecological, reproductive isolation, genetic drift, mating system, gene flow, and selection (Schaal *et al.*, 1998; Tripathi *et al.*, 2007; Thendral Hepsibha *et al.*, 2010). The genetic diversity greatly contributes to the sustainability of plant populations (Wang *et al.*, 2007) and its measurement is important to understand the events that occur in the population.

CONCLUSIONS

This is the first report on the mangaba species from northeast of Brazil, and our study is a contribution to the characterization of native genotypes. The genetic variation and the genetic relationships among wild mangaba were efficiently determined using RAPD markers. The discrimination of mangaba from Sergipe (Brazil) and identification of genotypes more genetically divergent may contribute to development of strategies for the implementation of conservation, breeding programs efforts and commercial exploitation.

ACKNOWLEDGEMENTS

We would like to thank to Maribeth Latvis (Florida Museum of Natural History - University of Florida, USA) for helping us with English proofreading.

REFERENCES

Almeida, S.P., Proença, C.E.B., Sano, S.M., Ribeiro, J.F. 1998. Cerrado: espécies vegetais úteis. Embrapa, Brazil.

Brown, A.H.D., Weir, B.S. 1983. Measuring genetic variability in plant populations, in: Tanksley, S.D., Orton, T.J. (eds). Isozymes in Plant Genetics and Breeding, Part A. Elsevier Science, The Netherlands. pp 219-239.

Bussell, J.D. 1999. The distribution of random amplified polymorphic DNA (RAPD) diversity amongst populations of *Isotoma petraea* (Lobeliaceae). Molecular Ecolology. 8: 775-789.

Caixeta, R.P., Carvalho, D., Rosado, S.C.S., Trugilho, P.F. 2003. Variações genéticas em populações de Eucalyptus spp. detectadas por meio de marcadores moleculares. Revista Árvore. 27:357-363.

Caldas, L.S., Machado, L.L., Caldas, S.H., Campos, M.L., Caldas, J.A, Pharis, R.P., Pereira-Neto, A.B. 2009. Growth-active gibberellins overcome the very slow shoot growth of *Hancornia speciosa*, an important fruit tree from the Brazilian ''Cerrado''. Trees. 23:1229–1235.

Cardoso, M.A., Provan, J., Powell, W., Ferreira, P.C.G., Oliveira, D.E. 1998. High genetic differentiation among remnant population of the endangered *Caesalpinea echinata* Lam. (Leguminosae - Caesalpinioideae). Molecular Ecolology. 7: 601–608.

Cruz, C.D. 2006. Programa Genes - Análise multivariada e simulação. 1. ed. Editora UFV, Viçosa.

Darrault, R.O., Schlindwein, C. 2005. Limited fruit production in *Hancornia speciosa* (Apocynaceae) and pollination by nocturnal and diurnal insects with long mouth parts. Biotropica. 37: 381–388.

Dawson, I.K.A.J., Simons, R., Waugh, W.P. 1995. Diversity and genetic differentiation among subpopulations of *Gliricidia sepium* revealed by PCR-based assays. Heredity. 74: 10-18.

Doyle, J.J., Doyle, J.L. (1991) Isolations of plant DNA from fresh tissue. Focus. 1:13-15.

Fritsch, P., Rieseberg, L.H. 1996. The use of Random Amplified Polymorphic DNA (RAPD) in conservation genetics. In: Smith, T.B., Wayne, R.K. (eds). Molecular Genetic Approaches in Conservation. Oxford University Press, USA. pp 54–73.

Ganga, R.M.D., Chaves, L.J., Naves, R.V. 2009. Parâmetros genéticos em progênies de *Hancornia speciosa* Gomes do Cerrado. Scientia Forestalis. 37:395-404.

Ghislain, M., Zhang, D., Fajardo, D., Huamann, Z., Hijmans, R.H. 1999. Marker assisted sampling of the cultivated Andean patato Solanumphureja collections using RAPD markers. Genetic Resources and Crop Evolution. 46:547–555.

Gupta, S., Srivastava, M., Mishra, G.P., Naik, P.K., Chauhan, R.S., Tiwari, S.K., Kumar, M., Singh, S. 2008. Analogy of ISSR and RAPD markers for comparative analysis of genetic diversity among different *Jatropha curcas* genotypes. African Journal of Biotechnology. 7: 4230-4243.

Jaccard, P. 1908. Nouvelles recherches sur la distribution florale. Soc Vaud Sci Nat 44: 223-270.

Kageyama, P.Y., Sebbenn, A.M., Ribas, L.A., Gandara, F.B., Castellen, M., Perecim, M.B., Vencovky, R. 2003. Diversidade genética em espécies arbóreas tropicais de diferentes estágios sucessionais por marcadores genéticos. Scientia Forestalis. 64:93-107.

Kumar, R.V., Tripathi, Y.K., Shukla, P., Ahlawat, S.P., Gupta, V.K. 2009. Genetic diversity and relationships among germplasm of *Jatropha curcas* L. revealed by RAPDs. Trees. 23:1075-1079.

Kruskal, J.B. 1964. Multidimensional scaling by optimizing goodness of fit to a no metric hypothesis. Psychometrika. 29: 1-27.

Lacerda, D.R., Acedo, M.D.P., Lemos Filho, J.P., Lovato, M.B. 2001. Genetic diversity and structure of natural populations of *Plathymenia reticulate* (Mimosoideae), a tropical tree from the Brazilian Cerrado. Molecular Ecolology. 10: 1143–1152.

Loveless, M.D., Hamrick, J.L. 1984. Ecological determinants of genetic structure in plant populations. Annual Review of Ecology, Evolution, and Systematics. 15: 65-95.

Lynch, M., Milligan, B.G. 1994. Analysis of population genetic structure with RAPD markers. Molecular Ecolology. 3: 91-99.

Maguire, T.L., Peakall, R., Saenger, P. 2002. Comparative analysis of genetic diversity in the mangrove species *Avicennia marina* (Forsk.) Vierh. (Avicenniaceae) detected by AFLPs and SSRs. Theoretical and Applied Genetics. 104:388-398.

Maina, P.K., Okoth, S., Njoroge, C. N., Monda, E. 2009. Genetic relatedness among *Fusarium* species isolated from Taita Taveta region, Kenya. Tropical and Subtropical Agroecosystems.11: 337 – 345.

Manly, B.F.J. 1997. Randomization, bootstrap and Monte Carlo methods in Biology. Chapman & Hall, London.

Mayer, N.A., Lemos, E.G.M., Pereira, F.M., Wickert, E. 2008. Caracterização de três genótipos de umezeiro (*Prunus mume* Sieb. et Zucc.) por marcadores RAPD. Revista Brasileira de Fruticultura. 30:1045-1050.

Mello Moraes, T., Rodrigues, C.M., Kushima, H., Bauab, T.M., Villegas, W., Pellizzon, C.H., Souza Brito, A.R.M., Hiruma-Lima, C.A. 2008. *Hancornia speciosa*: indications of gastroprotective, healisng and anti-helicobacter pylori actions. Journal of Ethnopharmacology. 120: 161-168.

Moura, N.F., Chaves, L.J., Vencovsky, R., Zucchi, M.I., Pinheiro, J.B., Morais, L.K., Moura, M.F. 2005. Seleção de marcadores RAPD para o estudo da estrutura genética de populações de *Hancornia speciosa* Gomez. Bioscience Journal. 21: 119-125.

Nei, M. 1987. Molecular Evolutionary Genetics. Columbia University Press, New York.

Nunes, A.D., Bianchi, V.J., Fachinello, J.C., Carvalho, A.Z., Cardoso, G. 2008. Caracterização molecular de butiazeiro por marcadores RAPD. Revista Brasileira de Fruticultura. c 30:702-707

Oliveira, M.S.P., Amorim, E.P., Santos, J.B., Ferreira, D.F. 2007. Diversidade genética entre acessos de açaizeiro baseada em marcadores RAPD. Ciência e Agrotecnologia. 31:1645-1653.

Rossetto, M., Weaver, P.K., Dixon, K.W. 1995. Use of RAPD analysis in devising conservation strategies for the rare and endangered *Grevillea scapigera* (Proteaceae). Molecular Ecolology. 4:321– 329.

Santana, G.C., Mann, R.S., Ferreira, R.A., Gois, I.B., Oliveira, A.S., Boari, A.J., Carvalho, S.V.A. 2008. Diversidade genética de indivíduos de *Enterolobium contortisiliquum* Vell .Morong., por meio de marcadores RAPD no Baixo Rio São Francisco. Revista Árvore. 32: 427-433.

Schaal, B.A., Hayworth, D.A., Olsen, K.M., Rauscher, J.T. Smith, W.A. 1998. Phylogeographic studies in plants: problems and prospects. Molecular Ecology. 7: 465-474.

Sheng, Y., Zheng, W.H., Pei, K.Q., Ma, K.P. 2004. Population genetic structure of a dominant desert tree, *Haloxylon ammodendron* (Chenopodiaceae), in the southeast Gurbantunggut Desert detected by RAPD and ISSR markers. Acta Botanica Sinica. 46:675-681.

Silva, S.D.A., Antunes, L.D.C., Anthonisen, D.G., Lemões, J.S., Gonçalves, E.D. 2008. Caracterização de genótipos de mirtilo utilizando marcadores moleculares. Revista Brasileira de Fruticultura. 30: 180-184.

Silveira, D.G., Amorim, E.P., Jesus, O.N., Souza, F.V.D., Pestana, K.N.P., Santos, V.J., Santana, J.R.F. 2009. Variabilidade genética de populações naturais de caroá. Pesquisa Agropecuária Brasileira. 44:283-290.

Slow Food. 2011. Produtos do Brasil na Arca. Arca do Gosto. Slow Food International Association. http://www.slowfoodbrasil.com/content/view/326/59/

Thendral, H.B., Premalakshmi, V., Sekar, T. 2010. Genetic diversity in *Azima tetracantha* (Lam) assessed through RAPD analysis. Indian Journal of Science and Technology. 3:170-173.

Tripathi, Y.K., Gurha, P., Ghosh, D., Kumar, R.V., Goswami, H.K., Prakash, V. 2007. Determination of phylogenetic relationship among *Isoetes* species using random primers. Turkish Journal of Botany. 31:1–6.

Wang, W., Chen, L., Yang, P., Hou, L., He, C., Gu, Z., Liu, Z. 2007. Assessing genetic diversity of populations of top mouth culter (*Culteral burnus*) in China using AFLP markers. Biochemical Systematics and Ecology. 35:662–669.

Williams, J.G.K., Kubelik, A.R., Livak, K.J., Rafalski, J.A., Tingey, S.V. (1990) DNA polymorphism amplified by arbitrary primers are useful as genetic markers. Nucleic Acids Research. 18: 6531-6535.

Zarek, M. 2009. RAPD analysis of genetic structure in four natural populations of *Taxus baccata* from southern Poland. Acta Biologica Cracoviensia Series Botanica. 51:67–75.

Zhao, K., Zhou, M., Chen, L., Zhang, D., Robert, G.W. 2007. Genetic Diversity and Discrimination of *Chimonanthuspraecox* (L.) Link Germplasm Using ISSR and RAPD Markers. Hortscience. 42:1144–1148.

Zhao, W.G., Zhang, J.Q., Wangi, Y.H., Chen, T.T., Yin, Y., Huang, Y.P., Pan, Y., Yang, Y. 2006.. Analysis of Genetic Diversity in Wild Populations of Mulberry from Western Part of Northeast China Determined by ISSR Markers. Genetics and Molecular Biology. 17:196-203.

Žukauskien, J., Paulauskas, A., Daubaras, R., Česonie, L. 2009. Plant morphology and RAPD marker characterizations of *Vaccinium Oxycoccus* Lithuanian populations. AgronomijasVēstis. 12:142-147

2

Effect of Grazing *Cratylia Argentea* associated with *Brachiaria Brizantha*-Toledo on Quality Pasture and Weight Gain in Holstein × Zebu Heifers

Melvin González Arcia[1], Braulio Valles de la Mora[2]*, Miguel Ángel Alonso Díaz[2], Epigmenio Castillo Gallegos[2], Eliazar Ocaña Zavaleta[2] and Jesús Jarillo Rodríguez[2]

[1]Universidad Autónoma de Zacatecas, Unidad Académica de Medicina Veterinaria y Zootecnia. Km 31.5 Carretera Zacatecas-Fresnillo, El Cordovel, Enrique Estrada, Zacatecas, México.
[2]Universidad Nacional Autónoma de México, Facultad de Medicina Veterinaria y Zootecnia, Centro de Enseñanza, Investigación y Extensión en Ganadería Tropical. Km 7.5 Carretera Federal Martínez de la Torre-Tlapacoyan, CP 93600, Martínez de la Torre, Veracruz, México.
**Correponding Author: braulio_36@hotmail.com*

SUMMARY

The relationship between grazing a high quality shrub legume, blood serum urea (BSU) and daily weight gain (DWG) in F_1 Holstein x Zebu heifers was evaluated. Two pastures were compared: 1) high crude protein diet (HCP): *Cratylia argentea* + *Brachiaria brizantha* Toledo grass, and 2) low crude protein diet (LCP): *Brachiaria brizantha* Toledo grass alone. Grazing of each treatment lasted 164 days. Dry matter yield (DMY), crude protein (CP), neutral detergent fiber (NDF), acid detergent fiber (ADF) and lignin (LIG) were evaluated in both pastures. Before grazing, DMY at HCP and LCP was 1922 ± 405 and 770 ± 405 kg ha^{-1}, respectively, and CP was 8.3 ± 0.85 and 8.7 ± 0.85 %, respectively. The NDF, ADF and LIG of Toledo grass at HCP and LCP were similar, averaging 73.1 ± 0.92, 42.7 ± 1.2 and 13.5 ± 0.7 %, respectively, while in *C. argentea* averages at HCP and LCP were 51.6 ± 0.58 % (NDF), 32.5 ± 0.85 % (ADF) and 16.7 ± 1.3 % (LIG). The BSU was 11.01 ± 0.35 (HCP) and 7.87 ± 0.35 (LCP) mg dL^{-1}. Average DWG in HCP and LCP was 839 ± 258 and 580 ± 278 g day^{-1}, respectively. In conclusion, in F_1 Holstein x Zebu heifers DWG can reach over 800 g day^{-1} when grazing a high CP diet.

Key words: *Cratylia argentea*; *Brachiaria brizantha* Toledo grass; blood serum urea; crude protein.

INTRODUCTION

In tropical and subtropical conditions, cattle graze mainly on native pastures, which are poor in quality and have marked seasonality in their growth. The major constrains in cattle production systems are the scarcity and the poor quality of feed, in terms of low crude protein (CP) level and digestibility, mainly during the dry season. This poor quality of the feed results in low live weight and/or milk production, poor body condition and unsatisfactory reproductive performance. To enhance ruminant productivity during critical seasons, it is common to supply expensive commercial concentrates. However, the current world food crisis is making people to reconsider the use of alternative feeds for animal production. Makkar (2003) stated that because fodder trees and shrubs are not an important source of food for humans, they can be a significant source of nutrients for ruminants in biodiverse environments.

Tropical trees and shrubs are important sources of protein for ruminants (Alonso-Díaz *et al.*, 2010), but their high levels of secondary metabolites and/or their high fiber contents and/or their high level of CP are often considered as factors that might limit their use as ruminant food (Buttler, 1998; Ben Salem *et al.*, 2005; Sandoval-Castro *et al.*, 2005). *Cratylia argentea* is a shrub legume native to Latin America, which has high CP level, low to moderate concentration of tannins, is drought-tolerant, and in tropical regions is used as foliage to feed cattle (Lascano *et al.*, 2002). Andersson *et al.* (2006) have reported CP content from 184 to 237 g kg^{-1} in this legume, which reveals its high quality.

Studies using a cut and carry system have reported that inclusion of *C. argentea* in animal diets improves their productivity (Ibrahim *et al.*, 2001; Lascano and Plazas, 2003; Sánchez and Ledin, 2006). However, there is a lack of information about the use of this legume through direct grazing/browsing by heifers during the dry season. This information might be important in those regions where the workforce is expensive. Indeed, it has been widely demonstrated that the use of forage legumes associated with grasses improves the productivity of ruminants (Pereira *et al.*, 2009; Manríquez-Mendoza *et al.*, 2011; Paciullo *et al.*, 2011), because of an increased forage quality (Deinum *et al.*, 1996; Yamamoto *et al.*, 2007) and soil fertility (Power *et al.*, 2003).

When a high CP pasture is consumed by ruminants, the efficiency of nitrogen use is a concern. It has been shown that high concentrations of serum urea nitrogen may affect reproductive function and cause decreased pregnancy rates (Chapa *et al.*, 2001). According to Butler (1998), the excess of ruminally degradable protein or ruminally undegradable protein can contribute to reduced fertility in lactating cows. Dietary protein utilization and its associated effects on ovarian or uterine physiology have been monitored with urea nitrogen in plasma or milk. To this respect, Butler (1998) indicated that concentrations of urea nitrogen in plasma above 19 mg dL^{-1} have been associated with reduced fertility in dairy cows. These considerations are important, because raising of heifers is often the second or third largest expense in dairy systems.

We hypothesized that heifers grazing a high quality pasture will achieve greater weight gain, and that their blood serum urea concentrations will not reach toxic levels that could risk their future reproductive performance. Therefore, the objective of this study was to evaluate the effect of the content of CP in the diet in heifers grazing a grass-legume association (*Brachiaria brizantha* Toledo – *C. argentea*; high CP diet) or a single grass (*B. brizantha* Toledo; low CP diet) pasture, on blood serum urea levels and daily weight gain, in tropical conditions.

MATERIALS AND METHODS

Experimental site

The study was conducted in Atzalan, Veracruz, Mexico, located at Lat. 20°02'N, Long. 97°06'W, at 114 m above sea level, in a wet-dry tropical climate. The study had a duration of 164 days (12/19/2010 to 05/31/2011), time that corresponded to the dry season. The period of habituation to pastures was 20 days. Sixteen F_1 Holstein x Zebu heifers, 11 month-old, with average initial body weight of 237 ± 37 kg, were divided in two groups of eight heifers each, and assigned to one of two grazing treatments: 1) an associated grass-legume (*B. brizantha* Toledo – *C. argentea*) pasture or high CP (HCP; grass 8.3 %, legume 22.0 %) level; and 2) grass (*B. brizantha* Toledo) alone pasture or low CP (LCP, 8.7 %) level. *Cratylia argentea* had a density of 570 shrubs ha^{-1}, and *B. brizantha* Toledo (associated or alone) was planted using 8 kg of seeds ha^{-1}. The legume was planted within rows every 2.5 m, with a distance between rows of 3 m, and the grass was sown between each row. The age of the legume and grass at the beginning of the experiment was 14 and 6 months, respectively. The grazing area for each treatment was 2.21 ha, divided in six paddocks of 3863 m^2 each. Because there are no previous experiences with grazing of *C. argentea*, we determined the stocking rate below the carrying capacity of *B. brizantha* Toledo grass, that is usually 2.0 AU ha^{-1} (AU = animal unit = one animal weighing 450 kg) in the dry season (Euclides *et al.*, 2001). Therefore, the initial stocking rate in both groups of heifers was 1.9 AU ha^{-1}. Each group grazed the corresponding treatment in an intensive rotational system, in a fixed cycle of 4 (grazing) and 20 (rest) days.

Quantitative and qualitative forage analysis

Throughout the experimental period, three (fixed) paddocks from the six of each treatment were sampled before and after grazing, in order to estimate dry matter yield (kg ha^{-1}) and daily growth rate (kg ha^{-1} day^{-1}), using the comparative yield method (Haydock and Shaw, 1975), with a set of 100 reference quadrats (50 x 50 cm) for each component corresponding to the specific pasture. These components were leaves and stems for *B. brizantha* Toledo, and only leaves for *C. argentea*. Samples from these evaluations were used to determine the nitrogen concentration, expressed as CP (N % x 6.25; AOAC, 1990), neutral detergent fiber (%), acid detergent fiber (%) (Goering and Van Soest, 1970), and lignin (%) (Robertson and Van Soest, 1981), using the Ankom filter bags technique.

Urea nitrogen determination

Throughout the experimental period, blood samples were obtained every seven days from eight heifers (four per treatment). Heifers were taken from the pasture at 0700 h and blood samples were collected via coccygeal venipuncture as aliquots of 3.5 mL, which were centrifuged at 1500 rpm/15 min and stored at -10 °C until analysis. Serum urea nitrogen was determined using an IDEXX Dry-Slide technology (VetTest® Chemistry Analyzer). Heifers were returned to their respective pasture following each blood sample collection.

Daily weight gain was recorded every 28 days, after a 12 h-fast. A trace mineral salt block (12 % P) was available in both groups of heifers for *ad libitum* consumption. Sanitary measures such as vaccination and deworming were applied as rutine procedures for all animals.

Statistical analysis

A completely randomized design was applied, using the animals as experimental units for daily weight gain and serum urea levels. Statistical significances were obtained with the F test, without the need of multiple range tests, having only two treatments. Also, repeated measurements data procedure was made using Proc Mixed (SAS/STAT,1995). An analysis of structures of covariance was performed in order to detect tendencies of correlation between serum urea - CP pasture variables.

RESULTS AND DISCUSSION

Dry matter yield and daily growth rate

Results of dry matter yield (DMY, kg ha^{-1}; averages ± SE) before grazing in HCP and LCP, during the period January-May 2011, were 1922 ± 405 and 770 ± 405 kg ha^{-1}, respectively, whereas after grazing they were 734 ± 242 and 413 ± 242, respectively (Figure 1). No statistical differences were found between treatments, neither before or after grazing. A tendency of decrease in forage allowance was observed troughout the experimental period as a result of transition from winter to the dry period, as well as by effect of an increasing stocking rate, from 1.9 to 2.3 AU ha^{-1}.

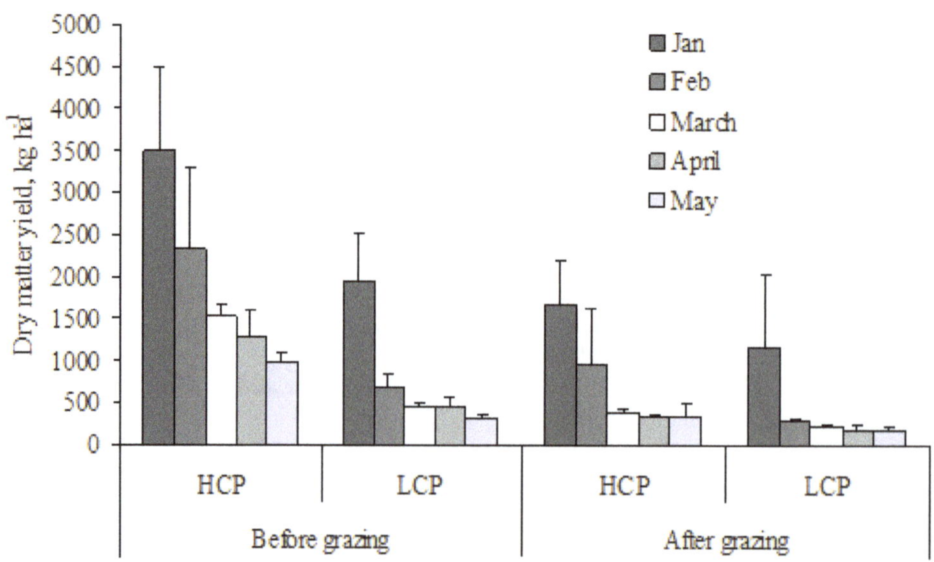

Figure 1. Dry matter yield (DMY, kg ha^{-1}) of high (*Brachiaria brizantha* Toledo - *Cratylia argentea*) and low (*B. brizantha*) crude protein pastures, before and after grazing (January-May, 2011) by Holstein x Zebu heifers.

Literature on grazing of *C. argentea* showed no effect in the subject, therefore, the discussion is focused in experiments with different species and or management of this association. In Colombia, Aparicio *et al.* (2002) reported values from an experiment using six cows/treatment (average weight 425 kg) with *B. decumbens* alone, or including restricted grazing (2 h d) of *C. argentea*. In the association, leaves of *B. decumbens* and *C. argentea* yielded 2153 and 7036 kg DM ha^{-1}, respectively. These values are higher than those reported here, but it should be noted that, in the Colombian experiment, the forage legume was used in restricted grazing; therefore, in the case of *Cratylia*, the use by the animals was less intense. Dry matter yield from the associated grass was 55 % higher than the grass alone. This agrees with reports of other authors (Alpizar, 1987; Kass, 1994), who have found that grasses associated with a N_2-fixing legume yield more forage than the same grass growing alone.

In Mexico, Valles *et al.* (2010) found that a mixture of native (*Paspalum* and *Axonopus* species, mainly) and introduced (*Cynodon nlemfuensis*, *Brachiaria arrecta*) pastures grazed by cows in a rotational system of leaders (48 cows) and followers (120 cows), in a grazing cycle of 1 to 3 days of use and 36 to 60 days of rest, yielded during dry season an average of 1977 ± 72 and 1468 ± 78 kg DM ha^{-1} before and after grazing. These values are higher than those reported here for LCP pasture, where the rest period was shorter.

Daily growth rate values were 22.8 ± 3.1 and 10.3 ± 3.1 kg DM ha^{-1} d^{-1}, for HCP and LCP, respectively (P ≤ 0.05) (Figure 2), representing the last value a daily contribution of only 45 % with respect to HCP. In both cases, this variable decreased as the experimental period progressed, and this fact could be attributed to the gradually increased grazing pressure over the pastures as the animals gained weight. The value of LCP was similar to that reported by Ramírez (1974), of daily growth rate for *Cynodon plectostachyus* of 10.4 kg DM d^{-1}, with occupation period of four days.

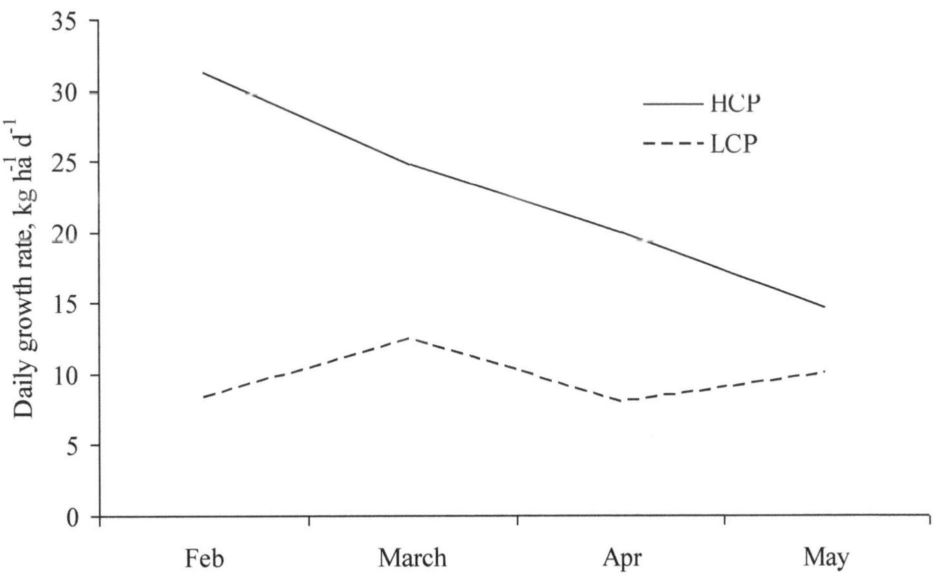

Figure 2. Daily growth rate (kg ha^{-1} d^{-1}) of high (HCP, *Brachiaria brizantha* Toledo - *Cratylia argentea*, range: 15.1 to 13.7 %) and low (LCP, *B. brizantha*, range: 8.7 to 7.4 %) crude protein pastures (February-May, 2011).

Quality variables in pasture botanical components

Crude protein

Results of CP (%) in the botanical components of the pastures are shown in Table 1. Crude protein values for *B. brizantha* Toledo grass alone and associated with *C. argentea* were similar before and after grazing (P ≥ 0.05), although a little decrease was observed in CP values after grazing. In the case of *C. argentea*, CP values were higher than in *B. brizantha* Toledo before and after grazing. Crude protein values for *B. brizantha* Toledo associated or alone were slightly higher than those reported by Paciullo *et al.* (2011) in Brazil, who evaluated the characteristics of the pasture and performance of dairy heifers in a silvopastoral system established with *B. decumbens* grass in combination with four tree species (*Acacia mangium*, *A. angustissima*, *Mimosa artesiana* and *Eucalyptus grandis*; 105 trees ha^{-1}) and in a monoculture system of *B. decumbens*. They found CP concentrations of 7.2 to 7.5 % in a pre-grazing sampling, with pasture management of 7 days in grazing and 45 days of rest, during the dry season. It is likely that these differences in favor of the *B. brizantha* Toledo - *C. argentea* pasture were given because the rest period for this pasture was shorter (20 days) than for *B. decumbens*, and the heifers consumed younger leaves and stems, and therefore better quality in terms of CP.

Table 1. Crude protein (mean ± SE) in *Brachiaria brizantha* Toledo grass alone or associated with *Cratylia argentea*, and *Cratylia argentea*, before and after grazing.

Comparisons between pasture components	Crude protein %	
	Before grazing	After grazing
B. brizantha Toledo grass associated with *C. argentea*	8.3 ± 0.85[NS]	7.6 ± 0.84[NS]
B. brizantha Toledo grass alone	8.7 ± 0.85	7.4 ± 0.63
B. brizantha Toledo grass associated with *C. argentea*	8.3 ± 0.76[b1]	7.6 ± 0.84[b]
vs.		
C. argentea associated with *B. brizantha* Toledo grass	22.0 ± 0.76[a]	19.9 ± 1.53 [a]

NS = Not significant.
[1]Within each column, different literals are statistically different at $P \leq 0.05$.

Neutral detergent fiber (NDF), acid detergent fiber (ADF) and lignin

Statistical comparisons between plant components in LCP and HCP before grazing were made. The comparisons between *B. brizantha* Toledo grass (alone or associated) showed similar values ($P \geq 0.05$) for these quality variables (Table 2), contrary to what was observed in the HCP plant components where, in all cases, statistical differences were found ($P \leq 0.05$). Table 2 also shows the values after grazing, but for this case no statistical procedures were conducted; however, as with the values obtained before grazing, similar tendencies were observed between plant components.

Table 2. Neutral detergent fiber (NDF), acid detergent fiber (ADF) and lignin (mean ± SE) in *Brachiaria brizantha* Toledo grass as low crude protein (LCP) pasture, or associated with *Cratylia argentea* as high crude protein (HCP) pasture, before and after grazing by Holstein x Zebu heifers.

Grazing	Plant comparisons	NDF	ADF	Lignin
		%	%	%
Before	*B. brizantha* Toledo (HCP)	73.2 ± 0.93	42.8 ± 1.26	13.3 ± 0.71
	vs.	NS	NS	NS
	B. brizantha Toledo (LCP)	73.1 ± 0.91	42.6 ± 1.23	13.8 ± 0.69
	B. brizantha Toledo (HCP)	73.3 ± 0.60[a]	42.9 ± 0.88[a]	13.4 ± 1.27[b]
	vs.			
	Cratylia argentea (HCP)	51.6 ± 0.58[b]	32.5 ± 0.85[b]	16.7 ± 1.30[a]
After[1]	*B. brizantha* Toledo (LCP)	74.89 ± 2.8	45.01 ± 2.5	14.21 ± 3.5
	B. brizantha Toledo (HCP)	74.6 ± 6.3	46.0 ± 4.2	14.3 ± 3.6
	Cratylia (HCP)	52.2 ± 4.0	33.5 ± 2.7	16.7 ± 5.3

NS= Not significant.
[a,b]Different letters within each column/comparison are statistically different ($P \leq 0.05$).
[1]No statistical comparisons were made.

In Veracruz, Mexico, Manríquez-Mendoza *et al.* (2011) assessed nutricional quality in a silvopastoral system with *Guazuma ulmifolia* trees mixed with *B. brizantha* (Insurgente grass) and *Digitaria eriantha* during the winter, dry and rainy seasons. They reported, for the dry season, values of 73.6 %, 40.6 % and 6.6 % for NDF, ADF and lignin, respectively, which were similar for NDF, slightly lower for ADF and lower for lignin in comparison with the ones reported in this study. The values of *B. brizantha* obtained in the present study were lower than those indicated by Costa *et al.* (2009) in Brazil, after cutting every 30 days three cultivars (Marandú, Xaraes and MG-4) with no fertilization, with values for NDF and

ADF of 68.8 to 70.6 % and 37.2 to 37.9 %, respectively. In the present study, in *C. argentea* the values for NDF and ADF were slightly lower than those reported by Franco (1997), who evaluated NDF and ADF in this legume at 2, 3 and 4 month-old, and obtained values of 55.6 to 57.2 % and 33.8 to 36.1 % for NDF and ADF, respectively. This difference was probably because in this study, *C. argentea* was sampled at an earlier age, it is 28 (before grazing) to 32 (after grazing) days. However, in the case of lignin, the differences were greater, since they reported values in a range of 8.7 to 10.9 %, which are much smaller than those obtained in this study. In Nicaragua, Sánchez and Ledin (2006) offered 3 month-old *C. argentea* to Criollo Reyna cows, and values for NDF and ADF were 60 % and 32.5 %, respectively, which were higher than those from the present study.

Blood serum urea levels

Blood serum urea nitrogen was significantly higher ($P \leq 0.05$) in heifers grazing the HCP pasture (11.01 ± 0.35 mg dL^{-1}) with respect to heifers grazing the LCP pasture (7.87 ± 0.35 mg dL^{-1}). The concentration of serum urea in all but the first sampling was higher in HCP (Figure 3). In the fourth sampling date, both treatments showed similar serum urea nitrogen concentrations. In other samples, the HCP treatment exceeded ($P \leq 0.05$) to LCP at different rates. Also, the interaction treatment x sampling was statistically significant ($P \leq 0.05$). Urea levels in blood serum had ups and downs throughout the study, but the heifers in HCP always showed higher values for this variable (11.3 ± 21 mg dL^{-1}) compared to heifers in LCP (8.4 ± 3.4 mg dL^{-1}).

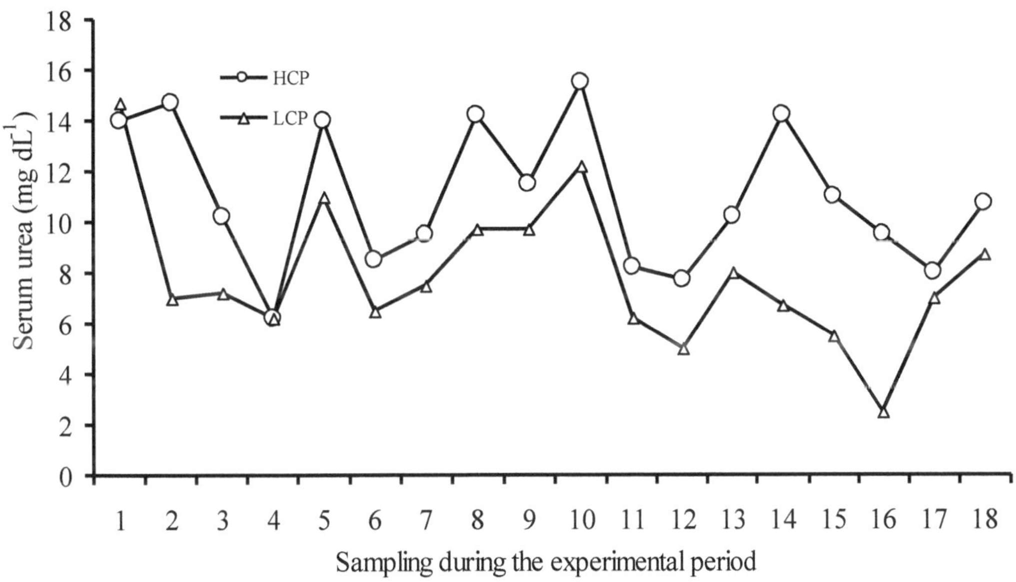

Figure 3. Blood serum urea concentration in Holstein x Zebu heifers grazing pastures with a high (HCP, *Brachiaria brizantha* Toledo - *Cratylia argentea*) or low (LCP, *Brachiaria brizantha* Toledo) crude protein levels, sampled every seven days.

Because no information was found on grazing systems using *C. argentea*, the discussion is based on pastures and/or similar diets. Carroll *et al.* (1988) compared the effect of 13 and 20 % CP rations (50 % grass-legume silage: 50 % corn silage; and 60 % concentrate [DM basis]) on the reproductive performance of early lactating dairy cows. They observed that plasma urea nitrogen increased rapidly in cows on the 20 % CP diet, maintaining a 10 mg dL^{-1} superiority over the other group fed 13 % CP, after the fourth week (of 14) of the experiment. However, these authors concluded that no relationship was found between reproduction and CP level in the ration. Previously, similar results were obtained by Barton *et al.* (1996) in a study conducted to determine the impact of dietary CP (13 % vs. 20 %), parity (first vs. second lactation or later), and breed (Holstein vs. Jersey) on the reproductive efficiency of dairy cows. Plasma urea nitrogen concentrations were influenced by diet (8.6 vs. 21 mg dL^{-1}, 13 and 20 % CP, respectively), parity and breed. Ferguson *et al.* (1993) mentioned values of urea in blood serum of cattle ranging from 10 to > 20 mg dL^{-1}, and established a negative relationship when these concentrations exceed 20 mg dL^{-1}, resulting in lower conception rates. Concentrations found here are lower than those reported by Butler (1998) in Holstein cows fed 50:50 % forage-commercial concentrate (18.9 ± 0.3 mg dL^{-1}), and by Aparicio *et al.* (2002) in cows grazing *Brachiaria decumbens* grass with restricted access (2 h day^{-1}) to *C. argentea* with 18.2 mg dL^{-1}.

The records of blood serum urea for each of the four heifers that were sampled by treatment throughout the experimental period are shown in Figure 4. In average, heifers grazing the grass-legume treatment had serum urea nitrogen concentration 33.3 % higher (10.8 ± 0.6 mg dL^{-1}) than heifers grazing the *B. brizantha* Toledo grass alone (7.2 ± 0.2 mg dL^{-1}), showing in all cases a consistent difference between treatments. Blood serum urea concentration was above 20 mg dL^{-1} in both treatment groups, although the HCP group reflected the effect of the diet with higher CP content in the pasture. These values in both treatments were lower than those found by Hammon *et al.* (2005), feeding Holstein dairy cows with a diet containing alfalfa hay, alfalfa silage, and corn silage, and 50 % concentrate formulated to provide at least 20 % CP. They recorded blood urea nitrogen in a range of 17.0 ± 0.34 to 22.36 ± 0.44 mg dL^{-1}.

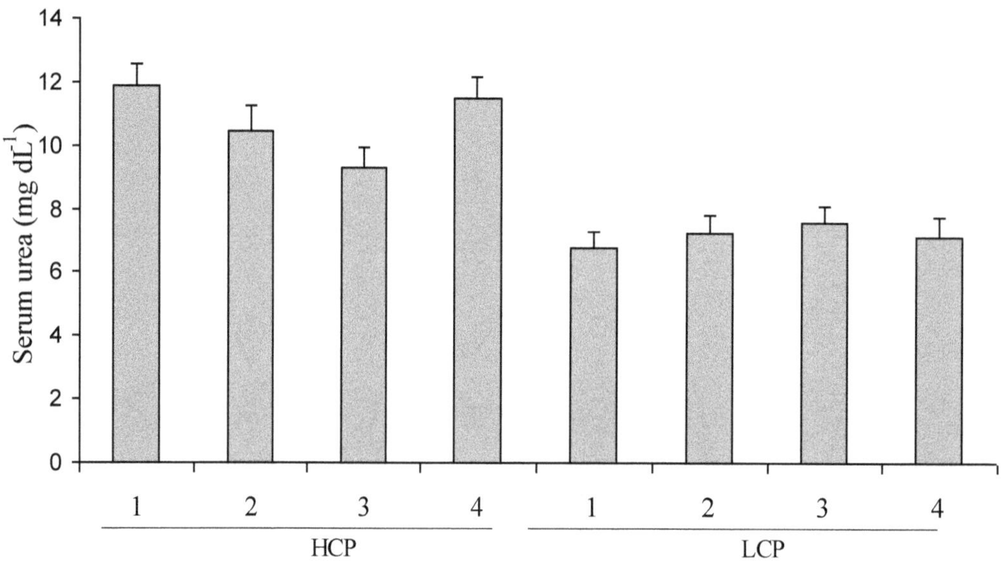

Figure 4. Serum urea concentration (mg dL^{-1}) in each of the four heifers sampled (1 to 4), grazing in pastures with high (*Brachiaria brizantha* Toledo - *Cratylia argentea*) and low (*B. brizantha* Toledo) crude protein levels. Lines upon bars are standard error of means.

Looking for a model consistent with data of relationship serum urea - CP in pastures, an analysis of structures of covariance was performed in order to detect tendencies of correlation between these variables (Table 3). The analisys showed that the first order autoregresive (AR) model produced the best results, considering the statisticals of fitting.

Table 3. Covariance structures for repeated measures analysis of serum urea nitrogen in heifers grazing *Brachiaria brizantha* Toledo grass alone or the association *B. brizantha* Toledo - *Cratylia argentea.*

Fit Statistics[1]	Covariance structure						
	CS	CSH [2]	UN [2,3]	AR(1)	ARH(1)	ARMA(1,1)	HF[4]
-2 RLL	470.7	------	------	468.1	438.7	467.2	------
AIC	476.7	------	------	474.1	478.7	475.2	------
AICC	477.0	------	------	474.3	488.4	475.6	------
BIC	477.0	------	------	474.3	480.3	475.5	------

[1] "Smaller is better."
[2] "Stopped because of infinite likelihood."
[3] "Unable to make hessian positive definite."
[4] "Did not converge."

Weight gain by heifers

Heifers that grazed the HCP pasture averaged a daily weight gain of 839 ± 258 g, while for the LCP pasture weight gain was 580 ± 278 g ($P \leq 0.05$; Table 4). The weight change for the period 02/02/2011 to 05/25/2011 in HCP was from 235 ± 39 kg to 327 ± 53 kg, compared to LCP from 231 ± 36 kg to 293 ± 47 kg. These changes accounted for a cumulative weight gain during the experimental period, of 92 and 62 kg by animal, respectively (Figure 5).

Table 4. Daily weight gain (g day^{-1}) of heifers grazing in pastures with high (HCP, *Brachiaria brizantha* Toledo - *Cratylia argentea*) and low (LCP, *B. brizantha* Toledo) crude protein diet during the 2011 dry season (February-May) in Atzalan, Veracruz, Mexico.

Weighing	HCP diet		LCP diet		Diff	Pr>|t|	
	DWG[1]	SE[2]	DWG	SE			
Feb	592	0.080	328	0.088	0.263	0.0866	NS
March	993	0.065	690	0.092	0.303	0.0302	*
Apr	781	0.051	754	0.059	0.027	0.7326	NS
May	951	0.085	522	0.077	0.429	0.0026	**

[1]Daily weight gain.
[2]Standard error.
NS= Not significant. * = $P < 0.05$; ** = $P < 0.01$.

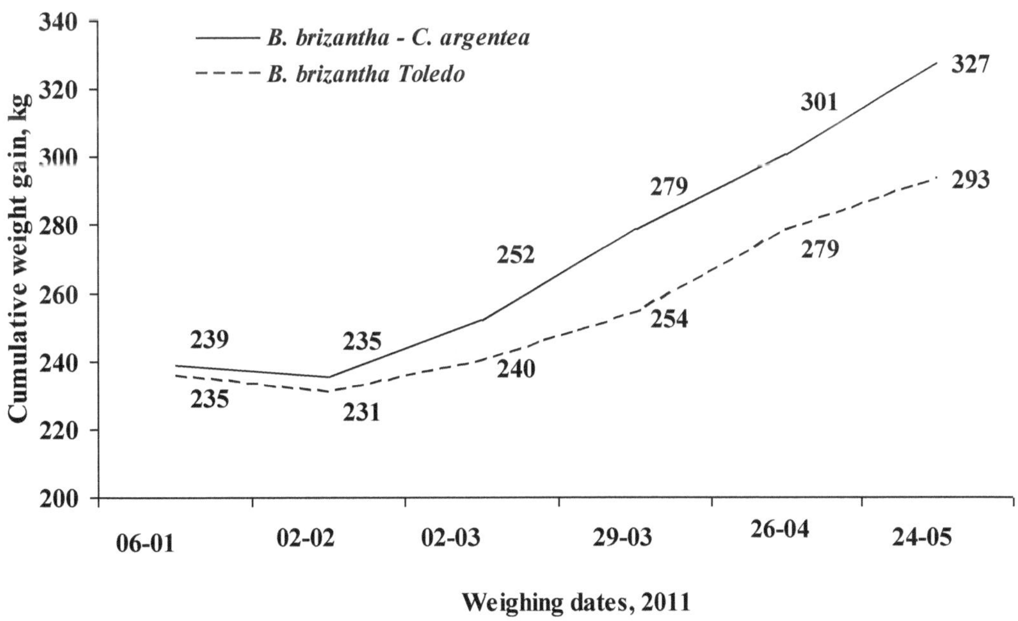

Figure 5. Cumulative weight gain (kg) in Holstein x Zebu heifers grazing in pastures with high (*Brachiaria brizantha* Toledo - *Cratylia argentea*) and low (*B. brizantha* Toledo) crude protein levels.

The daily weight gains reported here were higher than those obtained by Aranda *et al.* (2001) in a 153-day grazing experiment with heifers (six heads ha^{-1}), in the humid tropic of Tabasco, Mexico. They evaluated the grazing of African stargrass (*Cynodon plechtostachyus*) alone (CP 8.9 %), supplemented with sugarcane (CP 10.3 %), sugarcane without urea 1 % (CP 4.2%), or sugarcane with urea plus a protein supplement (CP 23.4 %). Heifers receiving the protein supplement had the highest daily gain (528 g d^{-1}) compared to the other treatments (control group 0.320 kg day^{-1}; sugarcane without urea 0.327 kg day^{-1};

sugarcane plus urea 0.366 kg day^{-1}). The experiment of Paciullo *et al.* (2011) that evaluated a silvopastoral system (*B. decumbens* associated with four species of trees) and a monoculture (*B. decumbens*), during three years, reported that the lowest body weight gains were observed during the dry season of the three years, averaging 355 (silvopastoral system) and 338 (monoculture) g day^{-1}, which can be attributed to the lower CP (7.3 % as average) content of the pasture, as a consequence of the decreased rainfall and increased air temperature.

Our results suggest that grazing *C. argentea* associated with *B. brizantha* Toledo promoted weight gain higher than expected in the dry season, attributing these results to the high quality of the mixed pasture.

Coefficients of simple correlation (R^2) between blood serum urea levels and daily weight gain were, comparing HCP, LCP and HCP+LCP vs. daily weight gains: 0.20, 0.29 and 0.01, respectively, indicating a poor relationship between these variables.

CONCLUSION

Considering the greater forage yields, the higher quality of the components of the association especially in terms of CP, the fact that blood serum urea concentrations never reached the toxic levels reported in the literature (> 20 mg dL^{-1}), and the best performance in terms of daily weight gain by heifers grazing a high CP pasture, the association *B. brizantha* Toledo - *C. argentea* can be a desirable alternative to improve tropical animal production systems based on grazing.

ACKNOWLEDGEMENT

The authors acknowledge the economic support to this experiment provided by DGAPA-UNAM through the PAPIIT Project No. IN202410.

REFERENCES

Alonso-Díaz, M.A., Torres-Acosta, J.F., Sandoval-Castro, C.A., Capetillo-Leal, C.M. 2010. Polyphenolic compounds of nutraceuticals trees and the variability of their biological activity measured by two methods. Tropical and Subtropical Agroecosystems. 12:649-656.

Andersson, M.S., Peters, M., Schultze-Kraft, R., Franco, L.H., Lascano, C.E. 2006. Phenological, agronomic and forage quality diversity among germplasm accessions of the tropical legume shrub *Cratylia argentea*. Journal of Agricultural Science. 144:237-248.

AOAC. 1990. Association of Official Analytical Chemists. Official methods of analysis. 15th ed. Washington, D.C.

Ben Salem, H., Saghrouni, L., Nefzaoui, A. 2005. Attempts to deactivate tannins in fodder shrubs with physical and chemical treatments. Animal Feed Science and Technology. 122:109-121.

Alpizar, L. 1987. Results from the CATIE "Central Experiment": Pasture and shade tree associations. In: Advances in Agroforestry Research., Proceedings. Beer, J., Fassbender, H.W., Heuveldop, J. (eds.). CATIE. Turrialba, Costa Rica. 1-11 September, 1985. pp. 209-214.

Aparicio, R., Lascano, C., Avila, P. 2002. Utilización de *Cratylia argentea* madura y joven como banco de proteína por vacas de leche. Revista Científica. 12(2 Suplemento):595-598.

Aranda, E., Mendoza, G.D., García-Bojalil, C., Castrejon, F. 2001. Growth of heifers grazing stargrass complemented with sugar cane, urea and a protein supplement. Livestock Production Science. 71:201-206.

Barton, B.A., Rosario, H.A., Anderson, G.W., Grindle, B.P., Carroll, D.J. 1996. Effects of dietary crude protein, breed, parity, and health status on the fertility of dairy cows. Journal of Dairy Science. 79:2225-2236.

Butler, W.R. 1998. Review: Effect of protein nutrition on ovarian and uterine physiology in dairy cattle. Journal of Dairy Science. 81:2533-2539.

Carroll, D.J., Barton, B.A., Anderson, G.W., Smith, R.D. 1998. Influence of protein intake and feeding strategy on reproductive performance of dairy cows. Journal of Dairy Science. 71:3470-3481.

Chapa, A.M., McCormick, M.E., Fernandez, J.M., French, D.D., Ward, J.D., Beatty, J.F. 2001. Supplemental dietary protein for grazing dairy Cows: Reproduction, condition loss, plasma metabolites, and insulin. Journal of Dairy Science. 84:908-916.

Costa, K.A. de P., Pereira, O.I., Faquin, V., Pereira, S.G., da Costa. S.E. 2009. Producao de massa seca e nutricao nitrogenada de cultivares de *Brachiaria brizantha* (A. Rich) Stapf sob doses de nitrogenio. Ciência e Agrotecnologia, Lavras. 33:1578-1585.

Deinum, B., Sulastri, R.D., Seinab, M.H.J., Maassen A. 1996. Effects of light intensity on growth, anatomy and forage quality of two tropical grasses (*Brachiaria brizantha* and *Panicum maximum* var. Trichoglume). Netherlands Journal of Agriculture Science. 44:111-124.

Euclides, V.P.B., do Valle, C.B., Macedo, M.C.M., Oliveira M.P. 2001. Evaluation of *Brachiaria brizantha* ecotypes under grazing in small plots. Proceedings XIX IGC 2001. Sao Paulo, Brazil.

Ferguson, J.D., Galligan, D.T., Blanchard, T., Reeves M. 1993. Serum urea nitrogen and conception rate: The usefulness of test information. Journal of Dairy Science. 76:3742-3746.

Franco, M.H. 1997. Evaluación de la calidad nutricional de *Cratylia argentea* como suplemento en el sistema de producción doble propósito en el trópico subhúmedo de Costa Rica. Tesis Magister Scientiae. Centro Agronómico Tropical de Investigación y Enseñanza (CATIE), Turrialba, Costa Rica.

Goering, H.K., Van Soest, P.J. 1970. Forage fiber analisis (apparatus, reagents, procedures and some applications). Agricultural Handbook 379. ARS, USDA. Washington, DC.

Hammon, D.S., Holyoak, G.R., Dhiman, T.R. 2005. Association between blood plasma urea nitrogen levels and reproductive fluid urea nitrogen and ammonia concentrations in early lactation dairy cows. Animal Reproduction Science. 86:195-204.

Haydock, K.P., Shaw, N.H. 1975. The comparative yield method for estimating the dry matter yield of pasture. Australian Journal of Experimental Agriculture and Animal Husbandry. 15:663-670.

Ibrahim, M., Franco, M., Pezo, D., Cameron, A. 2001. Promoting intake of *Cratylia argentea* as a dry season supplement for cattle grazing *Hyparrhenia rufa* in the subhumid tropics. Agroforestry Systems. 51:167-175.

Kass, D.L. 1994. *Erythrina* Species - Pantropical multipurpose tree legumes. In: Gutteridge, R.C. and Shelton, H.M. (eds.). Forage Tree Legumes in Tropical Agriculture. CAB International, UK. pp. 84-96.

Lascano, C., Rincón, A., Bueno, G., Argel, P.J. 2002 Cultivar veranera (*Cratylia argentea* (Desvaux) O. Kuntze) Leguminosa arbustiva de usos múltiples para zonas con períodos prolongados de sequía en Colombia. Corpoica, Ministerio de Agricultura y Desarrollo Rural y CIAT. Villavicencio, Colombia.

Lascano, C., Plazas, C.H. 2003. Utilidad de la leguminosa semiarbustiva *Cratylia argentea* en sistemas de ganado doble propósito del Piedemonte Llanero: Validación y difusión. CIAT, PRONATTA, Ministerio de Agricultura y Desarrollo Rural. Bogotá, Colombia.

Manríquez-Mendoza, L.Y., López-Ortiz, S., Olguín-Palacios, C., Pérez-Hernández, P., Díaz-Rivera, P., López-Tecpoyotl, Z.G. 2011. Productivity of a silvopastoral system under intensive mixed species grazing by cattle and sheep. Tropical and Subtropical Agroecosystems. 13:573-584.

Makkar, H.P.S. 2003. Effects and fate of tannins in ruminant animals, adaptation to tannins, and strategies to overcome detrimental effects of feeding tannin-rich feeds. Small Ruminant Research. 49:241.

Paciullo, D.S.C., Castro, C.R.T., Gomide, C.A.M., Mauricio, R.M., Pires, M.F.A., Muller, M.D. Xavier D.F. 2011. Performance of dairy heifers in a silvopastoral system, Livestock Science. 141:166-172.

Pereira, J.M.M., Tarré, R.M., Macedo, R., Rezende, C.P., Alves, B.J.R., Urquiaga, S. Boddey, R.M. 2009. Productivity of Brachiaria humidicola pastures in the Atlantic forest region of Brazil as affected by stocking rate and the presence of a forage legume. Nutrient Cycling in Agroecosystems. 83:179-196.

Power, I.L., Thorrold, B.S., Balks, M.S. 2003. Soil properties and nitrogen availability in silvopastoral plantings in *Acacia melanoxilon* in North Island, New Zealand. Agroforestry Systems. 57:225-237.

Ramírez, P.A. 1974. Efecto del ciclo de uso, la presión de pastoreo y la fertilización nitrogenada en la producción de praderas de pasto estrella (*Cynodon plectostachyus* (K. Schum) Pilger). Tesis Magister Scientiae. Instituto Interamericano de Ciencias Agrícolas (IICA, FAO). Turrialba, Costa Rica.

Robertson, J.B., Van Soest, P.J. 1981. The detergent system of analysis and its application to human foods. In: James, W.P.T. and

Theander, O. (eds.). The Analysis of Dietary Fiber in Foods. Marcel Dekker, New York. pp. 123-158.

Sánchez, N.R.S., Ledin, I. 2006. Effect of feeding different levels of foliage from *Cratylia argentea* to creole dairy cows on intake, digestibility, milk production and milk composition. Tropical Animal Health and Production. 38:343-351.

Sandoval-Castro, C.A., Lizarraga-Sanchez, H.L., Solorio-Sanchez, F.J. 2005. Assessment of tree fodder preference by cattle using chemical composition, *in vitro* gas production and *in situ* degradability. Animal Feed Science and Technology. 123-124:277-289.

SAS/STAT. 1995. SAS/STAT, Guide for personal computer (Release 6.12). SAS Institute, Inc., Cary, North Carolina, USA.

Valles, B., Castillo, E., Barragán, J., Jarillo, J., Ocaña, E. 2010. Dinámica de una pastura mixta bajo apacentamiento intensivo en el trópico húmedo veracruzano. Avances en Investigación Agropecuaria (México, Segunda Época). 14:3-21.

Yamamoto, W., Dewi, I.A., Ibrahim, M. 2007. Effects of silvopastoral areas on milk production at dual-purpose cattle farms at the semi-humid old agricultural frontier in central Nicaragua. Agricultural Systems. 94:368-375.

3

Proton-Pumping and Nitrogen Metabolism in Rice under Humid Tropic Conditions

Rodrigo Grazinoli Garrido[1*], Fabiola de S.R.G. Garrido[2], Sonia Regina de Souza[3], Manlio Silvestre Fernandes[4]

[1]*Institute of Forensic Genetic Research, DGPTC, PCERJ, Brazil*
Scholl of Health Science an Environment, UCB
E-mail: grazinoli.garrido@gmail.com
[2]*Environmental and Administrative Science Department, ITR, UFRRJ, Brazil*
[3]*Chemistry Department, ICE, UFRRJ, Brazil*
[4]*Soils Department, IA, UFRRJ, Brazil*
**Corresponding author*

SUMMARY

In rice plants, it is important to recognize N metabolism efficiency parameters in order to design strategies for improvements, increasing grain protein and enhancing people nutrition quality. Two rice varieties grown under humid tropic conditions were evaluated: Piaui, a landrace adapted to low N availability, and reduced light supply, and IAC-47, an improved variety. The assay was carried out in controlled conditions of 24°C and 12/12 hour light/dark periods (200 $\mu E.m^{-2}.s^{-1}$) using a growth chamber, simulating humid tropic environment. Rice plants were grown under 0.1 and 1.0 mM NH_4^+-N, in a Hoagland & Arnon nutrient solution, pH 5.5, up to 26 days after germination. The activity of PM H^+-ATPases in the plasma membrane, vacuole VH^+-ATPases, H^+-PPases as well as GS, GOGAT, and GDH were determined. PM H^+-ATPase presented higher activity in IAC-47 roots. On the other hand, Piaui roots showed an enhanced microsomal protein content and VH^+-ATPases activity, which apparently allows this variety to absorb N as well as the improved one. Increases in the cytosolic ammonium would result in high GDH deamination in Piaui variety at 1.0 mM NH_4^+-N. Meanwhile, GS increases were observed in IAC-47 shoots, together with GOGAT increases at same treatment.

Keywords: *Oryza sativa;* Ammonium; V-ATPase; P-ATPase; V-PPase; GS; GOGAT; GDH.

INTRODUCTION

About half planet population, more than three billion of people depend on rice for their nutrition (IRRI, 2012). Asia alone produces and consumes 90% of all world rice production. Out of the Asian continent, Brazil represents the greatest cereal producer (Pereira et al., 2005).

Particularly in Humid Tropic, rice growing starts when rainy summer begins, between December and July. It is characterized by low light and not so high temperatures (Agritempo, 2010). Tropical environmental conditions produce strongly acid soils with low natural fertility and frequently with high aluminum saturation. Under these conditions nitric-N is lixiviated from soil and, ammonium-N may raise its concentration by nitrification inhibition (Rodrigues and Garrido, 2005). Then, NH_4^+ can be the only mineral N source available for plants (Britto et al., 2001).

Ammonium assimilation is specially important since NH_4^+ use is faster and economic than NO_3^- assimilation, because ammonium by-pass two steps of nitrate reduction, however toxicity symptoms are common (Holzschuh et al., 2009). In flooded rice it was observed that up to three weeks old, NH_4^+ absorption and use is higher than NO_3^- (Kronzucker et al., 2000).

Actually, it is well known that ammonium uptake is carried out by two transporter systems: the hight affinity (HATS) system is active at low external NH_4^+ concentrations, and a low affinity (LATS) system is constitutive and is active above 1mM ammonium in the soil solution (Wang et al., 1994; Kuumar et al., 2003).

Ammonium influx is reduced due to PM H^+-ATPases inhibition. This influx is charge-dependent, but it is not directly affected by external pH variation (Howitt and Udvardi, 2000). In HATS, rice presents a light level N regulation: it decreases with external NH_4^+ increasing, and increases with prolonged light periods (Kumar et al., 2003).

The proton-motive force for this process is generated by PM H^+-ATPases. Beyond plasmalem, H^+-ATPases are widely distributed in cell membranes, as vacuole (VH^+-ATPases), endoplasmatic reticulum and vesicles (Palmgren, 2001). In vacuole, it is also known pyrophosphate dependent enzyme (H^+-PPases). Only in extreme metabolic stress, where PPi becomes the principal energy donor, H^+-PPases would be greater than VH^+-ATPases activity (Santos et al., 2009; Ramos et al., 2005).

The aim of the study is to prospect parameters of N absorption and assimilation efficiency, specially in rice landscape variety that enhances grain protein levels, despite environmental conditions. It allows to design strategies for improvements, contributing to enhance human nutrition quality. For both, two rice varieties were evaluated: Piaui, adapted to Humid Tropic conditions, and IAC-47, an improved variety. Piaui presents N assimilation efficiency, while the improved variety is efficient in N absorption (Ferraz Jr et al., 1997).

Two rice varieties were cultivated under Humid Tropic conditions, low NH_4^+-N levels and weak light, up to 26 DAG. Ammonium absorption and assimilation, PM H^+-ATPases, V H^+-ATPases and H^+-PPases as well as glutamine synthetase (GS), glutamate synthase (GOGAT) and glutamate dehydrogenase (GDH) kinetics, pH variation and NH_4^+ depletion in nutrition solution are elucidated.

MATERIAL AND METHODS

The study was carried out at the Universidade Federal Rural do Rio de Janeiro, Seropédica campus, Brazil, during 2007. Two rice (*Oryza sativa,* L.) varieties were used: Piaui and IAC-47. Rice plants were cultivated in a growth chamber, at 24°C of average temperature and with 12 hour light/dark periods (200 $\mu E\ m^{-2}\ s^{-1}$). The assay was designed with varieties and 2 treatments (0.1 and 1.0 mM NH_4^+-N, pH 5.5) and 3 replications. Replications are constituted of blocks with 5 repetitions and 5 plants each for extraction and enzyme assays. Means were compared using T test.

Rice plants were grown in nutrient solutions until 26 days after germination (DAG). A modified Hoagland & Arnon (1950) solution was used at ¼ of its ionic strength. At 4 DAG it was changed to ½ of its ionic strength and kept so up to 8 DAG. After that, plants were grown in a full strength solution up to the final harvest at 26 DAG.

Absorption NH_4^+ kinetic assay was carried out in plants which received nutrient solution without ammonium at 24 DAG. After 48 hours without N (26 DAG), it was reestablished 0.1 and 1.0 mM NH_4^+-N, pH 5.5 nutrient solutions. Variation in pH and NH_4^+ depletion in nutrient solution were followed during 72 hours. Ammonium was measured according Mitchell (1972).

For proton-pumping activities (PM H^+-ATPases, V H^+-ATPases and H^+-PPases) determination, tonoplast and plasma membrane vesicles were extracted by cell fractionation (Yoshida et al., 1983, Ramos et al., 2005, Garrido et al., 2008). At 26 DAG, plants were harvested, washed in distillated water, dried and 9 grams of roots was grounded in an ice cold mortar and pestle, with 25 mL of a buffer solution (0.25 M sorbitol; glycerol 10%; 50 mM TRIS-acetate; 250 mM sucrose; 2 mM EDTA; 2 mM EGTA; 5 mM DTT; 2 mM 2-mercaptoethanol; 1 mM PMSF; 0.5% BSA; 1% PVP 40T; 150 mM KCl; 100 mM choline chlorate; pH 8.0). The grounded material was filtered through 4 layers of cheesecloth. The extract was centrifuged at 3.600 x g for 10 minutes, at 4° C. The supernatant was taken and centrifuged at 10.000xg for 10 minutes, at 4° C. The new supernatant was centrifuged at 105.000 x g for 45 minutes at 4°C. Pellet was ressuspended in 1.5 mL of ressuspending buffer, pH 7.5 (30 mM TRIS-HCl; glicerol 15%; 1 mM EGTA, 1 mM EDTA, pH 7.5; 2 mM $MgCl_2$; 2 mM DTT; 1 mM PMSF). Total protein was determined by Bradford (1976) in purified fraction. It was stored in 0.5 ml samples and, after liquid N_2 frozen, kept at -20°C until used for the proton-pumps activity.

Hydrolytic proton-pump activities were measured by inorganic phosphate (Pi) releasing (Fiske, Subarrow, 1925). Proton pumping assays occurred as described by Yan et al. (2002), with minor modifications. Thc basic reaction mediums' contained KCl 1 M; 0.1 M $MgSO_4$; 40 mM ATP-BTP (0.1 M sodium pyrophosphate); 0.1 M Na_2MoO_4; 1% Brij 58; 0.1M NaN_3; 40 μg.mL^{-1} protein. After 45 min at 30°C, reaction was stopped by cool acid solution 1 mL (2% H_2SO_4; 3% SDS; 0.7% $(NH_4)MoO_4$). Buffers were modified according the proto-pump type: PM H^+-ATPase 0.5 M MOPS-BTP, pH 6.5; V H^+-ATPases 0.5 M HEPES-BTP, pH 7.5; H^+-PPase 0.5 M MOPS-Imidazole, pH 7.2. The controlled mediuns were assayed with/without 200 nM Bafilomicine; 1.0 M KNO_3; 10 mM Na_3VO_4; 100 mM KCl; 2.5mM NaF, respectively, according the inhibition profile; 40 mM ATP-BTP (0.1 M sodium pyrophosphate); and protein. Then, 50 μL ascorbic acid was added. Ten minutes afterward, 1.45 mL of citrate reagent containing 4% sodium citrate and 2% acetic acid was added to prevent ATP/PPi acid hydrolysis.

After absorption analysis of ammonium assimilation pattern for both varieties was determined at the same environmental conditions. For this, ammonium assimilatory enzymes: Glutamine Sinthetase (GS), Glutamate Sinthase (GOGAT) and Glutamate Dehydrogenase (GDH) (amination and deamination) were assayed

Ammonium assimilation enzymes GS, GOGAT and GDH were purified from plants harvested at 26 DAG. One gram of shoots and roots was grounded in liquid N_2 with a mortar and pestle, with 4 mL of extraction buffer (imidazole-HCl 0.5 M, pH 5.0). The grounded material was filtered and the cooled extract (1.5 mL) was centrifuged at 15.000 x g, for 15 min (0°C). The supernatant was used in enzymatic assays. Enzymes activity was measured by spectrum absorption variation according Bergersen (1980).

RESULTS AND DISCUSSION

Under both NH_4^+-N (0.1 and 1.0 mM) levels and weak light (200 μE.m^{-2}.s^{-1}), Piaui plants presented higher protein concentration in microsomal fraction (Table 1). Garrido et al. (2008) showed similar result when the same variety was cultivated with NO_3^- pulses. These conditions would enable reduction in protein transcription, like that of ammonium transporters (Kumar et al. 2003). This apparent contradiction highlights the discussion of Piaui morphophysiologic adaptations to Humid Tropic conditions (Ferraz Jr. et al., 2001, Garrido et al., 2008).

Table 1. Protein Purified (μg. g^{-1} MF) in root microsomal fraction of Piaui and IAC-47 rice plants cultivated according to Material and Methods. Average of 5 pots with 5 plants each.

Rice Varieties	NH_4^+-N levels (mM)	Protein (μg. g^{-1} FW)
Piaui	0.1	1230,54± 38,08
	1.0	1308,67± 63,88
IAC-47	0.1	544,38±113,84
	1.0	901,61±159,72

IAC-47 plants showed higher PM H^+-ATPases hydrolytic activity in both treatments (Table 2). Similar profile is observed under nitrate nutrition (Santos et al., 2009, Garrido et al., 2009). Ammonium transport across a uniport is maintained due to proton-pumps activity that established a ΔΨ (Howitt and Udvardi, 2000). Despite this, it was not reflected in solution acidification and ammonium depletion (Fig. 1).

Table 2. ATP and PPi (μmol Pi.mg^{-1} Ptn. h^{-1}) hydrolysis activity by PM H^+-ATPase, V H^+-ATPase, and H^+-PPase in root microsomal fraction of Piaui and IAC-47 rice plants cultivated according to Material and Methods. Average of 5 pots with 5 plants each.

Rice Varieties	NH_4^+-N levels (mM)	Enzyme Activity (μmol Pi.mg^{-1} Ptn. h^{-1})		
		PM H^+-ATPase	V H^+-ATPase	H^+-Ppase
Piaui	0.1	1.28 ± 1.13	8.05 ± 1.94	2.17 ± 0.57
	1.0	0.67 ± 0.33	8.72 ± 2.02	3.04 ± 0.45
IAC-47	0.1	4.93 ± 0.93	1.61 ± 0.86	2.72 ± 0.41
	1.0	4.96 ± 1.60	1.41 ± 0.47	3.30 ± 0.79

When V H^+-ATPases activity was analyzed, it showed a completely distinct pattern in Piaui plants (Table 2). Garrido et al. (2008) observed that V H^+-ATPases activity was higher in Piaui than in IAC-47, when plant received nitrate pulse. Santos et al. (2009) observed this pattern in Piaui plants under NO_3^--N starvation. So, it was noticed as a varietal characteristic, independently from treatments (Table 2).

Oppositely to what happens to nitrate, vacuolar ammonium stock would be more transitory and not allowed by acid pH (Howitt and Udivar, 2000). High VH^+-ATPases activities found in Piaui would diminish ammonium movement through tonoplast, maintaining its cytosolic levels for continuous assimilation even under low concentration.

Otherwise, IAC-47 plants, improved for higher N nutrition, would be able to maintain and stock ammoniun in the tonoplast when necessary. High Piaui tonoplast pumps activity would also compensate for low activity in plasmalem allowing for cytoplasm alkalization and probably keeping the $\Delta\Psi$ (not measured) in relation to apoplasm. This finding could explain the high N-ammonium and N-amino observed in Piaui plants by Garrido, et al. (2008)

H^+-PPases activity was not different between varieties or between treatments (Table 2). In general, plants use H^+-PPases in a parallel to V H^+-ATPases. H^+-PPases would only be higher than V H^+-ATPases in case of extreme energetic stress, when PPi would act as the principal cell energy donor (Ramos et al., 2005).

Nutrient solution acidification by plants after 26th DAG (Figure 1A) was enhanced in higher NH_4^+-N concentration. Acidification happens due to positive charge equivalents exudation against excessive cation absorption, avoiding membrane depolarization and maintaining $\Delta\Psi$ and ΔpH necessaries for HATS (Howitt and Udivardi, 2000).

Ammonium depletion followed nutritive solution acidification (Figure 1A and B). The similar slopes observed for Piaui and IAC-47 would be explained by different adaptative mechanisms: higher protein membrane content in Piaui (Table 1), probably electrogenic pumps, and higher specific PM H^+-ATPase activity in IAC-47 (Table 2).

Figure 1B showed that NH_4^+ was exhausted in 0.1 mM solution during the earliest four hours. During this time, the curve slope was very similar for both nutritional levels. This fact would point out to the same ammonium transporter class, although, after 72h, both varieties in 1.0 mM treatment showed 0.4 mM ammonium in nutritive solution. Thus, under these nutritional levels there were transporters with different thresholds.

Piaui plants cultivated under 1.0 mM NH_4^+-N showed higher amination activity by glutamate dehydrogenase (GDH) in shoots (Table 3). Nevertheless, according to Labboun et al. (2009), studying tobacco in a real time analysis of NH_4^+ metabolism, GDH activity does not present considerable level *in vivo*, even when GDH genes are overexpressed.

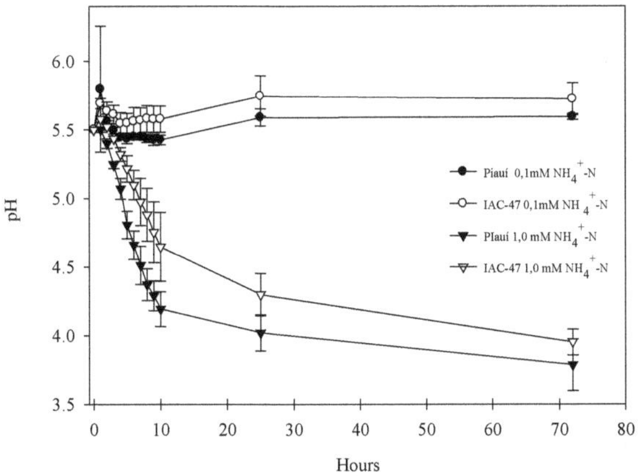

A

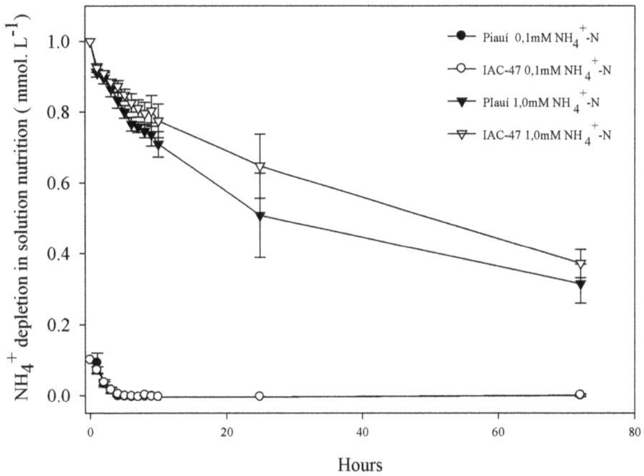

B

Figure 1. pH (A) and NH_4^+ (mmol.L^{-1}) depletion (B) in nutritive solution later than 26th DAG by Piaui and IAC-47 rice plants cultivated according to Material and Methods. Average of 5 pots with 5 plants each.

Table 3. Glutamate dehidrogenase amination (nmol NADH min^{-1} g^{-1}FW) and deamination (nmol NAD^+cons. min^{-1} g^{-1}FW); glutamate synthase (nmol NADH min^{-1} g^{-1}FW), and glutamine synthetase (nmol γ-glutamil min^{-1} g^{-1}FW) in shoots and roots of Piaui and IAC-47 rice plants cultivated according to Material and Methods. Average of 5 repetitions in purified material of 3 blocks with 3 pots and 5 plants each.

Rice Varieties	**NH_4^+-N levels (mM)**	**ENZYMATIC ACTIVITY**			
		Glutamate Dehidrogenase		**Glutamate Synthase**	**Glutamine Synthetase**
		Amination (nmol NADH min^{-1} g^{-1}FW)	Deamination (nmol NAD^+cons. min^{-1} g^{-1}FW)	(nmol NADH min^{-1} g^{-1}FW)	(nmol γ-glutamil min^{-1} g^{-1}FW)
		Shoots			
Piaui	0.1	342,33 ± 153.43	1.66 ± 0.27	0.00 ± 0.00	36.90 ± 4.38
	1.0	462,78 ± 185.02	3.05 ± 0.39	7.14 ± 4.76	57.81 ± 13.61
IAC-47	0.1	416,07 ± 165.03	2.76 ± 0.25	0.00 ± 0.00	93.45 ± 23.09
	1.0	136,88 ± 113.70	2.11 ± 0.04	29.13 ± 12.52	136.16 ± 30.40
		Roots			
Piaui	0.1	459.11 ± 206.16	19.61 ± 2.88	20.57 ± 4.67	29.40 ± 1.70
	1.0	499.24 ± 204.23	19.79 ± 1.71	25.37 ± 16.36	43.94 ± 14.87
IAC-47	0.1	537.85 ± 219.70	20.73 ± 2.73	15.94 ± 6.48	69.80 ± 29.19
	1.0	403.34 ± 161.90	15.01 ± 0.99	24.31 ± 17.96	129.42 ± 3.36

Thus, Piaui plants that also presented low GS/GOGAT activity in shoots (table 3), would compensate the assimilation with a NH_4^+ level maintenance in cytosol by a low PM H^+-ATPases and high V H^+-ATPases (table 2). This condition allows cytosolic ammonium concentration to supply the slow and gradual absorption, as an adaptative performance for low N conditions.

Despite this, high GDH amination *in vitro* was not found in roots of Piaui plants and under the lowest nutrition levels (Table 3). This reduction would explain the lowest soluble sugar concentration that resulted in similar NH_4^+ levels in Piaui roots cultivated in these conditions (Garrido et al., 2008).

GDH deamination was higher in roots of both varieties. In shoots, this activity was highest in Piaui under 1.0 mM N-NH_4^+ (Table 3). It seemed that GDH activity would be related essentially to the N recycling at high ammonium and amino-N (Lancien et al., 2000, Dubois et al., 2003), as observed for Piaui shoots under the same treatment conditions (Garrido et al., 2008). IAC-47 Plants showed higher GDH deamination in roots in the 0.1 mM NH_4^+-N treatment (Table 3). Its improvement conditions under high N concentration would produce a plant with an intense catabolism under low N conditions (Ferraz Jr et al., 2001)

Aminative activity of GDH was always higher than GS (GS1 + GS2) (Table 3). It is known that higher GS activity in roots and in the top part of rice plants during senescence is due to GSr and GS1 isoforms, while in green tissues, GS2 is the principal isoform. In shoots, the minor GDH amination (IAC-47 1.0 mM NH_4^+-N) was followed by high GS activity. In general, GS and GDH amination showed an inverse profile, principally when comparing developing and senescent parts (Souza et al., 2002, Stitt et al., 2002).

GS activity varied with N concentration in both treatments in shoots and roots (Table 3). It was not a surprise, because this enzyme is in ammonium primary assimilation pathway in higher plants (Souza et al., 2002, Souza, Fernandes, 2006; 2006a). However, in regular N assimilation, GS carries out the amination pathway with GOGAT, forming glutamine. Thus, when GOGAT NADH dependent were analyzed in shoots of IAC-47 plants under 1.0 mM NH_4^+-N treatment, it has showed highest levels of this enzyme activity than Piaui (Table 3).

The result was not repeated in roots (Table 3). However, in both Piaui and IAC-47 roots treatments, GOGAT and GDH activities were inverse to each other. This could point out to an enzyme competition for substrate.

The major ammonium assimilation pathway in IAC-47 shoots was GS/GOGAT, this was also observed for NO_3^- nutrition (Santos et al., 2009). In contrast with it, Piaui plants carried on slowly, but continuously, assimilation under 1.0 mM N-NH_4^+. Distinct levels of GS/GOGAT pathway between varieties can be explained by ammonium and amino-N accumulated in shoots (Garrido et al., 2008, Dubois et al., 2003) and high N-amino levels that inhibits GS in Piaui plants (Garrido et al., 2008, Souza, Fernandes, 2006a, Labboun et al., 2009). In roots, assimilatory ammonium, difference between varieties did not exist

CONCLUSION

Biochemical different mechanisms for ammonium absorption and assimilation were carried out by Piaui and IAC-47 under humid tropic conditions. Piaui plants were able to maintain nutritive solution acidification and ammonium depletion pattern similar to IAC-47.

Piaui plants presented higher V H^+-ATPases activity that supplied ammonium concentration in cytosol for slow and continuous assimilation by GS/GOGAT. However, GDH deamination in shoots, 1.0 mM N-NH_4^+, was also maintained high, acting as N recycle pathway when there was high ammonium and amino-N levels.

Improved variety carried out high GS/GOGAT assimilation, a standard pathway, and it have absorbed ammonium thanks to higher PM H^+-ATPases activity.

Acknowledgement

We acknowledge to CNPq, CAPES and FAPERJ by grants.

REFERENCES

AGRITEMPO. 2010. Sistema de Monitoramento Agrometeorológico. Available at www.agritempo.gov.br. Accessed in 12 Aug 2010.

Bergersen, F. J. 1980. Enzymes involved in metabolism related to nitrogenase. In: Bergersen, F.J (ed) Methods for evaluating biological nitrogen fixation. Chichester [Eng.] ; New York : J. Wiley and Sons, p.279-305.

Bradford, M.M. 1976. Rapid and sensitive method for quantitation of microgram quantities of protein utilizing the principle of protein dye binding. Analytical Biochemistry. 72:248-254.

Britto, D.T., Glass, A.D.M., Kronzucker, H.J., Siddiqi, M.Y. 2001. Cytosolic Concentration and Transmembrane Fluxes of NH_4^+/NH_3. An Evaluation of Recent Proposals. Plant Physiology. 125:523-526.

Dubois, F., Tercé-Laforgue, T., Gonzalez-Moro, M-B, Estavillo, J-M., Sangwan, R.; Gallais, A., Hirel, B. 2003. Glutamate dehydrogenase in plants: is there a new story for an old enzyme? Plant. Physiology and Biochemistry. 41:p.565-576.

Ferraz JR., A.S.L., Souza, S.R., Fernandes, M.S., Rossiello, R.O.P. 1997. Eficiência no Uso de Nitrogênio para a Produção de Grão por Cultivares de Arroz. Pesquisa Agropecuária Brasileira. 32(4):435-442.

Ferraz JR., A.S.L., Souza, S.R., Stark, E.M.L.M., Fernandes, M.S. 2001. Crude Protein in Rice Grown in Different Environmental Conditions. Physiology and Molecular Biology of Plants, 7(2):149-157.

Fiske, C.F., Subarrow, Y. 1925. The colorimetric determination of phosphorus. Journal Of Biological Chemistry. 66:375-400.

Garrido, F.S.R.G., Garrido, R.G., Bucher, C.A., Souza, S.R., Fernandes, M.S. 2008. Rice Varieties Tonoplast and Plasma Membrane H^+-ATPases Differential Activities in Response to Nitrate Pulses. Journal of Biological Sciences. 8(1): 107-112.

Garrido, R.G., Garrido, F.S.R.G., Souza, S.R., Fernandes, M.S. 2008. Physiological and Morphological Adaptations in Two Rice Varieties Cultivated Under Ammonium and Light Deficiency. Journal of Biological Sciences. 8(1):113-118.

Hoagland, D.R., Arnon, D.I. 1950. The water-culture method for growing plants without soil. California Agricultural Experiment Station Bulletin347:1-32.

Holzschuh, M.J., Bohnen, H., Anghinoni, I., Meurer, E.J., Carmona, F.C., Costa, S.E.V.G.A. 2009. Resposta do Arroz Irrigado ao Suprimento de Amônio e Nitrato. Revista Brasileira. de Ciência doSolo. 33:1323-1331.

Howitt, S. M., Udvardi K.M. 2000. Structue, function and regulation of ammonium transporters in plants. Biochimica et Biophysica Acta. 1465:152-170.

IRRI. 2012. Rice basics. Available at www.irri.org. Accessed in 20 Jan 2012.

Kronzucker, H.J., Glass, A.D.M., Siddiqi, M.Y., Kirk, G.J.D. 2000. Comparative kinetic analysis om ammonium and nitrate acquisition by tropical lowland rice: implications for rice cultivation and yield potential. New Phytology. 145:471-476.

Kumar, A., Silim, S.N., Okamoto, M., Siddiqi, M.Y., Glass, A.D.M. 2003. Differential expression of three members of the *AMT1* gene family encoding putative high-affinity NH_4^+ transporters in roots of *Oryza sativa* subspecies *indica.* Plant Cell and Environment. 26:907-914.

Lancien, M., Gadal, P., Hodges, M. 2000. Enzyme Redundancy and the Importance of 2-Oxoglutarate in Higher Plant Ammonium Assimilation. Plant Physiology. 123:817-824.

Lobboun, S., Tercé-Laforgue, T., Roscher, A. et al. 2009. Resolving the Role of Plant Glutamate Dehydrogenase. I. in vivo Real Time Nuclear Magnetic Resonance Spectroscopy Experiments. Plant Cell Physiology. 50(10):1761-1773.

Mitchell, H.T. 1972. Microdetermination of nitrogen in plant tissue. Journal of Association Official Agriculture. 55:1-3.

Palmgren, M.G. 2001. Plant Plasma Membrane H^+ATPases: Powerhouses for Nutrition Uptake. Annual Review of Plant Physiology and Plant Molecular Biology. 52:817-845.

Pereira, D.P., Bandeira, D.L., Quincozes, E. da R.F. (Ed.) 2005. Cultivo do arroz irrigado no Brasil. Available at http:// sistemasdeproducao.cnptia.Embrapa.br/ FontesHTML/Arroz/ArrozIrrigadoBrasil Accessed in 20 Jan 2012.

Ramos, A.C., Martins, M.A.,Façanha, A.R. 2005. Atividade ATPásica e pirofosfatásica em microssomos de raízes de milho colonizadas com fungos micorrízicos arbusculares. Revista Brasileira de Ciência do Solo. 29:207-213.

Rodrigues, F.S.,Garrido, R.G. 2005. Fluxo Sazonal de NO_3^- no Trópico Úmido. Revista Científica Eletrônica de Agronomia. IV(8):01-09.

Santos, L.A.; Bucher, C.A.; Souza, S.R.,Fernandes, M.S. 2009. Effects of Nitrogen Stress on Proton-Pumping and Nitrogen Metabolism in Rice. Journal of Plant Nutrition. 32:549-562.

Santos, M.A., Bucher, C.A., Stark, E.M.L.M., Fernandes, M.S., Souza, S.R. 2009. Effects of Nitrate Availability in Nutrient Solution on Nitrogen Uptake and Enzymatic Activity of Two Rice Cultivars. Bragantia. 68(1):215-220.

Souza, S.R., Fernandes, M.S. 2006. Nitrogen-Acquisition by Plants in a Sustainable Environmet. In: Singh, R.P., Jaiwal, P.K. Biotechnological Approaches to Improve Nitrogen Use Efficiency in Plants. Studium Press, LLC, USA., pp. 41-62.

Souza, S.R., Fernandes, M.S. 2006a. Nitrogênio In: Fernandes, M.S. (Ed.). Nutrição Mineral de Plantas.Sociedade Brasileira de Ciência do Solo,Brasil.pp. 215-252.

Souza, S.R., Stark, E.M.L.M., Fernandes, M.S. 2002. Enzimas de Assimilação de Nitrogênio em Plants. ed. Artware Projetos Culturais. 55p.

Stitt, M., Müller, C., Matt, P., Gibon, Y., Carillo, P. Morcuende, R., Scheible, W-R., Krapp, A. 2002. Steps towards an integrated view of nitrogen metabolism. J. Exp. Bot. (2002) 53 (370): 959-970.

Wang, M.Y., Glass, A.D.M., Shaff, J.E., Kochian, L.V. 1994. Ammonium Uptake by Rice Roots. III. Electrophysiology. Plant Physiology, 104:899-906.

Yan, F., Zhu, Y. Muller, C.,Zorb, C., Schubert, S.. 2002. Adaptation of H-pumping and plasma membrane h+atpase activity in proteoid roots of white lupin under phosphate deficiency. Plant Physiology. 129:50–63.

Yoshida, S., Uemura, M., Niki, T., Sakai, A., Gusta, L.V. 1983. Partition of Membrane Particles in Aqueous Two-Polymer Phase System and Its Practical Use for Purification of Plasma Membranes from Plants. Plant Physiology. 72:105-114.

4

Phenols and Flavonoids Concentration and Fungistatic Activity of Wood and Bark of Five Common Tropical Species

Artemio Carrillo-Parra[1*], Martha Rosales[2], Christian Wehenkel[3], Rahim Foroughbakhch[4], Humberto González[1] and Fortunato Garza[1]

Universidad Autónoma de Nuevo León, Facultad de Ciencias Forestales
Email: artemio.carrillopr@uanl.edu.mx
[2]Centro Interdisciplinario de Investigación para el Desarrollo Integral Regional
Unidad Durango. Instituto Politécnico Nacional
[3] Instituto de Silvicultura e Industria de la Madera.
Universidad Juárez del Estado de Durango
[4] Universidad Autónoma de Nuevo León, Facultad de Ciencias Biológicas
**Corresponding author*

SUMMARY

This research determine the total phenol and flavonoids content as well as the fungistatic activity of hot water wood sawdust and bark extracts on *Coniophora puteana* and *Trametes versicolor*. Extracts tested were taken from *Condalia hookeri* M.C. Johnst., *Ebenopsis ebano* (Berl.) Britton et Rose, *Helietta parvifolia* (*Gray*) Benth, *Leucaena leucocephala* (Lam.) de Wit, and *Prosopis laevigata* (Humb. et Bonpl. ex Willd). The extraction was developed with soxhlet aparats, phenol concentration was determined with Folin-Ciocalteu method and flavonoide was determined by Heimler procedure. Phenol concentration ranged between 50±11 to 827±23 mg g^{-1} and flavonoids content between 15±2 to 708±30 mg g^{-1}. All extracts tested inhibited growth of the fungal species. The highest inhibition effect (88%±1) occurred on *C. puteana* with *L. leucocephala* wood sawdust extracts at 2 mg ml^{-1}, this species also reduced the growth on *T. versicolor* by 75%±12 when used at 10 mg ml^{-1}, wood sawdust extracts of *H. parvifolia* and *C. hookerii* at same concentration reduced the growth of *T. versicolor* in 43%±3 and 40%±4 respectively. Inhibition of bark extracts of *E. ebano* was 84±5 and 80±7 % for *H. parvifolia*. A negative relationship between growing inhibitory activity and the content of total phenolics in the extracts was obtained.

Key words: Fungistatic effect; growing inhibition; extracts; native species.

INTRODUCTION

Thornscrub Shrubland vegetation grows naturally under harsh environmental conditions. Temperature ranges from 20 to 45°C and annual precipitation is less than 500 mm (Návar *et al*., 1999). The vegetation growing in this area is known as common tropical species and covers approximately 125,000 – 200,000 Km^2. It is characterized by shrubs and tree species with short stems with diameters of 0.15 m – 0.20 m (Rzedowski, 2006). Low economic input is obtained by traditional agricultural, free cattle management, and forest harvesting (Reid *et al*., 1990; Villalón and Carrillo, 2010). Some shrubs and tree species have demonstrated high natural resistance against fungi and other microorganism present in soil (Wolf and Perales, 1985; Carrillo, 2010; Carrillo *et al*., 2011). The natural durability of tree species is defined as the resistance to biological degradation and is attributed to several factors, including wood anatomic structure, certain physical properties and, above all, accumulation of extracts (Eaton, 1993; Schultz, 1995; Haupt *et al*., 2003; Eyles *et al*., 2003; Aloui *et al*., 2004). Wood extracts are present in heartwood, mostly in the parenchyma cells, but are also found in vessels, fibers, specialized cells, and substances with low molecular weight from 2 – 10 % of dry mass (Hillis, 1971; Mantanis, 1995). These substances are produced during the wood maturation process when cambium tissues form differentiated sapwood and heartwood cells. Trees growing in an environment with high UV or high temperatures as well as trees which have been injured mechanically by insects, or have suffered a microbial attack are rich in substances such as polyphenols, since these are part of defence mechanisms (Rowell, 2005; Kawamura *et al*., 2011).

Polyphenols are known to have strong antioxidant and antibacterial effects which act as radical scavengers, biocides, and metal chelators (Pietarinen *et al*., 2006). According to Schultz and Nicholas (2002) the combination of organic biocide with metal chelating and/or antioxidant additives improves protection against fungi. This enhances the development of environmentally benign wood preservative substances. Determining the total amount of extracts and flavonoids in trees and shrubs in there Thornscrub Shrubland areas can help to increase their economic importance.

Information with regard to the natural durability of wood, wood quality, chemical wood composition and the potential of wood extracts of Thornscrub Shrubland vegetation to inhibit the growth of decay fungi is lacking (Wolf and Perales, 1985; Carrillo, 2010). The aims of this study are: To determine the total phenol and flavonoids content of hot water extracted solutions from wood sawdust and bark of five tree species known to have high natural durability; and to establish how extracts inhibit the growth of *Coniophora puteana* and *Trametes versicolor* fungi under laboratory conditions.

MATERIAL AND METHODS

Species selection, extraction and determination of phenol and flavonoids content

Five tree species native to Northeast Mexico were selected according to importance, abundance, distribution and use. There were randomly selected three specimens from *Condalia hookeri* M.C. Johnst., *Ebenopsis ebano* (Berl.) Barneby & Grimes, *Helietta parvifolia* (Gray) Benth, *Leucaena leucocephala* (Lam.) DeWit and *Prosopis laevigata* (Humb. & Bonpl. ex. Willd.) M.C. Johnst. Bark and heartwood chips from the outer heartwood of them fresh cut tree were obtained at 1.3 m from the tree base, they were then air conditioned for two months and milled to a 40-mesh sieve size.

Soxhlet flasks were used for extraction procedures (Demirbaş, 1991; Schwanninger, 2002). Three portions of 100 g with 10% moisture content of saw wood and bark were separately grounded and sieved at 500 μm and then put in Soxhlet flasks. Each flask was filled with 200 ml of distillate water. The water solution and sawdust were boiled for four hours at a minimum boiling rate of 24 cycles. Extracts were then isolated from the solvent in a vacuum rotary evaporator (Rotavapor Type R-114 Büchi) at 40°C.

The total phenols were measured using the Folin-Ciocalteu index method according to Waterman (1994). Hot water extracts (10 mg) were diluted in 10 ml of acetone/water (1:1 v/v). Six ml of water and 0.5 ml of Folin-Ciocalteu reagent (Sigma-Aldrich) were successively added to 0.1 ml of this solution. In addition, 1.5 ml of a 20% sodium carbonate solution and water were added to obtain 10 ml. A reaction took place within 120 min. at room temperature. Absorbance was measured at 760 nm. Calibration was done using a gallic acid solution (0.5, 0.4, 0.3, 0.2 and 0.1 mg ml^{-1} in acetone/water 1:1 v/v, $y= 0.0011x + 0.016$, $R^2 = 0.9984$). The results were expressed as the equivalent to milligrams of gallic acid per gram of extract (mg GAE g^{-1}).

The total flavonoid content was determined according to Heimler *et al*. (2005). Ten mg of extracts were diluted in 10 ml of acetone/water (1:1 v/v). A solution of 0.25 ml of the suitably diluted sample was added to 0.75 μl of a $NaNO_2$ (5% w/v) solution, as well as 0.150 ml of a freshly prepared $AlCl_3$ (10% w/v) solution, and 0.5 ml of 1 M NaOH solution. The final volume was adjusted to 2.5 ml with deionized water.

The mixture was allowed to stand for 5 min and the absorption measured at 510 nm against the same mixture without the sample. Calibration was done using (+)-catechin as standard, for which a calibration curve was obtained with solutions of 0.25, 0.2, 0.15, 0.1 and 0.05 mg mL^{-1} ($y = 0.0032x + 0.0127$, $R^2 = 0.998$). The results were expressed as the equivalent to milligrams of catechin per gram of extract (mg CEQ g^{-1}).

Growing inhibition effect

The growing inhibition effect of extracts dissolved in a water/acetone solution (1:1) at 2 mg ml^{-1} and 10 mg ml^{-1} concentrations on brown rot fungus *C. puteana* and white rot fungus *T. versicolor* was determined. This solution was stirred with a mixer (Heidolph Type Mr 3003 Control G.) for 10 min. Subsequently, the extracts previously diluted with the water/acetone solution were transferred to 140 mm Petri dishes providing 2 mg ml^{-1} and 10 mg ml^{-1} concentrations. Fifty ml of malt-agar medium at 60°C were then added to the Petri dish. Plates were inoculated by placing a 10 mm diameter mycelium plug in the centre of each plate. The plugs were obtained from the edge of a growing colony of each fungus. Three replicates for each extractive were used; five from the acetone/water control (only acetone and water in the medium) and five with the medium as a control treatment. Cultures were kept in an incubator at 25°C until the mycelium of the control reached the edges of the plates. The inhibition effect was determined at the end of that period by calculating the average of diametric growth in the two perpendicular directions. The data were used in the following formula (Gérardin, 2004; Neya *et al.*, 2004).

$$G_i = 100 \cdot (1 - \frac{d_i}{d_o})$$

Where:
Gi = Growing inhibition effect (%)
d_i= Average diameter of the culture in the presence of extracts (mm)
d_0= Average diameter of the control culture (mm)

Statistical analysis

The data obtained were statistically analyzed using the SAS program. Analysis of variance was carried out and statistically significant differences were set at a 95% confidence level. Split plot desing ANOVA was applied to determine whether the growing inhibition was significantly different between species and between extractive concentrations. Additionally, interactions of growing inhibition effect for each fungus were examined with respect to tree species extracts kind of extractive and concentration.

RESULTS AND DISCUSSION

Total phenols

A higher phenol concentration was found for *E. ebano*, *P. laevigata*, and *C. hookeri* wood sawdust with values of 827±23, 818±8 and 400±15 mg g^{-1} respectively (Figure 1). Concentration for *C. hookeri*, *H. parvifolia* and *L. leucocephala* in bark was 232±29, 165±0.2 and 115±9 mg g^{-1}, respectively. No data describing phenol concentration for these species is available; however, tannin concentration on hardwood was presented by Foroughbakhch *et al.* (2008) with values ranging from 56 mg g^{-1} for *P. glandulosa* to 172.5 mg g^{-1} in *C. hookeri*. Scalbert *et al.* (1989) reported lower total phenol values from 17 trees species with lower values than reported in this research: *Quercus rubur* (62.6 mg m^{-1}), *Castanea sativa* (53.4 mg m^{-1}) and *Q. petrea* (39.3 mg m^{-1}). Hărmănescu *et al.* (2008) found concentrations from 52 to 400 mg g^{-1} in medicinal plants from Romania. Sáenz-Esqueda *et al.* (2010) found phenols concentration ranging from 48.9 to 399.2 mg g^{-1} in catequin equivalents from needle extracts of *Pinus cooperi* and *Pinus teocote,* respectively. Rosales-Castro *et al.* (2009) reported values of phenols from 491 to 604 mg g^{-1} and flavonoid from 292 to 385 mg g^{-1} in *Pinus cooperi, Pinus engelmannii, Pinus leiophylla and Pinus teocote*.

Total flavonoids

The content of total flavonoids varied between wood sawdust and bark, with the wood sawdust showing higher values of 708±30, 572±17, and 41±4 mg g^{-1} for *P. laevigata*, *E. ebano* and *C. hookeri,* respectively. Bark extracts showed lower amounts 75±5, 47±4 and 32±6 mg g^{-1} for *H. parvifolia*, *C.* hookeri and *P. laevigata*, respectively (Figure 2).

Growing inhibition

ANOVA results (Table 1) show that the growing inhibition effect on *C. puteana* fungus was highly significant among tree species ($p<0.0001$, $F=26.5$) and significant among extracts ($p< 0.0157$, $F=6.28$). The growing inhibition effect is shown in Table 2. A higher inhibition for *C. puteana* growth was shown by *L. leucocephala* wood sawdust and bark extracts at both concentrations: levels from 68±0.6 to 88±0.1% for bark (2 mg ml^{-1}) and wood sawdust (2 mg ml^{-1}). Biocide activity for *L. leucocephala* was also demonstrated on seedlings (Aderibigbe *et al.*, 2011). The activity was attributed to the high concentration of chitinases. These enzymes break down glycosidic bonds in chitin which composes the cell walls of fungi and exoskeletal elements of some worms and arthropods (Kaomek *et al.*, 2003). Chitosan was also used on wood modification under experimental conditions (Larnøy *et al.*, 2006). *E. ebano* extracts from bark at 10 mg ml^{-1} also reduced the growth of *C. puetana* (84±5%).

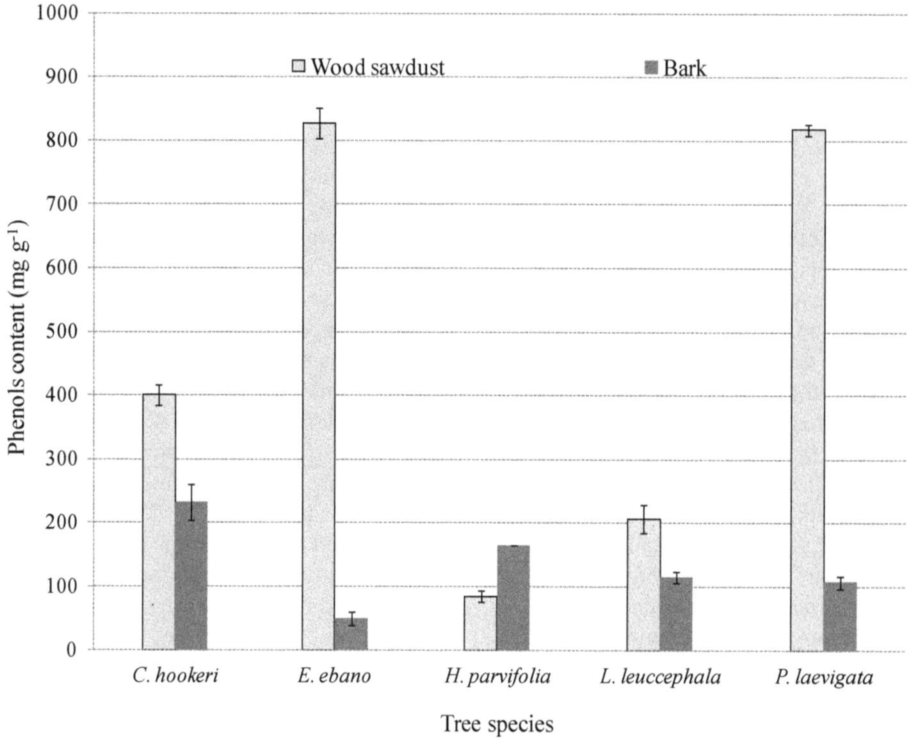

Figure 1. Total phenol concentration (mg g^{-1}) of sawdust wood and bark form five tree species.

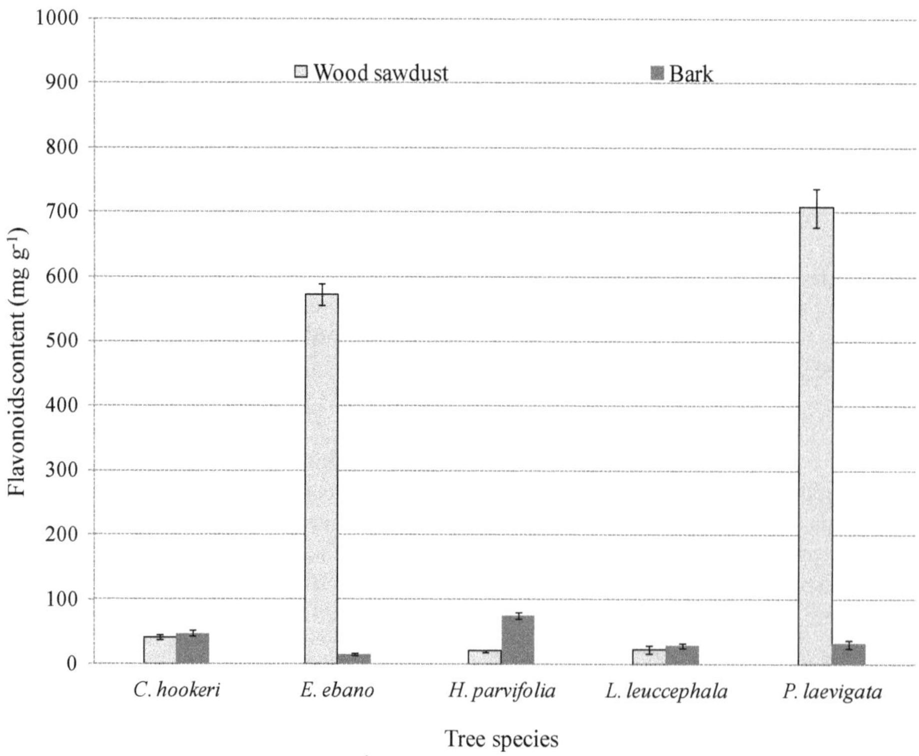

Figure 2. Flavonoids content (mg g^{-1}) of sawdust wood and bark form five tree species.

Table 1. Analysis of variance for growing inhibition of *C. puteana* fungi produced by wood and bark extracts from five tree species.

Source of variation	df	Sum of Squares	Mean Square	*F* value	*p* value
Wood species (WS)	4	33588.76	8397.19	26.50***	0.0001
Extracts (E)	1	1989.35	1989.35	6.28**	0.0157
WS*E	4	10246.63	2561.66	8.08***	0.0001
Concentration (C)	1	379.53	379.53	1.20NS	0.2792
E*C	1	12.57	12.57	0.04NS	0.843
Error	48	15209.96	316.87		

Table 2. Growing inhibition effect (%) on *C. puteana and T. versicolor* fungi by wood sawdust and bark extracts.

Tree species	*C. puteana*				*T. versicolor*			
	Wood sawdust mg ml^{-1}		Bark mg ml^{-1}		Wood sawdust mg ml^{-1}		Bark mg ml^{-1}	
	2	10	2	10	2	10	2	10
C. hookeri	55±1.3	9±5.1	3±0.3	1±0.2	3±0.0	40±4.4	4±2.0	2±2.1
E. ebano	16±0.4	45±13.8	56±2.4	84±5.1	2±2.3	15±5.7	2±0.4	30±1.5
H. parvifolia	5±1.4	47±5.8	44±3.7	80±6.5	22±1.7	43±12.8	21±0.8	49±3.3
L. leucocephala	88±0.1	80±5.9	68±0.6	87±0.2	37±1.6	76±11.8	36±1.0	45±0.5
P. laevigata	7±0.8	10±1.7	52±10.9	1±0.6	3±1.0	16±0.4	3±1.0	2±0.8

(±) standard deviation

Table 3 shows that the growing inhibition effect on *T. versicolor* was highly significant between species and concentration (p<0.0001, F=59.24), and significant between extracts (p< 0.0027, F=10.03). *T. versicolor* growth was reduced 76±12% by *L. leucocephala* wood sawdust extracts at 10 mg ml^{-1} concentration; it was followed by bark extracts at 10 mg ml^{-1} of *Helietta parvifolia* (49±3%). The wood of *H. parvifolia* has been classified with a high natural durability by Wolf and Perales (1985). In this research *P. laevigata* hot water extracts showed a low level of inhibition compared to the 82 and 90% values for *C. puteana* and *T. versicolor* presented by Carrillo *et al.* (2011). Juglone bark extract from *Juglans mandshurica* at concentrations ≥ 0.5 mg ml^{-1} displayed similar inhibition for *Aspergillus niger*, *Paecilomyces variotii*, *Trametes versicolor,* and *Gloeophyllum trabeum* (Dongmei *et al.*, 2009). A similar inhibition effect of up to 50% was found for water extracts and acetone extracts from *P. africana* heartwood diluted to concentrations of 1 mg ml^{-1}. The use of diethylether extracts as an inhibition agent affected the growth up to a level of 80% at 1 mg ml^{-1} (Gérardin, 2004.). Results from this research are somewhat similar to those studied by Huang *et al.* (2009). Here too no relationship was found between the amount of phenols and flavonoids and a growing inhibition effect on fungi.

CONCLUSION

Sawdust of *Prosopis laevigata* and *Ebenopsis ebano* has high concentrations of phenols and flavonoids. The fungistatic effect produced by sawdust and bark from species investigated is not related to high phenol and flavonoids yields. Wood sawdust and bark extracts of species such as *Helietta parvifolia* and *Condalia hookeri*, whose concentration of phenol and flavonoids were relatively low, showed high growth inhibition rates. Wood extracts from high natural durability tree species have a potential for use in wood protection. Still, further research is needed with regard to the identification of compounds present in wood sawdust and bark extracts and to their use in impregnating wood from a variety of tree species.

Table 3. Analysis of variance for growing inhibition on *T. versicolor* fungi produced by wood and bark extracts from five tree species.

Source of variation	Df	Sum of Squares	Mean Square	*F* value	*p* value
Wood species (WS)	4	15298.78	3824.69	59.24***	0.0001
Extracts (E)	1	647.33	647.33	10.03**	0.0027
WS*E	4	1544.52	386.13	5.98***	0.0005
Concentration (C)	1	4988.24	4988.24	77.26***	0.0001
E*C	1	580.88	580.88	9.00**	0.0043
Error	48	3099.24			

Acknowledgements

This study was supported by The Improvement of Professors Program (PROMEP) and The Scientific and Technological Research Support Program (PAICYT).

REFERENCES

Aderibigbe, A., Adetunji, O.A., Ayodele, O.M. 2011. Antimicrobial and pharmaceutical properties of the seed oil of *leucaena leucocephala* (LAM.) De Wit (Leguminosae). African Journal of Biomedical Research, 14, 63-68.

Aloui, F., Ayadi, N., Charrier, F., Charrier, B. 2004. Durability of European oak (*Quercus petraea* and *Quercus robur*) against white rot fungi (*Coriolus versicolor*): relations with phenol extractives. European Journal of Wood and Wood Products. 62(4):286-290.

Carrillo, A., Hapla, F., Mai, C., Garza-Ocañas, F. 2011. Durabilidad de la madera de *Prosopis laevigata* y efecto de sus extractos en hongos que degradan la madera. Madera y Bosques. 17(1) 7-21.

Carrillo, P.A., Garza, F., Bustamante, V., Sandoval, S.,Villalón, H. 2010. Natural durability of fence poles from fourteen semi-arid land species after six month exposition. In Internationar Research Group on Wood Protection, Biarritz, France.

Demirbaş, A. 1991. Analysis of beech wood fatty acids by supercritical acetone extraction. Wood Science and Technology. 25(5):365-370.

Dongmei, Y., Shuangyue, L., Shujun, L., Jian, L., Molong, S. Yan, J. 2009. Effect of juglone from Juglans mandshurica bark on the activity of wood decay fungi. Forest products journal. 59(9):79-83.

Eaton, R.A., Hale, M.D.C. 1993. Wood: decay, pests and protection. Chapman & Hall. London, New York. 546 pp.

Eyles, A., Davies, N.W., Mohammed, C. 2003. Novel detection of formylated phloroglucinol compounds (FPCs) in the wound wood of *Eucalyptus globulus* and *E. nitens*. Journal of Chemical Ecology. 29(4):881-898.

Foroughbakhch, R.P., Valdes, C.M.C., Alvarado, V.M.A., Cárdenas, A.M.L., Nuñez, G.M.A., Hernández, P.J.L. 2008. Contenido de taninos condensados en la madera de 11 especies del Matorral Tamaulipeco del Estado de Nuevo León International. Journal of Agriculture Environment and Biotechnology. 1(3):166-168.

Gérardin, P. 2004. Contribution of gums to natural durability of *Prosopis africana* heartwood. Holzforschung. 58:39-44.

Hărmănescu, M., Moisuc, A., Florina, R., Drăgan, S., Gregen, I. 2008. Total polyphenols content determination in complex matrix of medicinal plants from Romania by NIR spectroscopy. Bulletin UASVM, Agriculture. 65(1):123-128.

Haupt, M., Leithoff, H., Meier, D., Puls, J., Richter, H.G., Faix, O. 2003. Heartwood extractives and natural durability of plantation-grown teakwood (*Tectona grandis* L.)—a case study. European Journal of Wood and Wood Products. 61(6):473-474.

Heimler, D., Vignolini, P., Dini, M.G., Romani, A. 2005. Rapid tests to assess the antioxidant activity of Phaseolus vulgaris L. dry beans. Journal of Agriculture, Food and Chemestry. 53:3053–6.

Hillis, W. 1971. Distribution, properties and formation of some wood extractives. Wood Science and Technology. 5:272-289.

Huang, Z., Hashida, K., Makino, R., Kawamura, F., Shimizu, K., Kondo, R., Ohara, S. 2009. Evaluation of biological activities of extracts from 22 African tropical wood species. Journal of Wood Science. 55(3):225-229.

Kaomek, M., Mizuno, K., Fujimura, T., Sriyotha, P., Cairns, J.R. 2003. Cloning, expression, and characterization of an antifungal chitinase from *Leucaena leucocephala* de Wit. Bioscience biotechnology and Biochemistry. 67(4):667-676.

Kawamura, F., Ramle, S., Sulaiman, O., Hashim, R., Ohara, S. 2011. Antioxidant and antifungal activities of extracts from 15 selected hardwood species of Malaysian timber. European Journal of Wood and Wood Products. 69: 207–212

Larnøy, E., Dantz, S., Eikenes, M., Militz, H. 2006. Screening of properties of modified chitosan-treated wood. Wood Material Science and Engineering. 1(6):59-68.

Mantanis, G. 1995. Swelling of wood. Part III. Effect on temperature and extractives on rate and maximum swelling. Holzforschung. 49:239-241.

Návar, J., Charles, F., Jurado, E. 1999. Spatial variations of interception loss components by Tamaulipan thornscrub in northeastern Mexico. Forest Ecology and Management. 124(2-3):231-239.

Neya, B., Hakkou, M., Pétrissans, M., Gérardin, P. 2004. On the durability of *Burkea africana* heartwood: evidence of biocidal and hydrophobic properties responsible for durability. Annals of Forest Science 61(3):277-282.

Pietarinen, S., Willför, S., Ahotupa, M., Hemming, J., Holmbom, B. 2006. Knotwood and bark extracts: strong antioxidants from waste materials. Journal of Wood Science. 2(5):436-444.

Reid, N., Marroquín, J., Bayen-Munzen, P. 1990. Utilization of shrubs and trees for browse fuelwood and timber in the tamaulipan thurnscrub, northeastern Mexico. Forest Ecology and Management. 36:61-69.

Rosales-Castro, M., González-Laredo, R.F., Rocha-Guzmán, N.E., Gallegos-Infante, J.A., Peralta-Cruz, J., Karchesy, J.J. 2009. Evaluación química y capacidad antioxidante de extractos polifenólicos de cortezas de *Pinus cooperi*, *P. engelmannii*, *P. leiophylla* y *P. teocote*. Maderas y Bosques. 15(3):87-105.

Rowell, R. 2005. Handbook of wood chemistry and wood composites. Boca Raton, Fla. USA. 446 pp.

Rzedowski, J. 2006. Vegetación de México. Comisión Nacional para el Conocimiento y Uso de la Biodiversidad. http://www.biodiversidad.gob.mx/publicaciones/librosDig/pdf/VegetacionMx_Cont.pdf. Consulted: 18/05/2012.

Sáenz-Esqucda, M.Á., Rosales-Castro, M., Rocha-Guzmán, N.E., Gallegos-Infante, J.A., González-Laredo, R.F. 2010. Contenido fenólico y acción antioxidante de extractos de acículas de *Pinus cooperi*, *P. durangensis*, *P. engelmanii* y *P. teocote*. Madera y Bosques. 16(3):37-48.

Scalbert, A., Monties, B., Janin, G. 1989. Tannins in wood: comparison of different estimation methods. Journal of Agricultural and Food Chemistry. 37(5):1324-1329.

Schultz, T. 1995. Durability of Angiosperm heartwood: The importance of extractives. Holzforschung. 49:29-34.

Schultz, T.P., Nicholas, D.D. 2002. Development of environmentally-benign wood preservatives based on the combination of organic biocides with antioxidants and metal chelators. Phytochemistry. 61(5):555.

Schwanninger, M. 2002. Comparison of the classical wood extraction method using a soxhlet apparatus with an advanced extraction method. Holz als Roh- und Werkstoff. 60:343-346.

Villalón, H., Carrillo, A. 2010. Plantas productoras de leña y carbón. *In* De la lechuguilla a las biopelículas vegetales: las plantas útiles de Nuevo León, Alvarado-Vázquez, M.A., A. Rocha-Estrada, and S. Moreno-Limón (eds.), Monterrey, Nuevo León, México.

Waterman, P.G. 1994. Analysis of phenolic plant metabolites / Peter G. Waterman, Simon Mole. Blackwell Scientific Publications, Oxford.

Wolf, F., Perales, F. 1985. Durabilidad natural de la madera de algunas especies del matorral del noreste de México. Reporte científico No. 3. UANL.

5

Effects of Seven Different Mycorrhizal Inoculum *in Persea Americana* in Sterile and Non-Sterile Soil

Jacob Banuelos[1*], Dora Trejo[1], Liliana Lara[1], Mayra Gavito[2] and Yazmín Carreón[3]

[1]*Universidad Veracruzana, Facultad de Ciencias Agrícolas, Laboratorio de Organismos Benéficos. Circuito Gonzalo Aguirre Beltrán s.n., Zona Universitaria Xalapa, Veracruz. CP 91000, México.*
llacoob@gmail.com, (228)-8421700 ext. 11621
[2]*CIEco, UNAM, Antigua Carretera a Pátzcuaro No 8701, Col. Exhacienda de San José de La Huerta C.P. 58190, MoreliaMichoacán, México.*
[3] *Universidad Michoacana de San Nicolás Hidalgo, Av. Francisco J. Mujica S/N Ciudad Universitaria, C.P. 58030, Morelia, Michoacán, México.*
**Corresponding author*

SUMMARY

Some mycorrhizal fungal species could have certain compatibility and efficiency in the development of the host. Recently, avocado producers became interested in the use of arbuscular mycorrhizal fungi (AMF) for the production of plants in nurseries; therefore, it is important to revise the inoculants that are more efficient to this crop. The native microbiota could interfere in the establishment of introduced AMF. A 2 factorial experimental design was used. The first factor, AMF, had 8 levels, with 6 AMF species (*Rhizophagus fasciculatum, Gigaspora margarita, Claroideoglomus etunicatum, Pacispora scintillans, Rhizophagus intraradices, Acaulospora laevis*), a commercial consortia (MTZ-UV1) and a non-mycorrhizal control. The second factor, soil treatment, had 2 levels (sterile and not). Sterile soil treatments had a higher growth and the AMF inoculation increased the height, diameter, fresh and dry weight of the leaves. The inoculant *Rhizophagus fasciculatum* in sterile soil had tendency of the highest growth in most of the variables. On the other hand, *Pacispora scintillans* and *Acaulospora laevis* in not sterile soil decreased the plant growth. The results obtained showed possible plant-AMF compatibility, as well as the importance of the sterilization of the soil before AMF inoculation

Keywords: Desinfection; inoculation; Mycorhizal; microbiota; compatibility; efficiency.

INTRODUCCIÓN

Mexico is the avocado center of origin (Gutiérrez-Díez *et al*., 2009; Gutiérrez-Contreras *et al*., 2010) and more than 20 different species and three varieties (mexican, antillana, and guatemalteca) are known (Barrientos-Priego and López-López, 2000; Chen *et al.*, 2008). Some studies reports that the avocado is a mycotrophic crop that responds favorably to the mycorrhizal inoculation (Menge *et al.*, 1980; Alarcón, 1997; da Silveira *et al*., 2003; Bárcenas *et al*., 2010). This crop generally goes through a nursery phase (da Silveira *et al*., 2003), but without the porper management during its transplant, it is susceptible to pathogen attack (Graham, 2001; Morales-García, 2009). The soil sterilization and the reintroduction of arbuscular mycorrhizal fungi (AMF) could favor protection, the survival and increase the plant production.

The mechanisms by which AMF improve plant development and protection, could be related to the increase of nutrient absorption, allowing a better nutritional status and a adequate stress response (Smith *et al.*, 2010), as well as allowing a higher survival rate to transplant (Trejo *et al*., 2011). Although it has not been fully documented, the efficiency of the AMF could depend on the fungal species, host compatibility and biological; physical and chemical soil factors (Klironomos, 2000; González and Cuenca, 2008); as well as the diversity of native fungi or by the introduction of better-adapted fungi (Díaz and Honrubia, 1995; Gustafson and Casper, 2006). Smith and Giaginazzi, (1988) reported that the compatibility between fungal species and the host depends on the genotype, root exudates, root geometry and root hairs, as well as soil pH, humidity, texture, fertility and surrounding microorganisms.

Studies involving avocado plants inoculated with AM fungi have been explored (Haas and Menge, 1990; Jaizme-Vega and Azcón, 1995; da Silveira *et al.*, 2002), where they have proved that the avocado plant is a mycotrophic plant and therefore its interaction is of interest.

Some studies indicate that introduced AMF could be more effective than native ones (González and Cuenca, 2008). But other studies contrast, were the native AMF were more effective than the introduced ones (Requena *et al*., 2001). Therefore the identification of HMA is a prerequisite for efficient inoculation programs (Caravaca, *et al.*, 2005), because the compatibility level between AMF species and the host depends on the species involved (Smith and Read, 1997).

Although, specificity between fungi and host has not been reported, few researches have mentioned that some AMF could be more efficient in nutrient uptake than others, depending on the host (Klironomos, 2000; González and Cuenca, 2008; Trejo *et al.*, 2011).

Due to the high microorganisms diversity, that interact by different means (Linderman, 1992), the establishment of introduced microorganisms is affected, caused by competition among native populations (Garbaye, 1991)

Considering that Mexico is the origin center of the avocado and that AMF are associated to this plant, this paper therefore seeks to evaluate the response of avocado seedlings to various AMF inoculants in sterile soil and not.

MATERIALS AND METHODS

Experimental conditions

The experiment was developed under greenhouse conditions, at the facilities of the Laboratory of Beneficial Organisms, Faculty of Agricultural Sciences, Universidad Veracruzana. Avocado seeds (Criollo var.) were sown on black plastic bags (2 lt) 45 days after sown. Uniform plants of 15 cm height were selected for the experiment and at 75 days after being sown, the plants were inoculated with the mycorrhizal inoculants, composed mainly by soil containing spores and root fragments. The plants were irrigated to field capacity with tap water and then harvested after 165 days after AMF inoculation (DAI).

Substrate

A mixture of soil, sand and pumice (1:2:1 v/v) was autoclaved for one h for three days and aerated for six h each day. This mixture was characterized by a pH of 6.04, organic matter of 5.93%; N, 22 ppm; P, 4.4 ppm; K^+, 885 ppm; Ca^{2+}, 13,58 ppm; Mg^{2+}, 237 ppm; Fe, 6.7 ppm; Cu, 0.5 ppm; Zn, 1.2 ppm and Mn, 3.9 ppm.

AMF inoculation

Six monoxenic species were used, one comercial species, *Rhizophagus fasciculatum* (Micofert); four species provided by the Laboratory of Interaction plant-microorganism-environment of CIEco, UNAM, *Gigaspora margarita* (Gm), *Acaulospora laevis* (Al), *Claroideoglomus etunicatum* (Ce), *Pacispora scintillans* (Ps), *Rhizophagus intraradices* (Ri); and one isolate provided by the Department of Agroecology-Plant Pathology and Entomology, Aarhus University, Denmark. One of the inoculum used was a consortia (MTZ-UV) that consisted of 12 AMF species, *Acaulospora morrowiae, Acaulospora spinosa, Acaulospora scrobiculata, Funneliformis mosseae, Funneliformis geosporus, Gigaspora rosea, Gigaspora decipiens, Glomus macrocarpum, Glomus aggregatum, Rhizophagus intraradices, Scutellospora pellucida, Claroideoglomus etunicatum* provided by the Laboratory of Beneficial Organisms, Faculty of Agricultural Sciences, University of Veracruz. They were propagated through the modified Sieverding technique (Sieverding, 1991), using *Brachiaria decumbens* as a host plant in sterile sand as substrate, under greenhouse conditions during 5 months before its use.

Assessment of variables

Plants were harvested 165 DAI and physical variables (height, diameter, fresh and dry weight of root and shoot) and mycorrhizal colonization, by root clearance (Phillips and Hayman, 1970) and the intersection method (Giovannetti and Mosse, 1980) were recorded.

The experiment had a factorial design with two factors, soil sterilization (sterile and not) and mycorrhizal inoculation, with eight levels (with seven mycorrhizal inoculants and a control). Each treatment had three replicates. Data were subjected to a factorial analysis of variance with eight treatments followed by Fisher's least significant difference (LSD) test, with the statistical software STATGRAPHICS XV for Microsoft Windows.

RESULTS

The sterilization of the soil had significant ($\alpha \leq 0.05$) differences between treatments, expressed in an increased in the growth variables, compared to the not sterile soil (Table 1). The AMF inoculation had a significant effect on most of the plant growth variables, compared to the non-inoculated plants (Table 1). There was an interaction between the sterilization of the soil and the AMF inoculation, mainly on height, leaf area and shoot dry weight (Table 1).

Plants inoculated with *R. fasciculatum* displayed the highest increase on plant growth expressed in leaves number, especially when the soil was sterile (Figure 1a).

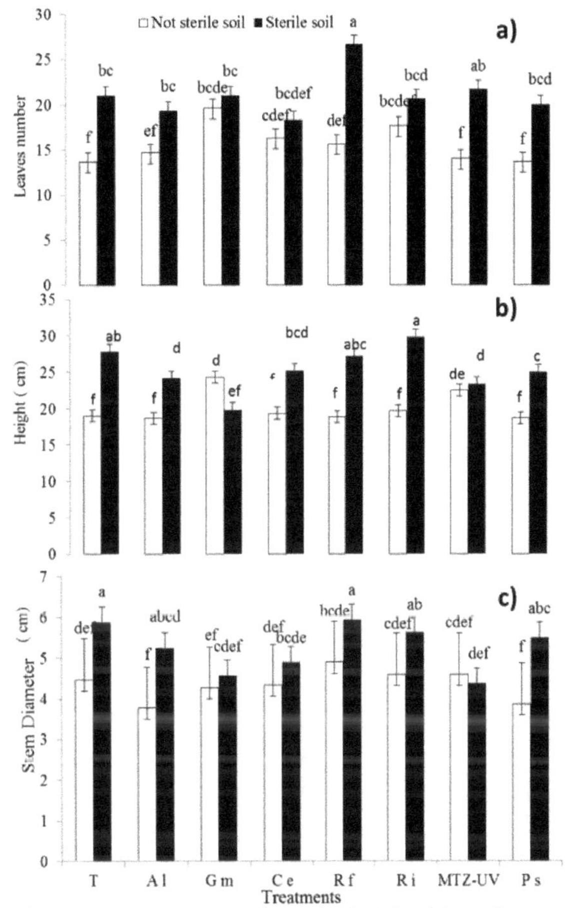

Figure 1. Leaves number a), height b) and stem diameter of avocado plants inoculated and not with seven different mycorrhizal inocula in sterile soil and not. Bars with equal letters are statistically similar (LSD, $\alpha \leq 0.05$); n=3. T, control; Al, *A. leavis*; Gm, *G. mosseae*; Ce, *C. etunicatum*; Rf, *R. fasciculatum*; Ri, *R. intrarradices*; MTZ-UV; Ps, *P. scintillans*.

Table 1. ANOVA analysis of each variable between main factors.

Factor	Height	Diameter	Leaves number	Leaf area	Fresh weight leaf	Shoot dry weight	Fresh weight root	Shoot dry weight
Soil	***	***	***	**	***	***	*	**
AMF	*	*	NS	***	*	***	NS	NS
Soil+AMF	***	NS	NS	**	NS	*	NS	NS

Fisher ($\alpha \leq 0.05$);***, 0.001;**, 0.01; *, 0.05; NS, not significant; n=3.

Plants inoculated with *R. fasciculatum* and *R. intraradices* exhibit a slight increase in shoot fresh and dry weight, but was not significant (Figure 2a and b). However, plants inoculated with MTZ-UV, had a significant plant growth depression in sterile soil. In not sterile soil, there was not difference among treatments (Figures 2a and b).

The AMF colonization was observed to be higher in sterile soil compared to not sterile only in *A. leavis* and *P. scintillans* treatments (Figure 3). The treatments with the highest AMF root colonization were *A. leavis, R. fasciculatum* and *P. scintillans* in sterile soil (Figure 3), and the treatment with the lowes was *C. etunicatum*.

For the not sterile soil, there were not significant differences among treatments (Figure 3).

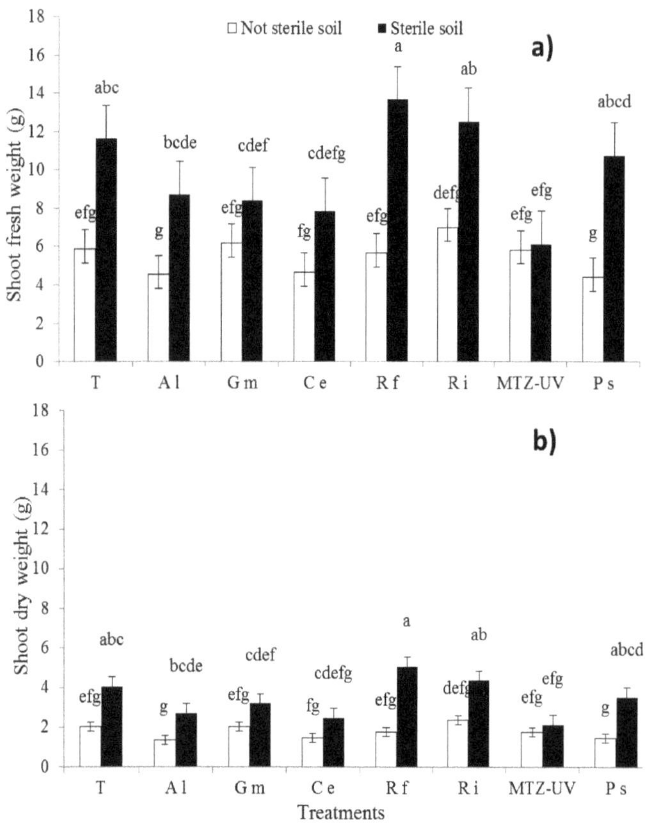

Figure 2. Shoot fresh a) and dry b) weight of avocado plants inoculated and not with 7 different mycorrhizal inocula in sterile soil and not. Bars with equal letters are statistically similar (LSD, $\alpha \leq 0.05$); n=3. T, control; Al, *A. leavis*; Gm, *G. mosseae*; Ce, *C. etunicatum*; Rf, *R. fasciculatum*; Ri, *R. intrarradices*; MTZ-UV; Ps, *P. scintillans*.

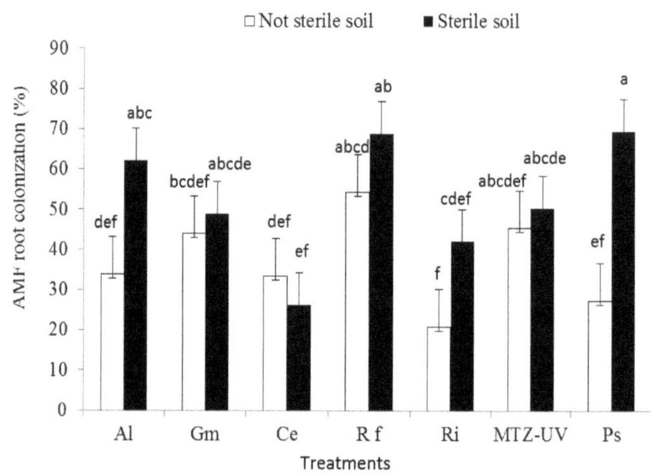

Figure 3. AMF root colonization of avocado plants inoculated with 7 different mycorrhizal inocula in sterile soil and not. Bars with equal letters are statistically similar (LSD, $\alpha \leq 0.05$); n=3. T, control; Al, *A. leavis*; Gm, *G. mosseae*; Ce, *C. etunicatum*; Rf, *R. fasciculatum*; Ri, *R. intrarradices*; MTZ-UV; Ps, *P. scintillans*.

DISCUSSION

Some of the species used in the experiments display a better response that could be translated as compatibility with the host, special with *R. fasciculatum* , that presented the highest values among the variables evaluated, but whith a very short difference between the other treatmens. In the other hand the plants inoculated with *P. scintillans*, exhibited a greater relative difference between the response of the plant unders sterile and non-sterile soil conditions in most of the variables.

In the present experiment we found that the soil sterilization promoted an increase on the plant growth, in accordance with De Deyn *et al.* (2004) and Al-Khaliel (2010). Such result could be explained by absence of competition for resources in the soil by other microorganisms (Calvet *et al.*, 2002) or, as found by Miransari *et al.* (2009) in wheat, the increase of nutrient uptake in sterile soils by its solubilization. The sterilization promoted a greater colonization percentage, compared with non-sterile soil. Shuch effect could be explained by the low competition for space and nutrients (Graham, 2001; Calvet *et al.*, 2002).

These findings are different of what Ingham *et al.* (1985), Callaway *et al.* (2004) and He and Cui (2009) found, where they observed a decrease of plant biomass in sterile soil. According to He and Cui (2009), perhaps this could be related to changes in the pH and soil nutrient availability. As some microorganisms still remain unknown; and therefore, their role in nutrient cycle is not understand, it is complicated to know how they influence plant growth (Crecchio *et al.*, 2004).

The results of this studio could point a functional complementarity between the *R. fasciculatum* and the avocado var. Criollo. Mainly, due to the greater plant growth response of such interaction, especially on sterile soil. Such effect may suggest a better relationship in terms of efficient between avocado plants (var. Criollo) and this inoculum. Similarly, Da Silveria, *et al.* (2003) observed in avocado plants var. Carmen a better plant response of two inoculants (*G. etunicatum* and *S. hetorogama*) compared to other inocula tested. Besides, the availability of the nutrients could have been greater due to low competition for resources under sterile conditions, facilitating the disponibility of the nutrients to the AMF.

A few inoculants in his experiment had a detrimental effect on the plant growth (*C. etunicatum*), but some studies have considered the differential effect of the AMF (Gustafson *et al.*, 2006; Toussaint *et al.*, 2007).

The low response of plant growth to the AMF inoculation in non-sterile soil could be due to competition of space (Graham, 2001) and nutrients (Calvet *et al.*, 2002).

Therefore, this result could give us indication of the effectiveness of species like *R. fasciculatum*, above others because it displays a tendency that could be significant if the experiment continued for a few months more. But when comparing the relative difference between the response of an inoculum under sterile and not sterile soil conditions, the AM fungi *P. scintillans* could be recommended. Therefore, it could indicate host compatibility under the conditions expressed previously. Further studios could be performed to unveil the mechanisms involved in this interaction under sterile conditions.

REFERENCES

Alarcón, A., Ferrera-Cerrato R. 1999. Manejo de la micorriza arbuscular en sistemas de propagación de plántulas frutícolas. TERRA Latinoamericana. 17:170-191.

Alarcón, A., Ferrera-Cerrato R. 2003. Aplicación de fósforo e inoculación de hongos micorrízicos arbusculares en el crecimiento y estado nutricional de *Citrus volkameriana* Tan and Pasq. TERRA Latinoamericana. 21:91-99.

Al-Khaliel, A.S. 2010. Effects of Arbuscular Mycorrhization in sterile and non-sterile soils. Tropical Life Sciences Research. 21:55–70.

Bárcenas, O.A.E., Molina, E.J, Huanosto, M.F., Aguirre P.S. 2003. Contenido de macro y micro elementos en hojas flor y fruto de aguacate "Hass" en la región de Uruapan Michoacán. Proceedings V World Avocado Congress. October Granada Spain 2003. pp. 365-371.

Bárcenas, O.A.E., Martínez, N.A., Aguirre, P.S. 2002. Fenología del Aguacate (*Persea americana* Mill) var. Hass en cuatro diferentes altitudes del municipio de Uruapan, Mich. Revista Divulga. CIC de la UMSNH. Morelia, Michoacán.

Bárcenas, O.A.E., Varela-Fregoso, L., Stürmer, S.L., Chávez-Bárcenas, A.T. 2011. Cátalogo de hongos micorrizógenos arbusculares de huertos de aguacate de Michoacán, México. Proceedings VII World Avocado Congress (ActasVII Congreso Mundial de Aguacate). September Cairns, Australia.

Barrientos-Priego, A.F., Borys, M.W., A. Ben-Ya'acov, L. López-López, M. Rubí-Arriaga, G. Bufler, Solis-Molina A. 1995. Progress of the study of the avocado genetic resources. III. Findings in the Mexican Gulf region. Proceedings III World Avocado Congress. Israel. pp. 211-216.

Barrientos-Priego, A.F., Borys, M.W., Escamilla-Prado, E., Ben-Ya'acov, A., de la Cruz-Torres, E., López-López, L. 1992. Study of avocado germoplasm resources, 1988-1990. IV. Findings in the Mexican Gulf region. Proceedings of II World Avocado Congress II. pp. 551-558.

Barrientos-Priego, A.F., López-López, L. 2000. Historia y genética del aguacate. In: Téliz D, González H, Rodríguez R, Dromundo R, (eds.). El aguacate y su manejo integrado. México D.F. Mundi-Prensa México. pp. 19–31.

Bellon, M.R., Perales, H. 2009. Diversidad y conservación de recursos genéticos en plantas cultivadas, en Capital natural de México, vol. II: Estado de conservación y tendencias de cambio. CONABIO, México, pp. 355-382.

Callaway, R.M., Thelen, G.C., Barth, S., Ramsey, P.W., Gannon, J.E. 2004. Soil fungi alter interactions between north American plant species and the exotic invader *Centaurea maculosa* in the field. Ecology. 85:1062-1071.

Calvet, C., Estaun, V., Camprubi, A., Pinochet, J. 2002. Interactions in rhizosphere. In: Sharma, A.K., Johri, B.N. (eds.). Arbuscular Mycorrhizae, Interactions in Plants, Rhizosphere and Soils. Oxford and IBH Publishing Co. Pvt. Ltd. pp. 3-69.

Caravaca, F., Alguacil, M.M., Barea, J.M., Roldán, A. 2005. Survival of inocula and native AM fungi species associated with shrubs in a degraded mediterranean ecosystem. Soil Biology and Biochemestry. 37:227-233.

Chen, H., Morrell P.L., Ashworth, V.E.T.M., de la Cruz, M., Clegg, M.T. 2008. Tracing the Geographic Origins of Major Avocado Cultivars. Journal of Heredity. 100:56–65.

Crecchio, C., A. Gelsomino, R. Ambrosoli, J.L. Minati, Ruggiero, P. 2004. Functional and molecular responses of soil microbial communities under differing soil management practices. Soil Biology and Biochemistry. 36:1873-1883.

da Silveira, S.V., de Souza, P.V.D., Koller, O.C. 2002. Effect of arbuscular mycorrhizal fungi on growth of avocado. Pesquisa Agropecuaria Brasileira. 37:1597-1604.

da Silveira, S.V., de Souza, P.V.D., Koller, O.C., Schwarz, S.F. 2003. Elementos minerales y carbohidratos en plantones de aguacate 'Carmen' inoculados con micorrizas arbusculares. Proceedings V World Avocado Congress. November Granada Spain. pp. 415-420.

de Deyn, G.B., Raaijmakers, C.E., van der Putten, W.H. 2004. Plant community development is affected by nutrients and soil biota. Journal of Ecology. 92:824–834.

Díaz, G., Honrubia, M. 1995. Effect of native and introduced arbuscular mycorrhizal fungi on growth and nutrient uptake of *Lygeum spartum* and *Anthyllis cytisoides*. Biologia Plantarum. 37:121–129.

Dos Santos, R., Girardi, C.G., Pescador, R., Stürmer, S.L. 2010. Effects of arbuscular mycorrhizal fungi and phosphorus fertilization on post vitro growth of micropropagated. Revista Brasileira de Ciência do Solo. 34:765-771.

Garbaye, J. 1991. Biological Interactions in the mycorrhizosphere. Experimentia. 47:370-375.

García, J.D. 2003. El aguacate ecológico aspectos técnicos, sociales y medio ambientales de su cultivo en Andalucía. Proceedings V World Avocado Congress. October, Granada Spain. pp. 707-712.

Giovannetti, M., Mosse, B. 1980. An evaluation of techniques for measuring vesicular arbuscular mycorrhizal infection in roots. New Phytologist. 84: 489-500.

González, M., Cuenca, G. 2008. Respuesta de plantas de plátano (*Musa* AAB cv.Hartón) a la inoculación con hongos micorrízicos arbusculares nativos e introducidos, bajo condiciones de campo. Revista de la Facultad de Agronomía (LUZ). 25:470-495.

Graham, J.H. 2001. What Do Root Pathogens See in Mycorrhizas? New Phytologist. 149:357–359.

Gustafson, D.J., Casper, B.B. 2006. Differential host plant performance as a function of soil arbuscular mycorrhizal fungal communities: experimentally manipulating co-occurring Glomus species. Plant Ecology. 183:257–263.

Gutiérrez-Contreras, M., Lara-Chávez, Ma., Guillén-Andrade, H., Chávez-Bárcenas, A.T. 2010. Agroecología de la franja aguacatera en Michoacán, México. Interciencia. 35:647-653.

Gutiérrez-Díez, A., Martínez-de la Cerda, J., García-Zambrano, E.A., Iracheta, L., Ocampo Morales, J.D., Cerda-Hurtado, I.M. 2009. Estudio de diversidad genética del aguacate nativo en Nuevo León, México. Revista Fitotecnia Mexicana. 32:9-18 .

Haas, J.H., Menge, J.A. 1990. VA-mycorrhizal fungi and soil characteristics in avocado *(Persea americana* Mill.) orchard soils. Plant and Soil. 127:207-212.

He, W.M, Cui, Q.G. 2009. Manipulation of soil biota in ecological research. Web Ecology. 9:68–71.

Ingham, R.E., Trofymow, J.A., Inham, E.R., Coleman, D.C. 1985. Interactions of bacteria fungi and their nematode grazers: Effects of nutrient cycling and plant growth. Ecological Monographs. 55:119-140.

Jaizme-Vega, M.C., Azcón, R. 1995. Responses of some tropical and subtropical cultures to endomycorrhizal fungi. Mycorrhiza. 5:213-217.

Klironomos, J. 2000. Host-specificity and functional diversity among arbuscular mycorrhizal fungi. In: Bell C.R., Brylinsky M., Johnson-Green P., (eds.). Microbial biosystems: new frontiers, 8th International Symposium on Microbial Ecology. Halifax, Canada: Atlantic Canada Society for Microbial Ecology. pp. 845–851.

Linderman, R.G. 1992. Vesicular-Arbuscular Mycorrhizae and Soil Microbial Interactions. In : Miccorrhizae in Sustainable Agriculture, G.J. Bethlenfalvay and Linderman, R.G. (eds.). ASA Special publication. pp. 45-70.

Menge, J.A., Michael, R., Johnson, E., Zentmyer, R. 1978. Mycorrhizal fungi increase growth and reduce transplant injury in avocado. California Agriculture. 32:6-7.

Menge, J.A., Larue, J., Labanauskas. C.K., Johnson, E. 1980. The Effect of Two Mycorrhizal Fungi upon Growth and Nutrition of Avocado Seedlings Grown with Six Fertilizer Treatments. American Society for Horticultural Science. 105:400-404.

Miransari, M., Bahrami H.A., Rejali F., Malakouti, MJ. 2009. Effects of arbuscular mycorrhiza, soil sterilization, and soil compaction on wheat (*Triticum aestivum* L.) nutrients uptake. Soil and Tillage Research. 104:48–55.

Morales-García, J.L. 2009. Enfermedades de importancia económica en el cultivo de aguacate en Michoacán, México. III Congreso latinoamericano del aguacate. November 2009, Medellín, Colombia. pp. 15-31.

Phillips, J.M., Hayman, D.S. 1970. Improved procedures for clearing roots and staining parasitic and vesicular-arbuscular mycorrhizal fungi for rapid assessment of infection. Transactions of the British Mycological Society. 55:158-161.

Sieverding, E. 1991. Vesicular-Arbuscular Mycorrhiza Management in Tropical Agrosystems. Friedland, Germany. TZ-Verlagsgesellschaft mbH für Technische Zusammenarbeit.

Smith, S.E., Gianinazzi-Peearson, V. 1988. Physiological interactions between symbionts in vesicular-arbuscular mycorrhizal plants. Annual Review of Plant Physiology and Plant Molecular Biology. 39:221-224.

Smith, S.E., Facelli, E., Pope S., Smith, F.A. 2010. Plant performance in stressful environments: interpreting new and established knowledge of the roles of arbuscular mycorrhizas. Plant and Soil. 326:3–20.

Toussaint, J., Smith, F., Smith, S. 2007. Arbuscular mycorrhizal fungi can induce the production of phytochemicals in sweet basil irrespective of phosphorus nutrition. Mycorrhiza. 17:291–297.

Trejo, D., Ferrera-Cerrato, R., García, R., Varela, L., Lara, L., Alarcón, A. 2011. Effectiveness of native arbuscular mycorrhizal fungi consortia on coffee plants under greenhouse and field conditions. Revista Chilena de Historia Natural. 84:23-31.

Violi, H.A., Treseder, K.K., Menge, J.A., Wright, S.F., Lovatt. C.J. 2007. Density dependence and interspecific interactions between arbuscular mycorrhizal fungi mediated plant growth, glomalin production, and sporulation. Canadian Journal of Botany. 85:63-75.

Xu, L., Ravnskov, S., Larsen, J., Nilsson, R.H., Nicolaisen, M. 2011. Soil fungal community structure along a soil health gradient in pea fields examined using deep amplicon sequencing. Soil Biology and Biochemistry. 46:26-32.

Xu, X., Qin G., Tian, S. 2008. Effect of microbial biocontrol agents on alleviating oxidative damage of peach fruit subjected to fungal pathogen. International Journal of Food Microbiology.126:153–15.

6

Antimicrobial Activity and Chemical Composition of the Essential Oils of *Malvaviscus Arboreus* Cav, *Pimenta Dioica* (L.) Merr., *Byrsonima Crassifolia* (L.) Kunth and *Psidium Guajava* L.

Daniel Alejandro Vazquez-Cahuich[1], Judith Espinosa Moreno[1*], Dora Centurion Hidalgo[1], Jose Rodolfo Velazquez Martinez[1], Rocio Borges-Argaez[2] and Mirbella Caceres Farfan[2].

[1]División Académica de Ciencias Agropecuarias de la Universidad Juárez Autónoma de Tabasco Km. 25+2 de la Carretera Villahermosa-Teapa, R/a La Huasteca 2ª Sección.
Villahermosa, Tabasco, México.
*e-mails: alexvc14@hotmail.com, *juespinosa@hotmail.com, dora.centurion@ujat.mx, rborges@cicy.mx,*
[2]Unidad de Biotecnología,Centro de Investigación Científica de Yucatán, Calle 43 No. 130 Chuburná de Hidalgo, C.P. 97200, Mérida, Yucatán, Mexico.
**Corresponding author*

SUMMARY

The essential oils of leaves of *Malvaviscus arboreus* Cav., *Pimenta dioica* (L.) Merr., *Byrsonima crassifolia* (L.) Kunth and *Psidium guajava* L., obtained by hydrodistillation were evaluated as potential antimicrobial agents against *Staphylococcus aureus* (ATCC 25923), *Salmonella typhimurium* (ATCC 4028) and *Bacillus cereus* (ATCC 11778). Disk diffusion, microdilution and bioautography methods were used for antimicrobial activity evaluation; subsequently, the possitive extracts were analyzed by gas chromatography coupled to a mass spectrometry detector (GC-MS) to obtain the chemical profile. The essential oils of the four species showed antimicrobial activity by the agar diffusion method against the three strains of microorganisms. As for the bioautography, it was found that the component at a frontal reference (Rf) 0.42 from fraction four of the *P. dioica* essential oil, inhibited the growth of *B. cereus*, *S. typhimurium* and *S. aureus*. In the microdilution test, *B. cereus* was the most susceptible microorganism to all essential oils tested. Eugenol was the major component of *P. dioica* and *P. guajava* with abundance percentages of 94.86% and 33.84%, respectively.

Keywords: Essential oils; antimicrobial activity; *Malvaviscus arboreus* Cav.; *Pimenta dioica* (L.) Merr.; *Byrsonima crassifolia* (L.) Kunth; *Psidium guajava*.

INTRODUCTION

Essential oils from aromatic and medicinal plants have been known to possess a broad range of biological activities, being antimicrobial one of its main medicinal properties reported (Viuda-Martos *et al.*, 2008; O'Bryan *et al.*, 2008). These activities are strongly associated by their chemical composition which is determined by the environmental and agronomic conditions as well as the genotype of each plant (Marotti *et al.*, 1993; Cosentino *et al.*, 1999). The interest in the application of essential oils to pathogenic microorganisms control has been increased in recent years (Rojas-Grau *et al.*, 2006; Oussalah *et al.*, 2004; Gutierrez *et al.*, 2008). Essential oils are part of plant metabolites and they are usually composed of volatile terpenes. These compounds can be located in different parts of the plant such as flowers, seeds, leaves and fruits. They constitute 0.1% to 1% of plant dry weight, and they are usually less dense than water, with a high refractive index (Lopez, 2004). The concentration and composition of the essential oil metabolites varies due to environmental factors (Toncer *et al.*, 2010; Gazim *et al.*, 2010), soil nutrients (Powell and Raffa, 1999), the part of the plant used (Pala-Paul *et al.*, 2005), the drying conditions (Mejia *et al.*, 2007), the method of extraction (Quintero *et al.*, 2004), among other factors.

Malvaviscus arboreus Cav, *Pimenta dioica* (L.) Merr., *Byrsonima crassifolia* (L.) Kunth and *Psidium guajava* L are four plants with a traditional use as condiment and medicinal purposes. *M. arboreus* (tulipancillo, in Spanish) has been used to treat whooping cough, as an urinary antiseptic and as a diuretic; *P. guajava* is used to treat gastrointestinal diseases, chills and stomach pain (UNAM, 2009). On the other hand, *P. dioica* is used traditionally as anti-hemetic (Germosén-Robineau, 2005) whereas *B. crassifolia* is used against skin infections and gastrointestinal disorders, it is also used as digestive, emenagoge, febrifuge and the bark infusion is used as an antidiarrheal, to treat inflammation, to treat scabies and as wound healing (Centurion *et al.*, 2004; Gonzalez, 2007). Although there are reports of traditional medicinal use of these species, there are no reports on the study of their antimicrobial activity. Therefore, the objective of this study was to evaluate the antimicrobial activity of essential oils from the leaves of *M. arboreus*, *P. dioica*, *B. crassifolia* and *P. guajava* against *Staphylococcus aureus*, *Salmonella typhimurium* and *Bacillus cereus*, as well as the chemical profile obtained by GC-MS.

MATERIALS AND METHODS

Plant material

Plant species were obtained from a local market called "Jose Maria Pino Suarez" in Villahermosa, Tabasco, Mexico on September of 2010. The leaves were dried at room temperature and protected from light. Then, samples were ground in a hammer mill and sieved through a mesh number 60 (particle size of 0.250 mm). The sieved product was packaged in polyethylene bags stored in PET containers in a dry place for further analysis.

Essential oil extraction

A sample of 500 g of plant material was extracted by hydrodistillation in a Clevenger type equipment for 7 hours at 100 °C and the essential oil was obtained by condensation from the oil/water mixture. Hexane was added to separate the essential oil from water, and the oil was placed in amber vials and stored at 4 °C.

Gas chromatography-mass spectrometric (GC-MS) analysis

Essential oil components obtained by steam distillation were determined on an Agilent Technologies 6890N Gas Chromatography apparatus coupled with a mass selective detector 5975B, using the following chromatographic conditions: split injection of 1 μL of 10 μg of sample in hexane; HP5 MS phenyl column (30 m x 0.25 μm), flow rate 1 mL/min (helium as carrier gas); samples were analyzed with the column held initially at 60 °C for a minute after injection, then increased to 260 °C with a gradient of 10 °C/min heating program. Identification of components in the extract was performed by computer searches in commercial references libraries. The fragmentation patterns of the mass spectra were compared with those from the NIST05 libraries (Pala-Paul *et al.*, 2005, Pala-Paul *et al.* 2007, Raseetha *et al.*, 2009).

Bacterial strains

Cultures of *Staphylococcus aureus* (collection number 25923), *Salmonella typhimorium* (4028) and *Bacillus cereus* (11778) were obtained commercially on the American Type Culture Collection (ATCC) and were maintained on tryptose soy agar (TSA) at 4 °C.

Agar diffusion method

Agar diffusion method was performed according to the procedure described by Moulari *et al.* (2006). An inoculum of each bacterial strain (1×10^6 bacteria mL^{-1}) was applied on a homogeneous dispersion in Mueller Hinton medium plates. The essential oil samples were applied on Whatman filter paper disks at a concentration of 10 mg mL^{-1}. Each disk, containing the sample to be tested, was brought into contact with an inoculated medium and, after incubation for 24 h at 37 °C, the diameter of the clear zone around the disk (inhibition diameter) was measured.

Bioautography on TLC plates

Direct bioautography with *S. aureus*, *S. typhimorium* and *B. cereus* was performed on silica gel GF254 TLC Aluminum plates. Briefly, 10 µl of each extract solution (20 µg mL^{-1}) were applied on silica gel plates separately. TLC plates were development using *n*-hexane-acetone (80:20 v/v) as solvent system. The plates were then dried under an extraction hood. After drying for complete removal of solvent, the TLC plates were placed in 9 cm x 9 cm sterile square Petri dishes, to which an inoculum of the bacteria (1 x 10^6 bacteria mL^{-1}) in TSA were added. The medium with the inoculum was distributed over the developed TLC plates uniformly. After media solidification, the plates were incubated at 37°C for 24 h. Finally, the plates with a solution of 3,5-triphenyl-2H-tetrazolium chloride (MTT). The inhibition zones were revealed and visualized by spraying the plates with MTT solution followed by their incubation at 37 °C for 1 hour. Active compounds in the extract were detected as clear spot against a dark red background due to the transformation of MTT in its reduction form, formazan (Moulari *et al.*, 2006).

Statistical design

For antimicrobial activity by agar diffusion method, we conducted a completely randomized design with a 4x3 factorial distribution: four plant species (*M. arboreus*, *P. dioica*, *B. crassifolia* and *P. guajava*) and three microorganisms (*Staphylococcus aureus*, *Salmonella typhimurium* and *Bacillus cereus*) with a total of 12 treatments and three replicates of each one were performed.

Determination of minimum inhibitory concentration

The extracts showing an inhibition zone in the bioautography assay were chosen to test the minimum inhibitory concentration (MIC) with the agar dilution method reported by Andrews *et al.* (2001). Additionally, the minimum bactericidal inhibition was measured. The test microorganisms (*S. aureus, S. typhimurium* y *B. cereus*) were grown for 24 h before the bioassay in MH broth at 37°C. Then, the optical density of the culture was adjusted (1 x 10^8 cell ml^{-1}) using the 0.5 Mac Farland turbidity standard. Microdilution tests were performed in a sterile 96-well micro-plate. All extracts were initially tested at 20 mg ml^{-1} and were serially diluted two-fold to 0.0156 mg ml^{-1}, after which 100 µg of bacterial culture (final concentration 1 x 10^6 cell ml^{-1}) were added to each well. The antibiotic amikacin was included as a positive control in each assay. Two additional wells were used as culture medium control without inoculum and as a growth control with the inoculum. The microplates were incubated overnight at 37°C. To indicate bacterial growth, 100 µl of a solution of 1% MTT was added to each well and the plates were incubated at 37°C for one hour. Bacterial growth in the well was indicated by a red color, whereas clear wells indicated growth inhibition by the essential oil. The minimum bactericide concentration (MBC) was determined by making a replication with a 96 teeth brush on a MH medium plate, incubating it at 37 °C for 24 h and observing the presence or absence of growth (Taylor *et al.*, 1983).

RESULTS AND DISCUSSION

The essential oils of *P. dioica*, *B. crassifolia*, *P. guajava* and *M. arboreus* showed antimicrobial activity in the agar diffusion method (Table 1) for both *S. aureus* (8.00 mm to 17.06 mm) and *B. cereus* (13.46 mm to 18.16 mm). Also, *S. typhimurium* showed sensitivity to *M. arboreus* and *P. dioica* essential oils similar to the amikacina control. These results coincide with the ones found by Alzamora *et al.* (2001) with the *Eucalyptus globulus* (Myrtaceae family) essential oil obtained by steam distillation with inhibition zones of antibacterial activity of 13, 12 and 12 mm on *S. aureus, S. typhi* and *S. typhimurium*, respectively.

On the other hand, Sahadeo y Vilas (2011) reported that the *P. dioica* essential oil presented an inhibition zone of 19 y 35 mm for *B. cereus* and *S. aureus*, respectively. The inhibition zones obtained in the present study: 13 and 17 mm for Gram positive (*B. cereus* and *S. aureus*) and 19 mm for Gram negative (*S. typhimurium*) were bigger (Table 1) than the ones found by Girova *et al.* (2010) for *P. dioica* essential oil that inhibited Gram positive bacteria growth with inhibition zones of 12-13.7 mm and 10-11 mm for Gram negative bacteria. It has been reported antibacterial activity of extracts of Malpighiaceae family species on *S. aureus, E. faecalis* (Suffrendin *et al.,* 2004) and *Byrsonima crassifolia* root extracts presented antimicrobial activity on *S. aureus, E. coli* and *S. typhi* (Martinez-Vasquez *et al.,* 1999).

Table 1. Antimicrobial activity of essential oils using the agar diffusion method.

Essential oil (50 µg mL^{-1})	Inhibition zone (mm)		
	S. aureus	*S. typhimurium*	*B. cereus*
P. dioica	17.06^{c} ±2.90	19.46^{c} ±5.10	13.46^{c} ±2.20
B. crassifolia	15.30^{c} ±5.20	0.00	18.16^{c} ±3.20
P. guajava	8.00^{a} ±1.70	0.00	13.80^{b} ±1.20
M. arboreus	10.00^{a} ±2.00	21.80^{c} ±0.80	14.20^{b} ±1.50
(Amikacine 1 µg mL^{-1})	23.83^{d} ±2.70	22.83^{d} ±2.70	28.96^{d} ±2.20

Data are expressed as mean ± standard deviation (n=3). Different letters indicate statistically significant differences (Tukey, α=0.05).

The bioauthography results confirm that the fraction 4 of the leaf essential oil *of P. dioica* have metabolites with antimicrobial activity, with a R_f of 0.42 cm, against the three studied microorganisms (*B. cereus*, *S. typhimurium*, *S. aureus*) meanwhile the fraction 3 presented such activity on *S. aureus* with a R_f of 0.34 (Table 2). *M. arboreus* essential oil presented one inhibition zone on *S. aureus* growth without definition of specific fractions (Figure 1). Moreover, the essential oil of *P. guajava* presented three compounds (fractions 2, 3 and 4), with inhibition R_f of 0.5, 0.56 y 0.68 cm, respectively. On the other hand, essentials oils from *B. crassifolia* and *M. arboreus* leaves showed activity against *S. aureus* without defined fractions. Borges-Argaez *et al.* (2010) reported that the fraction 4 of *P. dioica* leaf essential oil obtained by steam distillation, Rf of 0.5 cm, inhibited *S. aureus* growth, data similar to the one obtained in the present study. Reports were not found in the literature related to antimicrobial activity of the other species.

Table 2. Essential oil active fractions and their frontal references (R_f in cm).

Leaf essential oils	*S. aureus*	*S. typhimurium*	*B. cereus*
P. dioica	Fraction 3 and 4 R_f : 0.34 and 0.42	Fraction 4 R_f : 0.42	Fraction 4 R_f : 0.42
M. arboreus	A	NA	NA
B. crassifolia	A	NA	NA
P. guajava	Fraction 2, 3 and 4 R_f :0.5, 0.56 and 0.68	NA	NA

A= Active, NA= Not active

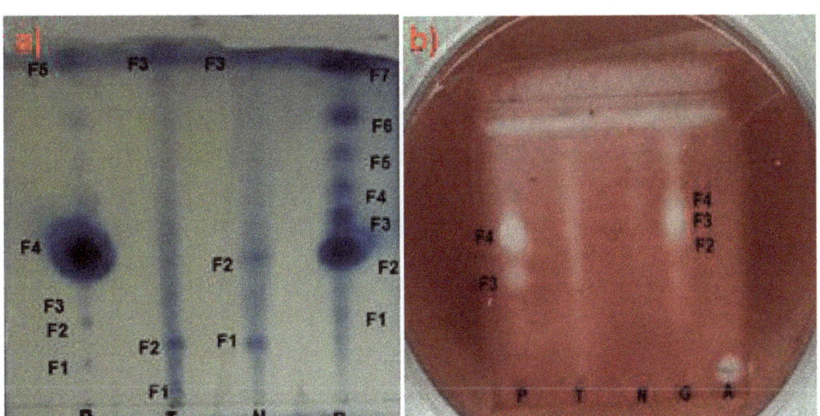

Figure 1. a) Chromatographic plate revelated with phosphomolibdic acid; b) Chromatographic plate of bioautography bioassay. (P: *Pimenta dioica*, T: *Malvaviscus arboreus*, N: *Byrsonima crassifolia*, G: *Psidium guajava*, A: Amikacina).

The most sensitive microorganism to the presence of *P. dioica*, *P. guajava*, *B. crassifolia* and *M. arboreus* essentials oils was *B. cereus* (Table 3). *P. guajava* essential oil presented the lowest MIC (5 mg mL^{-1}) followed of *P. dioica* (10 mg mL^{-1}), *B. crassifolia* and *M. arboreus* (both at 20 mg mL^{-1}). In the other hand *P. dioica* (10 mg mL^{-1}), *B. crassifolia* (20 mg mL^{-1}) and *M. arboreus* (20 mg mL^{-1}) inhibited *S. aureus* growth, meanwhile *S. typhimurium* was the least sensitive microorganism, although *P. dioica* and *P. guajava* essential oils inhibited its growth at concentrations of 20 and 10 mg mL^{-1}, respectively.

Table 3. Minimal Inhibitory Concentration (mg mL^{-1}) of leaf essentials oils.

Essential oil	MIC (mg mL^{-1})		
	S. typhimurium	*S. aureus*	*B. cereus*
P. dioica	20	10	10
P. guajava	10	NA	5
B. crassifolia	NA	20	20
M. arboreus	NA	20	20

MIC= Minimal Inhibitory Concentration, NA= No activity.

It has been reported that the leaf *P. dioica* essential oil obtained by steam distillation presented a MIC of 6.25 mg mL^{-1} on *S. aureus* with the plate microdilution method (Borges-Argaez *et al*., 2010). In the present study, the MIC value was 10 mg mL^{-1} obtained on the same microorganism, probably due to that the essential oil was obtained by hydrodistillation and it had different concentration of some components with antimicrobial activity. Although MIC of *B. crassifolia* leaf essential oil reports have not been found, the result obtained in this paper (20 mg mL^{-1}) was expected because diverse species and plant structures of *Byrsonima* have antimicrobial activity, such as Rivero-Cruz *et al.* (2009) found that the dichloromethanic extract of *B. crassifolia* bark had activity with a MIC of 63 mg mL^{-1} on *S. aureus* using the microdilution method. Even though reports have not been found on *P. guajava* leaf essential oil, some reports exist that demonstrate antimicrobial activity of extracts obtained with other solvents and using the same method. Sanches *et al.* (2005) reported that the *P. guajava* extracts obtained by ethanol:water maceration of leaf, root and bark presented a MIC of 0.5, 0.25 y 0.12 mg mL^{-1}, respectively, on *S. aureus*. Barbieri *et al.* (2002) reported a MIC of 0.25 mg mL^{-1} on *S. aureus* with ethanol-water (90:10) extract of *P. guajava* leaves. Finally, data on *M. arboreus* essential oil were not found in the consulted literature.

Leaf *P. dioica* essential oil showed bactericidal activity on *S. aureus* and *B. cereus* at 20 and 2.5 mg mL^{-1}, respectively, meanwhile *S. typhimurium* required an essential oil concentration of 20 mg mL^{-1} presenting bacteriostatic activity (Table 4). On the other hand, *P. guajava* needed essential oil concentrations of 5 and 10 mg mL^{-1} in order to present bactericidal activity on *B. cereus* and *S. aureus* whereas the same essential oil showed bacteriostatic activity on *S. typhimurium* at 20 mg mL^{-1}, the maximum concentration assayed in this study. *B. crassifolia* essential oil was the least effective on *S. aureus*, *S. typhimurium* and *B. cereus* elimination because it had bacteriostatic activity at 20 mg mL^{-1} and *M. arboreus* required a lower concentration (5 mg mL^{-1}) for *B. cereus* elimination, although *S. typhimurium* elimination required the maximum concentration assayed (20 mg mL^{-1}) and *S. aureus* showed a bacteriostatic activity at the same concentration.

Table 4. Minimal Bactericidal Concentration (mg mL^{-1}) of essential oils.

Leaf essential oil	MBC (mg mL^{-1})		
	S. aureus	*S. typhimurium*	*B. cereus*
P. dioica	20 (Bactericidal)	20 (Bacteriostatic)	2.5 (Bactericidal)
P. guajava	10 (Bactericidal)	20 (Bacteriostatic)	5 (Bactericidal)
B. crassifolia	20 (Bacteriostatic)	20 (Bacteriostatic)	20 (Bacteriostatic)
M. arboreus	20 (Bacteriostatic)	20 (Bactericidal)	5 (Bactericidal)

The CG-MS analysis of essential oils from the leaves of the studied plant species is showed in Table 5. *P. dioica* essential oil had nine compounds with retention times from 6.61 to 30.20 min; some of these substances are eugenol with 94.86%, alpha-terpineol with 2.45% and cinnamic acid methyl ester with 1.06% as main compounds. *P. guajava* essential oil presented 13 substances such as eugenol (33.84%), caryophylene (7.08%), caryophylene oxide (7.02%), beta-bisabolene (4.62%), longiborneol (4.33%) and nerolidol (3.48%).

Rao *et al.* (2010) found eugenol as the mayor component (95 %) in leaf *P. dioica* essential oil, being this value similar to the one found in this study. On the other hand, Jirovetz *et al.* (2007) reported that this species, found in Jamaica, had eugenol (76.02%), alpha–selinene (1.04%), 4-terpineol (0.19%), alpha-terpineol (0.09%) and beta–selinene (0.52%). Moreover, Nascimiento *et al.* (2000) determined the antimicrobial activity, with the disc diffusion method, of secundary metabolites such as eugenol (positive on *S. aureus, S. choleraesuis, Bacillus subtilis* and *Candida albicans*), cinnamic acid (positive on *S. aureus, Enterobacter aerogenes* and *E. coli*). Joseph y Priya (2011) reported that eugenol is one of the main substances in *P. guajava* leaf, besides of alpha-pinene, limonene, menthol and volatile compouds such as β-cariofilene, β-bisabolene, cariofilene, β-copaenoe and eucalipte in lower percentages. Most of these components also formed part of the chemical composition of *P. guajava* essential oil used in this study. On the other hand, de Lima *et al.* (2010) reported selin-11-en-4-α-ol (22.19%), 1,8-cineol (12.83%), *trans*-cariophylene (8.89%), cariophylene oxide (9.09%), guaiene (6.74%), α-selinene (5.79%) and cadinol (8.18%) as main components found in the essential oil of the Brazilian species. For these reason, it is confirmed that ethnogeographic differences establish a variation between the composition and chemical substances abundance (Toncer *et al.*, 2010; Gazim *et al.*, 2010; Powell y Raffa, 1999; Pala-Paul *et al.*, 2005; Amna *et al.*, 2010; Mejia *et al.*, 2007; Quintero *et al.*, 2004).

It has been reported that gram positive bacteria, *L. monocytogenes* and *S. pyogenes*, were less sensitive to inhibition of eugenol, α-terpineol and γ-terpinene at concentration between 0.25 to 0.50% than the two gram negative bacteria, *P. vulgaris* and *E. coli* at 0.50 to 0.75% (Oyedemi *et al.*, 2009; Tippayatum and Chonhenchob, 2007; Dorman and Deans, 2000). Finally, in the *M. arboreus* and *B. crassifolia* essential oils were identified only two and three compounds, respectively, in minimal concentrations (>0.001%) that indicate that both oils are poor in aromatic compounds. It is important to mention that reports were not found on chemical composition of these species in the consulted literature.

Table 5. Abundance (%) of mayor substances of essential oils of studied plant species.

Substance	*P. dioca*	*P. guajava*	*M. arboreus*	*B. crassifolia*	Retention time (min)
4-terpineol	>0.0001	-	-	-	11.62
Eucaliptol	>0.0001	-	-	-	6.6
Caryophylene	>0.0001	7.08	-	-	21.24
Cariyophylene oxide	0.72	7.02	-	-	27.51
α - caryophylene	-	1.25	-	-	24.20
Cinnamic acid methyl ester	1.06		-	-	20.10
Eugenol	94.86	33.84	>0.001	-	19.28
Selinene	-	4	-	-	
α - Selinene	>0.0001		-	-	
α-Terpineol	2.45	0.83	-	-	12.18
β - Selinene	>0.0001		-	-	23.84
Acoradiene	-	0.55	-	-	22.97
Cis α-bisabolene	-	2.02	-	-	24.65
β-bisabolene	-	4.62	-	-	24.89
Copaene	-	0.53	-	-	19.50
Longiborneol	-	4.33	-	-	30.25
Nerolidol	-	3.48	-	-	27.06
α-bisalobol	-	0.47	-	-	31.41
1-4- dichlorobencene	-	-	>0.001	>0.001	6.10
Curcumene	-	-	-	>0.001	30.16
Bencene 1-(1,5-dimethyl-4-hexenyl)-4-methyl	-	-	-	>0.001	23.82

(-) = Not present

CONCLUSIONS

Leaf essential oils of *P. dioca, P. guajava, M. arboreus* and *B. crassifolia* presented antimicrobial activity with the agar diffusion method against *S. aureus* and *B. cereus.* Essential oils of *P. dioica* and *P. guajava* were the most effective in *B. cereus* elimination, at a concentration of 2.5 mg mL^{-1}. Essential oil of *P. guajava* presented three active fractions for *S. aureus* growth inhibition. Fracti0on 4 of the *P. dioica* essential oil inhited *B. cereus, S. typhimurium, S. aureus* growth meanwhile fraction 3 was active againt *S. aureus.*
The main compound of *P. dioica* and *P. guajava* essential oils was eugenol, 94.86% and 33.84%, respectively. The present study allows knowing the potential antimicrobial activity of plant species that could have future applications as preservatives for perishable food such as meat or milk products.

Acknowledgement

To Fondo Mixto CONACYT - Gobierno del Estado de Tabasco for the research funding to the project titled "Caracterización de propiedades de antimicrobianos de extractos de plantas alimentarias tropicales" (FOMIX-2009-C17-120880).

REFERENCES

Amna, T., Deepika, S., Ganeshaiah, K.N., Shaanker, R.U., Puri, S., Qazi, N. 2010. Changes in the essential oil content and composition of *Origanum vulgare* L. during annual growth from Kumaon Himalaya. Current Science, 98(8): 1010-1012.

Andrews, J.M. 2001. Determination of minimum inhibitory concentration. Journal Antimicrobial Chemotherapy, 48(1): 5-16.

Alzamora, L., Morales, L., Armas, L., Fernández, G. 2001. Medicina tradicional en el Perú: actividad antimicrobiana *in vitro* de los aceites esenciales extraídos de algunas plantas aromáticas. Anales de la Facultad de Medicina, 62(2): 156-161.

Barbieri, F.H., Pessini, G.L., Rogério, N.S., García, D.A.C., Vataru, C.N., Dias Filho, B.P. 2002. Screening of some plants used in the Brazilian folk medicine for the treatment of infectious diseases. Memórias do Instituto Oswaldo Cruz, 97(7): 1-5.

Borges-Argaez, R., Cáceres, F.M., Reyes, C.E.D., Zarrabal, U.G.J., Espinosa, M.J., Centurión, H.D., Novelo, P.J.I., Rocha, U.J.A. 2010. Antimicrobial activities and composition of essential oils of *Pimienta dioca* L. Merr. and *Plectranthus amboinicus* Lour obtained by supercritical CO_2 extraction and steam distillation. II Iberoamerican Conference on Supercritical Fluids, PROSCIBA. Brazil.

Centurion, H.D., Cazares, C.J., Espinosa, M.J. 2004. Inventario de recursos fitogenéticos alimentarios de Tabasco. Colección José María Pino Suárez, Universidad Juárez Autónoma de Tabasco, México. pp: 65-155.

Cosentino, S., Tuberoso, C.IG., Pisano, B., Satta, M. Mascia, V., Arzedi, E., Palmas, F. 1999. *In vitro* antimicrobial activity and chemical composition of Sardinian Thymus essential oils. Letters in Applied Microbiology, 29: 130-135.

Dorman, H.J.D., Deans, S.G. 2000. Antimicrobial agents from plants: antibacterial activity of plant volatile oils. Journal of Applied Microbiology 88:308–316.

Gazim, Z. C., Amorim, A.C. L., Hovell, A.M.C., Moraes, C. R., Nascimento, I.A., Ferreira, G.A., Garcia, C. D.A. 2010. Seasonal Variation, Chemical Composition, and Analgesic and Antimicrobial Activities of the Essential Oil from Leaves of *Tetradenia riparia* (Hochst.) Codd in Southern Brazil. Molecules, 15(8): 5509-5524.

Germosén-Robineau, L. 2005. Farmacopea vegetal caribeña. Prosisa, Unión Europea. 485 pp

Girova, T., Gochev, V., Jirovetz, L, Buchbauer, G., Schmidt, E., Stoyanova, A. 2010. Antimicrobial activity of essential oils from spices against psychrotrophic food spoilage microorganisms. Second Balkan Conference on Biology Special. Biotechnol. & Biotechnol. 21-23 de Mayo, Plodvid, Bulgaria.

Gonzalez, J. 2007. Malpighiaceae Juss. Catálogo: OET-PVFLO-000487. Flore Digital de Palo Verde. Organización para Estudios Tropicales.

Gutierrez, J., Barry-Ryan, C., Bourke, P. 2008. The antimicrobial efficacy of plant essential oil combinations and interactions with food ingredients. International Journal of Food Microbiology 124: 91-97.

Jirovetz, L., Buchbauer, G., Stoilova, I., Krastanov, A., Stoyanova, A., Schmidt, E. 2007. Spice plants: Chemical composition and antioxidant properties of *Pimenta* Lindl. essential oils. Part 1: *Pimenta dioica* (L.) Merr. Leaf oil from Jamaica. Nutrition, 31(2): 55-62.

Joseph, B. and Priya, R.M. 2011. Phytochemical and biopharmaceutical of *Psidium guajava* (L.) essential oil: A review. Research Journal of Medicinal Plant, 5(4): 432-442.

Lima, R.K. de, Cardoso, M.G, Aparecida, A.M., Nascimento, E.A., de Morais, S.A.L, Lee, N.D. 2010. Composition of the essential oil from the leaves of three domestic varieties and one wild variety of the guava plant (*Psidium guajava* L., Myrtaceae). Brazilian Journal of Pharmacognosy, 20(1): 41-44.

Lopez, L. M. T. 2004. Los aceites esenciales, aplicaciones farmacológicas, cosméticas y alimentarias. Ámbito Farmacéutico, (23) 7: 88-91.

Marotti, M., Dellacecca, V., Piccaglia, R., Gioaovanelli, E. 1993. Agronomic and chemical evaluation of three varieties of *Foeniculum vulgare* Mill. Acta Horticulturae, 331: 63-69.

Martinez-Vasquez, M., Gonzalez-Esquinca, A.R., Cazares, L.L., Moreno, G.M.N., García-Argaez, A.N. 1999. Antimicrobial activity of *Birsonima crassifolia* (L.) H.B.K. Journal of Ethnopharmacology, 66: 79-82.

Mejia, O., Julio, M., Sanchez, M.S., Bonilla,C., Vanegas, P. 2007. Efecto de la altura y frecuencias de corte y secado en el rendimiento y calidad del aceite esencial de pronto alivio. Scientia et Technica, XIII (033): 253-255.

Moulari, B., Pellequer, Y., Lboutounne, H., Girad, C., Chaumont, J., Millet, J., Muyar, F. 2006. Isolation and in vitro antibacterial activity of astilbin, the bioactive flavonone from the leaves of *Harungana madagascariensis Lam. ex Poir.* (Hypericaceae). Ethnopharmacology, 106(2): 272-278.

Nascimiento, G.F.G., Locatelli, J., Freitas, C.P., Silva, L.G. 2000. Antibacterial activity extracts and phytochemical on antibiotic-resistant bacteria. Brazilian Journal of Microbiolgy, 31: 247-256.

O'Bryan, C. A., Candrall, P. G., Chalova, V. I., Ricke, S. C. 2008. Orange essential oils antimicrobial activities against *Salmonella* spp. Journal of Food Science, 73: M264-M267.

Oussalah, M., Caillet, S., Salmieri, S., Saucier, L., Lacroix, M. 2004. Antimicrobial and antioxidant effects of milk protein-based film containing essential oils for the preservation of whole beef muscle. Journal of Agricultural and Food Chemistry. 52: 5598-5605.

Oyedemi, S. O., Okoh, A. I., Mabinya, L. V., Pirochenva, G., Afolayan, A. J. 2009. The proposed mechanism of bactericidal action of eugenol, α-terpineol and g-terpinene against *Listeria monocytogenes, Streptococcus pyogenes, Proteus vulgaris* and *Escherichia coli.* African Journal of Biotechnology 8(7)0: 1280-1286.

Pala-Paul, J., Copeland, L.M., Brophy, J.J., Goldsack, R.J. 2007. Essential oil composition of *Eryngium rosulatum.* P.W. Michael ined.: A new undescribed species from eastern Australia. Biochemical Systematics and Ecology, 34(11): 796-801.

Pala-Paul, J., Perez-Alonzo, J.M., Velasco-Negueruela, A., Varade, J., Villa, A.M., Sanz, J., Brophy, J. J. 2005. Analysis of the essential oil composition from the different parts of *Erygium glaciale Boiss* from Spain. Journal of Chromatography, A1094(1-2): 179-182.

Powell, J.S. and Raffa, K.F. 1999. Sources of variation in concentration and composition of foliar monoterpenes in tamarack *(Larix laricina)* seedlings: roles of nutrient availability, time of season, and plant architecture. Journal of Chemical Ecology, 25(8): 1771-1797.

Quintero, C.A., González de Colmenares, N., Staschenko, E. 2004. Aceite esencial de las hojas de *Hyptis umbrosa* Salzm extraído por diferentes técnicas. Acta Científica Venezolana 55(2): 181-187.

Rao, P.S., Sheth, N.R., Jayaveera, K.N., Rao, S.K. 2010. Pharmacognostic standardisation of the leaves of *Pimenta dioica* Linn. International Journal of Pharmaceutical Sciences and Research, 1(9): 110-115.

Raseetha, V.S., Cheng, S.F., Chuah, C.H. 2009. Comparative Study of Volatile Compounds from Genus *Ocimum*. American Journal of Applied Sciences, 6(3): 523-528.

Rivero-Cruz, J.F., Sanchez-Nieto, S., Benitez, G., Casimiro, X., Ibarra-Alvarado, C., Rojas-Molina, A., Rivero-Cruz, B. 2009. Antibacterial compounds isolated from *Byrsonima crassifolia.* Revista Latinoamericana de Química, 37(2): 155-163.

Rojas-Grau, M.A., Avena-Bustillos, R.J., Friedman, M. 2006. Mechanical, barrier and antimicrobial properties of apple puree edible films containing plant essential oils. Journal of Agricultural and Food Chemistry 54: 9262-9267.

Sahadeo, D.P., Vilas, A.K. 2011. Antibacterial activity of some essential oils against food borne pathogen and spoilage bacteria. International Journal of Pharma and Bio Sciences, 2(3):B143- B150.

Sanches, N.R., García, C.D.A., Schiavini, M.S., Nakamura, C.N., Dias, F.B.P. 2005. An Evaluation of Antibacterial Activities of *Psidium guajava* (L.). Brazilian Archives of Biology and Technology, 48(39): 429-436.

Suffrendin, B.I., Sader, S.H., Goncalves, G.A., Reis, O.A., Gales, C.A., Varella, D.A., Younes, N.R. 2004. Screening of antibacterial extracts from plants native to the Brazilian Amazon Rain Forest and Atlantic Forest. Brazilian Journal of Medical and Biological Research, 37(3):379-384.

Taylor, P.C., Shoenknecht, F.D., Sherris, J.C., Linner, E.C. 1983. Determination of minimum bactericidal concentrations of oxacillin for *Staphylococcus aureus*: Influence and significance of technical factors. Antimicrobial Agents Chemotherapy, 23(1): 142-150.

Tippayatum, P., Chonhenchob V. 2007. Antibacterial activities of thymol, eugenol and nisin against some food spoilage bacteria. Kasetsart Journal (Nat. Sci.) 41: 319 – 323.

Toncer. O., Karaman, S., Diraz, E. 2010. An annual variation in essential oil composition of *Origanum syriacum* from Southeast Anatolia of Turkey. Journal of Medicinal Plants Research, 4(11): 1059-1064.

UNAM. 2009. Guayaba. En: Argueta, V.A. (Coordinador General). Atlas de las plantas de la medicina tradicional Mexicana. Tomos I, II, III. Instituto Nacional Indigenista, México. Consultado el 15 de octubre de 2010 en la página electrónica: http://www.medicinatradicionalmexicana.unam.mx/index.php.

Viuda-Martos, M., Ruiz-Navajas, Y., Fernandez-López, J., Perez-Alvarez, J. 2008. Antibacterial activity of lemmon (*Citrus lemon* L.), mandaron (*Citrus reticulata* L.), grapefruit (*Citrus paradisi* L.) and orange (*Citrus sinensis* L.) essential oils. Journal of Food Safety, 28: 567-576.

International Migration and Change in Land use in Bella Esperanza, Veracruz

Martha Elena Nava-Tablada* and Enrique Martínez-Camarillo

El Colegio de Veracruz. Carrillo Puerto # 26, Zona Centro. CP. 91000, Xalapa, Veracruz, México.
** Corresponding Author: menavata@yahoo.com.mx*

SUMMARY

An increase in international migration from the State of Veracruz, Mexico, in the decade of the 1990s, mostly occurred as a result of the agricultural crisis in the rural sector. The state coffee producing sector proved to be no exception to the impact of the recurring crises, caused by the fall in the price of coffee grain in the international market. Many coffee growers migrated to the USA to stave off their own economic collapse. This investigation aimed to analyze the relationship between the process of international migration and change in land use in the communal landholding of Bella Esperanza, Veracruz. For this purpose, historic documentation, interviews with 21 coffee producing families -some of which included migrants-, interviews with key informants and geographical information systems were all used. The main change in land use entailed the substitution of shaded coffee plantations for sugarcane monoculture, implying deforestation. Urban expansion was shown to be incipient, in spite of the significant amount of money transfers directed towards housing construction and the proximity of this "ejido" or communal landholding to the cities of Xalapa and Coatepec. These changes are mainly associated with the crisis involving this commodity, but they have been exacerbated by emigration of family members, who were once in charge of coffee production.

Key words: *Coffea arabica* L.; rural; communal landholding; deforestation; urbanization.

INTRODUCTION

In Veracruz, Mexico, international migration intensified during the decades of 1990 and 2000. The state went from 27th place in 1997 to 4th place in 2002, in terms of federal states that sent the greatest number of migrants to the USA. This growing participation in migration was also reflected in the remittances sent home, as in the year 2002 $200 million MXN were received, of which $130 million MXN were invested in the state rural areas (Pérez, 2003). According to figures provided by INEGI (2011) for 2005, Veracruz ranked fourth nationally (in absolute and relative numbers) in terms of Mexican population residing in the USA, as a total of 56 521 migrants were registered, accounting for 5.7 % of the national total for that year. Regarding the trend for remittances received in Veracruz, according to CONAPO (2010) in 1995, the state ranked 15th place nationally in terms of attracting remittances, totaling $76 million USD, which represented 2.1 % of all remittances sent home to Mexico. In 2009, it ranked sixth with a total of $1 294 million USD, representing 6.1 % of the national total.

The increase in international migration in Veracruz is especially common in rural areas, following the collapse of the agricultural sector, forcing a mass of rural producers to seek other economic alternatives outside their place of origin. The Veracruz coffee sector was no exception and because of the recurrent crises that led many growers to economic collapse, migration to the USA has become one of the most common economic strategies to be employed in coffee growing areas (Mestries, 2003).

The increase of international migration in coffee producing areas is a phenomenon that affects the productive, environmental, economic and socio-cultural aspects of these regions, hence the importance of studying it. The interest in investigating this massive migration from Veracruz is reflected in several studies (Anguiano, 2005; Córdova *et al*., 2008). However, the relatively recent increase in migration from Veracruz is constantly and rapidly expanding to include the entire state, and shows no signs of reversal in the short term. Because of this, there is a need to elucidate little studied aspects such as its effects on the rural productive sector, from where most international migrants originate. Concerning products such as coffee, which has been very important in the national and state economy, it is essential to understand the impacts that migration flows are having on the growing areas and particularly to assess how the world coffee crisis has adversely affected the level of development of coffee regions in Mexico. Currently, a number of problems have been observed, such as job loss, abandonment of plantations and severe environmental impact resulting from tree cutting in coffee plantations, leading to changes in cultivation and urbanization, as well as a high incidence of pests and diseases that affect grain quality, resulting in low yields and a drastic fall in living standards for rural people who depend on coffee production (Martinez, 2004).

In the light of the coffee crisis and increased migration, shaded coffee plantations are being abandoned and replaced mainly with cash-crop monocultures such as sugarcane and also with housing, as mentioned in a number of studies (Nava-Tablada, 2009). However, these studies fail to address in depth, the issue of land use change as a result of international migration. Thus, there is a need to focus on this aspect of research in detail, using case studies to provide greater insight and to reveal the magnitude of the processes of land use change in coffee regions while analyzing their impact on future development.

López and Bocco (2006) state that only limited research exists employing geographic information systems (GIS) as tools for analyzing the impact of international migration on land use and resulting changes in the landscape. This is due to the difficulties in accessing demographic, socioeconomic and biophysical information relating to specific geographic areas which refer to similar time periods, especially in the instance of small scale studies such as this one. These authors have carried out one of the few investigations in Mexico, which describes the implications of migration in terms of land coverage and change in land use in the Lake Cuitzeo basin in Michoacán, during the period 1970-2000, for which they used demographic and socioeconomic information and a GIS. Their main conclusions indicate the "transformation to waste land" of land abandoned by migrants that was previously used for agriculture, and the expansion of urban settlements. These changes are associated with processes that are "degrading" the territory with resulting damage to communities, towns and landscapes. However, for the analysis of migration, the authors relied solely on the statistics from the municipalities that comprise the lake and from these they made inferences relating to the spatial analysis of land use change, but there is a lack of fieldwork that would take into account more aspects in order to better understand the relationship between the two phenomena in question.

In this context, the objective of this study was to analyze the relationship between international migration and changes in land use in the communal landholding of Bella Esperanza, a municipality of Coatepec, in Veracruz, Mexico.

MATERIALS AND METHODS

The research was conducted in the Bella Esperanza communal landholding, in the town of Coatepec, Veracruz, Mexico, in an area where small producers with shaded coffee plantations predominate, in order to provide relevant information about international migration as people in this area appeared willing to answer the survey. Fieldwork was conducted in 2008 with expeditions to explore the problem and contact key informants. Afterwards, a survey was undertaken to record general information about all family members (both residents and migrants), describing details about migration and the perception of producers concerning the impact of migration on coffee production and the community. Units of study consisted of families engaged in shaded coffee production, with an account of international migration on the part of a family member. As there was no prior knowledge of the number of households with migrants (due to the lack of reliable statistical data), no sample size was set and identification of the households included in the study was carried out using the "Snow Ball" technique (Goodman, 1961). The number of families who completed the questionnaire was limited up to the moment when "saturation" occurred, i.e. when cases no longer contributed by adding new data and response patterns were repeated. Twenty-one questionnaires were applied, preferably being answered by the family member who made the decisions concerning coffee production. Information about migrants was provided by the families who answered the questions. Data were analyzed with descriptive statistics using the Microsoft Office Excel 2007 program. The interviews of key informants recorded the historical evolution of migration in the community, the coffee crisis and change in land use, from the perspective of coffee growers. There were two in-depth interviews with people who had experienced the problems dealt with here.

Geographic information systems were used to assess the changes in land use in the community. Firstly the collection, classification and systematization of base material was carried out, including cartographic information, orthophotos, aerial photographs, maps, statistical information and historical background. Once the collected information had been systematized, we proceeded to generate a base map of the study area for the purpose of carrying out a deductive analysis using information from the most recently published satellite image indicating land use classified by three categories: urban, forest (including shaded coffee plantation) and agricultural monoculture with sugarcane. This identification was also carried out physically, once land use had been determined by GPS (MAGELLAN Professional) and defined points in the field were identified using a Google Earth image to identify similarities in terms of current use. Subsequently, an area of polygons was defined in order to obtain the first product in KMZ format for exportation, and this was imported using the Global Mapper software. This allowed us to obtain geo-referenced data and Geodetic GEO80 UTM projection for subsequent comparative analysis with material obtained using ArcView software. This information was linked to databases containing census statistics.

The analysis of land use change was carried out using satellite imagery from Google Earth, as the high resolution makes it possible to distinguish the type of ground cover. For the study area, Google Earth provides two images from the years 2003 and 2007, which because of their high definition and absence of cloud cover, can be used for comparison. It should be stated that no images of the study area exist in Google Earth for previous years and the black and white orthophotos available from the National Institute of Statistics, Geography and Informatics (INEGI) do not have the resolution required for the identification of crop type; it is only possible to distinguish the type of ground cover. Aerial photographs of the study area for 1980 and 1995 were acquired from INEGI; the earliest of these could not be used due to poor definition and is only employed again in 1995 as a benchmark to compare land cover before and after the increase in migratory flow; according to the testimony of informants, it was in the middle of the 1990s when international migration began in Bella Esperanza. Neither was there any statistical data concerning land use or crop type for each communal landholding in databases or publications from the INEGI. The lack of statistical information or quality geographic imagery is a constraint pointed out by López and Bocco (2006) for this type of study.

Description of the study area

The Bella Esperanza communal land holding is located to the southeast of the municipality of Coatepec and forms part of an agricultural corridor where there are 15,635 ha of coffee plantation, representing 11 % of the total crop growing area in the state (SAGARPA, 2007). Likewise, the region has a culture of coffee production dating from the late nineteenth century and is recognized in the international market for the quality of its grain, resulting from the optimal conditions for cultivation, in terms of altitude, climate and soil. The Bella Esperanza community (latitude 19°26’01’’ N, longitude 96°51’59” W; 1016 meters above sea level) is located in the mountainous central area of Veracruz State (Figure 1). The location falls within the area of influence of the metropolitan area of Xalapa (state capital) and is near Coatepec, another important urban area and capital of the municipality with the same name (INEGI, 2006). The climate is (A)C(FM) semi-warm humid with abundant rainfall in summer and annual rainfall varying between 1500 and 2000 mm, and with temperatures ranging between 10 and 24 °C (INEGI, 2006).

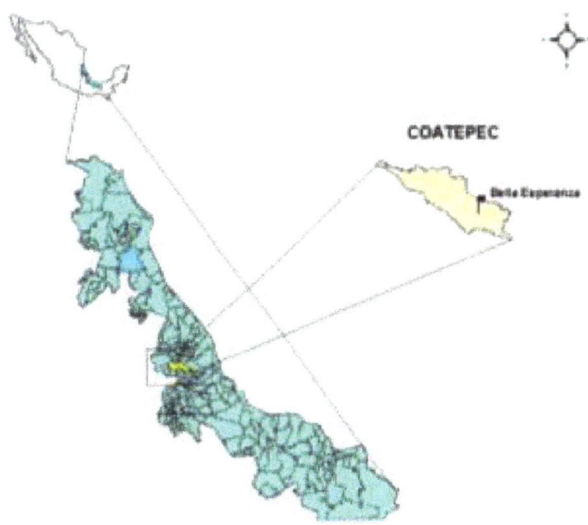

Figure 1. Location of study area, Bella Esperanza, Veracruz, Mexico.

Regarding land tenure, the communal landholding of Bella Esperanza obtained a final donation of 736.13 ha in 1931, which were distributed as follows: 680.84 ha as allotments, 11.38 ha as human settlement with title deeds, and 33.90 ha as area of communal land use (RAN, 2012). INEGI (2011) indicated that by 2010, the population of Bella Esperanza totaled 1618 inhabitants, 50.8 % men and 49.2 % women. Today there are a total of 425 homes, each with an average of four people per household. Concerning social issues; 33.55 % of the population has no right to medical services, 3.83 % of people aged 15 or older are illiterate and the average level of schooling is 7.73 years (INEGI, 2011). In 2005, CONAPO (2006) indicated that living standards are low.

Land use in the communal landholding is mainly agricultural (coffee and sugarcane). Given the characteristics of the soil, it is ideal for forestry; however, it has been primarily adapted for shade-grown coffee, which replaced the natural landscape of cloud forest and in the last years the coffee plantations have been being replaced by sugarcane production (INEGI, 2006). Due to the proximity of Bella Esperanza to the cities of Xalapa and Coatepec, it is common for residents to have jobs in the service sector in these cities, to which they commute daily. Concerning migration, INEGI (2006) for the year 2000 only registered four people who lived in another federal entity and one in the USA.

RESULTS AND DISCUSSION

Migration and change to coffee production in Bella Esperanza

International migration in Bella Esperanza initiates and increases in the 1990s influenced by nearby communities such as Tuzamapan (where migration to the USA has been going on for longer), and the fall in coffee prices left many people without work (coffee crisis). The survey recorded data from 115 persons who were members of the participating families, 76.5 % of these remained in the community and 23.5 % were active male migrants who resided permanently outside the community. The predominance of males among migrants coincides with that found in other case studies in Veracruz (Córdova *et al.*, 2008). The average age of migrants was 31.4 years, with a concentration in the age range of 16 to 30 (59.3 %) and 31 to 50 years (37 %). That is, 96.3 % consist of working age population who migrate as laborers seeking better job opportunities and wages, agreeing with observations by Ramírez and González (2006) referring to coffee communities in the state of Puebla.

Regarding the occupation of migrants at the place of destination, the majority (96.3 %) undertake activities in the secondary and tertiary sectors (builders, gardeners, employees in restaurants and shops) compared to a small percentage (3.7 %) employed in the agricultural sector. This is consistent with Anguiano (2005), who indicates that migration is mostly due to employment reasons, with the greatest number of migrants entering the secondary and tertiary sectors, because of their low rating as migrant labor, employed in activities that require no special training.

Concerning the destination of the migrants, 40.7 % were currently living in other Mexican states and 59.3 % were based in the USA, being the states of North Carolina and South Carolina the most frequent destinations. Apparently, the places for arrival are diversifying, consistent with Córdova *et al.* (2008).

Of the surveyed households, 95.2 % received remittances from the USA, destined to multiple uses, with the priority comprising household expenditure (food, clothing, medical and education expenses), home improvement and investment in agricultural production (purchase of land, inputs and payment of labor from outside the family). Remittances are spent primarily in household consumption and housing, but also function as a subsidy for replacing the missing work force to recreate the family production unit which is surviving precariously, but this has failed to re-instigate coffee production at a profitable level. Authors such as Yúnez-Naude (2002) agree that spending in family sustenance and improved housing are the main use for remittances, while investment in production is directed to hiring labor and inputs in order to replace the absent work force and subsidize production activity; however, in few cases does this represent a significant investment for reviving agriculture and making it profitable.

Of the migrants interviewed, 90.3 % indicated economic reasons (farming crisis, unemployment and lack of economic resources, and search for higher wages) as the motive for their first migration. There is consensus that economic motives are the most important (although not the only one) for the departure of migrants, especially when the employment and income situation in the place of origin is precarious (Aragon, 2006).

Of the households included in the study, 23.8 % mentioned that work has intensified for those who remain in the community (women, children and elderly), and 28.6 % indicated an increase in the hiring of laborers to replace the absent workforce, and these represent the main changes in the organization of family labor as a result of migration. This is consistent with the report by Martinez-Garza *et al.* (2010) for the dry tropical region in Central Veracruz, where because of the international migration of a member of the household, the labor force is reorganized to replace the laborer who has migrated.

Of the families interviewed, 23.8 % replaced coffee with sugarcane as a result of the low price of the coffee. It was apparent that this change to commercial monoculture, although incipient, has accelerated and is almost irreversible. One informant said: "Once the coffee plants have been cut down to plant sugarcane, it is nearly impossible to recover the plants" because of the high economic costs and the time it takes for a coffee plant to become productive (3 to 4 years). Likewise, the demolition of shade coffee plantations results in environmental degradation. Besides this, 9.5 % of households ceased planting coffee due to the production crisis, the low price in the market and the absence of family members to tend the coffee plants. In other words, the pattern of crop cultivation has been affected by migration, but the change relates more to effects caused by the crisis in the coffee sector. In this regard, Nava and Marroni (2003) state that in the Mixteca area of Puebla, migration led to the abandonment of crops due to the shortage of family labor (due to limited physical capacity to participate in diversified agriculture), but also because of the national crisis affecting the agricultural sector.

As for changes in production technology, the absence of members who were in charge of farming has led to a gradual abandonment of coffee plantations, many of which receive no agronomic management. Concurring with these results, Mestries (2003) recognizes that international migration encourages producers to abandon the work of fertilizing and weeding coffee plantations in the Xalapa-Coatepec region, causing a drastic drop in yields as well as pest infestation.

Families were also questioned about the effects on the community resulting from increased migration. It emerged that the positive changes relating to households economic improvement (improvement of the house, living standard and public services, and vehicle or land purchase) are the most commented (84.1 % of respondents). Contrastingly, 15.9 % of the comments reported changes considered negative by respondents (family breakdown, labor shortages, loss of rural culture, intensification of work for those who remain behind), as they refer to issues that point to the degradation of the material, social and cultural base of coffee production in the community. In this regard, Nava and Marroni (2003) argue that migration has contradictory effects, as in the family context remittances promote a substantial improvement in the quality of life and consumption, whereas in the community there is deterioration of the economic and cultural base that sustains farming.

Regarding the process of urbanization and the way it relates to the remittances sent to build houses, this aspect is largely perceived as positive; 50 % consider improving housing as a benefit for the family, while only one respondent (2.3 %) perceived the growth of the urban area as a negative effect, as it is transforming what was once farmland into housing areas. In this context, surveys in rural areas discovered that although remittances are invested in housing construction this has no significant effect on the growth of urban areas, as houses are not generally built on new ground, but rather the improvement is limited to remodeling and expanding the existing buildings.

Change in land use in Bella Esperanza

In this study, three forms of land use were identified: 1) forest (including shaded coffee plantations), 2) urban, and 3) cleared land (used for agriculture, mainly sugarcane). We identified two processes indicated in the survey and interviews: displacement of forest and shaded coffee replaced by sugarcane monoculture, along with the use of remittances for housing construction and a consequent saturation of the housing area (increased number of inhabitants per area), as well as a slight increase in urban area. Once uses had been established and identified, the polygons obtained were corroborated in the field using satellite images. Lastly, the percentages of surfaces covered by forestry-shaded coffee were obtained, in order to discover how they have evolved during the analyzed time period (1995-2003-2007), considering areas where tree cover has been removed as deforested (including both forest and shaded coffee plantations), destined for agriculture (mainly sugarcane).

Figure 2 indicates that in 1995 the communal landholding comprised a deforested area of 153.26 ha, and 210.59 ha in 2003, whereas a total area of 67.3 % retains its forestry or shaded coffee plantation. Meanwhile in 2007, 257.32 ha were identified as deforested with a decreased percentage of forest area and/or shade-grown coffee, which now represents 61 % of the total communal landholding area. That is, there is gradual loss in area of forest and shaded coffee, because in the period analyzed (1995-2007), 104.06 ha (14.13 % of total landholding area) had been deforested mainly for the introduction of sugarcane. Deforestation in the period 2003-2007 showed an annual rate of loss of forest cover of 1.74, and if it continues at this rate, it is to be expected that within five years the deforested area will represent 38 % of the total landholding area, and in 15 years, 45 %.

As for the relationship between urbanization and remittances (23.2 % of these are destined for housing construction), Figure 3 shows that in 1995 the urban area was 14.66 ha, in 2003 it was 15.60 ha, and in 2007 it increased to 17.37 ha, which represents an annual growth rate of 0.88. The slow growth of the urban area is in contrast with the saturation of the housing area (increased number of inhabitants per area), derived from the fact that new rooms were usually built to house the family of the children. Besides this, remittances are often used for remodeling the existing homes, but infrequently used to buy new land for housing.

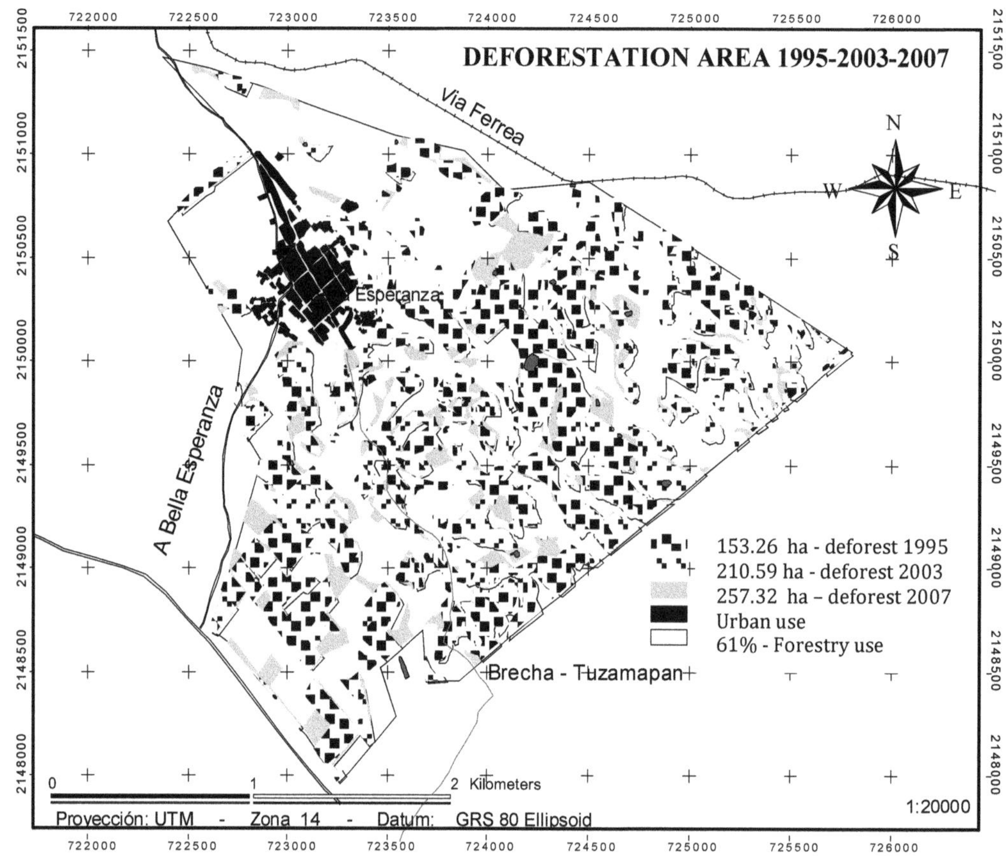

Figure 2. Deforestation in the Bella Esperanza communal landholding area, Veracruz, Mexico. 1995-2003-2007.

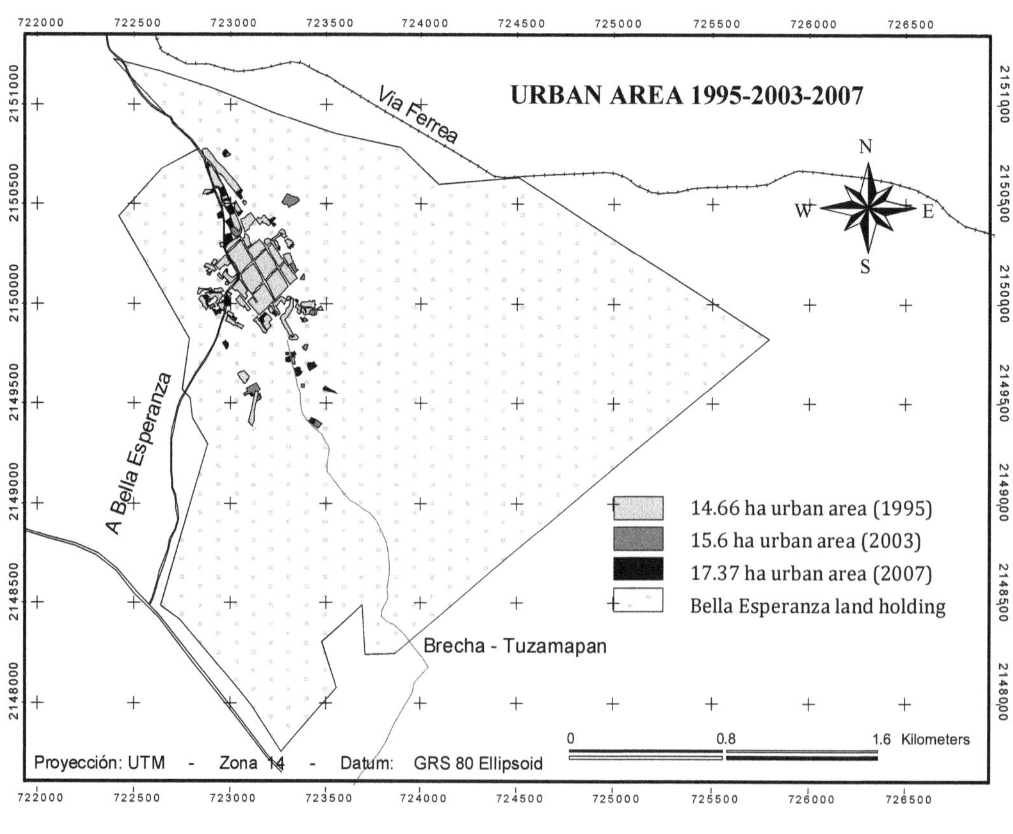

Figure 3. Growth of urban area in Bella Esperanza, Veracruz, Mexico. 1995-2003-2007.

Relationship between international migration and change in landuse

International migration in Bella Esperanza increased as a result of the production crisis caused by falling coffee prices in the international market, state withdrawal from the agricultural sector (privatization) and crisis in the Veracruz labor market (lack of jobs and low wages). The most important migratory flow is directed towards the USA, these migrants being generally working-age men, who leave for economic reasons and are inserted into low-skilled jobs in industry and services in urban areas of that country.

From 1990 onwards, Bella Esperanza experiences the desertion of working-age male population who emigrate to the USA following the first major coffee crisis (1990-1994), and international migration intensifies during the second coffee crisis (1998-2004) which was considered much more severe, so as reported in the survey, 80.6 % of migrants leave for the first time in the period between 2000-2008.

DISCUSSION

The other land use transformation of interest is the growth of the urban area in Bella Esperanza, detecting that this process is still in the early stages despite the large proportion of remittances that are used for housing construction (23.2 %), and the relative proximity of the communal landholding area to the cities of Xalapa and Coatepec. However, the apparent trends shown by neighboring communal landholding areas indicate that it is highly probable that conurbation of rural areas near these cities will occur and that housing developments will encroach upon agricultural land, affecting Bella Esperanza.

The results presented are consistent with those reported by Mestries (2003) in the central coffee region of Veracruz, where although most producers have not been willing or able to change their land use in the hope of an improvement in coffee prices, or for a lack of alternative products or resources, there has been a growing tendency to remove the coffee plantations in order to grow sugarcane. This causes serious environmental degradation, as trees that served as shade for coffee plantations are eradicated, contributing to loss of remnants of cloud forest. The increased dynamics in the land market are also notable, as more and more coffee plantations near the cities of Xalapa and Coatepec are being divided into lots to be sold as real estate.

Although the increase in international migration in Bella Esperanza parallels the increase in area planted with sugarcane, replacing shaded coffee plantations, this does not ensure that there is a direct correlation

The results referring to land use in Bella Esperanza, indicate that the main change is the replacement of shaded coffee plantations for sugarcane, which involves a process of deforestation. In this regard, although sugarcane is also facing a production and market crisis caused by the withdrawal of state subsidies, privatization of sugar mills and import of concentrated fructose, it still represents for the producers an alternative for family subsistence. Because sugar mills control almost the entire production process, the sugar growers earn a small but guaranteed amount of money, without having to invest much family labor or attention in the sugarcane plot. This applies even when sugar prices rise, as the production crisis is structural in origin and depends on international market fluctuations, so that prices experience periods of increase and recurrent falls. Similarly, rising prices are of little benefit to small producers, as most of the profits are for those who collect, process and market production (in this case the mill). A strong reason for opting to grow sugarcane (from the perspective of producers) is that this activity gives them access to Social Security, which implies health care and pension rights (Núñez, 2005). between these two variables, as the coffee crisis is an important factor for understanding these changes.

Thus, with restricted possibilities for maneuver in terms of the limited resources available, the fall in coffee prices, lack of access to credit and technical assistance, and limited management capacity, the coffee producers of Bella Esperanza responded with a short term perspective, in order to reduce the risk of financial failure in an uncertain situation. This has particularly promoted international migration, the adoption of commercial monoculture such as sugarcane production, representing an activity that despite all odds still provides them with economic and social benefit. However, this strategy implies high environmental costs (deforestation, land degradation, biodiversity loss and overexploitation of natural resources) that in the medium and long term can compromise their productive capacity and thus future development possibilities.

In other words, the decline of coffee production forces Bella Esperanza farmers to choose whether to risk trying new potentially feasible productive options which still have a small margin of profit (such as sugarcane), or to continue to invest their limited resources in coffee production, whose economic benefits are uncertain in the light of current national and international conditions, or to engage in emerging strategies such as emigrating to the USA ensuring the short-term survival of the family production unit, but with long-term disrupting effects, both in terms of the family production unit and the community.

CONCLUSION

The main change in land use in Bella Esperanza is the replacement of shaded coffee plantations, causing deforestation. Meanwhile, urban expansion is still incipient, despite the large proportion of remittances invested in housing construction (23.2 %) and the proximity of the communal landholding to the cities of Xalapa and Coatepec. Changing land use is more related to the crisis in the coffee sector than it is to migration, as the substitution of coffee for commercial monocultures such as sugarcane is mainly associated with the low price of coffee in the international market, but this transformation is exacerbated when families who once tended the coffee plantations migrate. This change is also influenced by government policies of state withdrawal, privatization of agriculture and the consequent increase in the economic vulnerability of small coffee producers. Although migration is not the cause of the decline in coffee production, it accelerates this process by decreasing family labor available for production of coffee, also influencing changes in the economic activities carried out by migrants who return to the community (they usually do not get involved in agricultural work again) taking up alternative jobs, as agriculture becomes less profitable.

Finally, the increase in international migration in Bella Esperanza, while coinciding with the increase in area planted with sugarcane gradually replacing the shaded coffee plantations, does not necessarily indicate that there is a direct relationship between these processes. The trigger for both phenomena is the coffee crisis, where international migration, together with changes in land use, constitutes interdependent phenomena that occur in parallel, reflecting new family production strategies for addressing the production crisis. Although the increase in international migration has an impact on land use, as it changes the dynamics inside the family farms, the coffee crisis is a more important factor for understanding these changes. In other words, the relationship between migration and change of land use cannot be simplified to a cause-effect relationship, as there are multiple effecting factors involved, making it complex.

ACKNOWLEDGEMENTS

The CONACYT Joint Fund – Government of the State of Veracruz funding provided for the project "Effects of migration on agricultural activity in four farming regions in the state of Veracruz" (code number 68277), which is a part of this investigation.

REFERENCES

Anguiano, T.M.E. 2005. Rumbo al norte: Nuevos destinos de la emigración veracruzana. Migraciones Internacionales. 3:82-110.

Aragón, G.C. 2006. Cafeticultura, inequidad y pobreza. In: Ramírez, V.B., Juárez, S.J.P., Cesín, V.A. (coord.). Productores Indígenas de Café de la Sierra Nororiente de Puebla. Problemas y Alternativas. Consejo Nacional de Ciencia y Tecnología, Colegio de Postgraduados. México. pp. 13-32.

CONAPO (Consejo Nacional de Población). 2006. Índices de Marginación 2005. www.conapo.gob.mx (Consulted: November 24, 2007).

CONAPO (Consejo Nacional de Población). 2010. Portal electrónico de CONAPO. www.conapo.gob.mx (Consulted: April 14, 2010).

Córdova, P.R., Núñez, M.C., Skerrit, G.D. 2008. Migración internacional, crisis agrícola y transformaciones culturales en la región central de Veracruz. Centro de Estudios Mexicanos y Centroamericanos. Universidad Veracruzana, Consejo Nacional de Ciencia y Tecnología. Ed. Plaza y Valdés. México.

Goodman, A.L. 1961. Snowball Sampling. The Annals of Mathematical Statistics. 32:148-170.

INEGI (Instituto Nacional de Estadística, Geografía e Informática). 2006. Conteo de Población y Vivienda 2005. Perfil Socio-demográfico, Veracruz de Ignacio de la Llave, Tabulaciones Básicas. INEGI. México.

INEGI (Instituto Nacional de Estadística, Geografía e Informática). 2011. Censo Nacional de Población y Vivienda 2010. www.inegi.org.mx/est/contenidos/proyectos/ccpv/cpv2010/default.aspx (Consulted: October 27, 2011).

López, E., Bocco, G. 2006. Patrones locales de cambios globales: efectos de la emigración en el cambio de uso de suelo en el centro de México. El caso de la cuenca del Lago de Cuitzeo. Scripta Nova X. 218:1-15.

Martínez, A.C. 2004. Transformación de la actividad cafetalera en los años noventa. In: Rubio, B. (coord.). El sector agropecuario mexicano frente al nuevo milenio. Universidad Nacional Autónoma de México. Ed. Plaza y Valdés. México. pp. 109-146.

Martínez-Garza, S.E., Nava-Tablada, M.E., Gallardo-López, F., Ruíz-Rosado O., Vázquez-García, V. 2010. Effects of international migration on the Acazónica and Hato de la Higuera agroecosystems in Veracruz, México. Tropical and Subtropical Agroecosystems. 12:1-10.

Mestries, B.F. 2003. Crisis cafetalera y migración internacional en Veracruz. Migraciones Internacionales. 2:121-148.

Nava, T.M.E., Marroni, G. 2003. El impacto de la migración en la actividad agropecuaria de Petlalcingo, Puebla. Agrociencia. 37:657-663.

Nava-Tablada, M.E. 2009. Efectos de la migración en la cafeticultura de la región de Coatepec, Veracruz. In: Galdaméz, G.J., Guevara, H.F., Soto, P.L., López, M.J., Vázquez, G.M. (comp.). Agricultura Sostenible Vol. 6. Universidad Autónoma de Chiapas. Sociedad Mexicana de Agricultura Sostenible, A.C.México. pp. 1-11.

Núñez, M.M.C. 2005. Ejido, Caña y Café. Política y Cultura Campesina en el Centro de Veracruz. Universidad Veracruzana. México.

Pérez, M.M. 2003. Las redes sociales en la migración emergente de Veracruz a los Estados Unidos. Migraciones Internacionales. 2:136-160.

Ramírez, V.B., González, R.A. 2006. La migración como respuesta de los campesinos ante la crisis del café: estudio en tres municipios del Estado de Puebla. Ra Ximhai. 2:319-341.

RAN (Registro Agrario Nacional). 2012. Padrón e Historial de Núcleos Agrarios (PHINA). phina.ran.gob.mx/phina2/ (Consulted: March 26, 2012).

SAGARPA (Secretaría de Agricultura, Ganadería, Desarrollo Rural, Pesca y Alimentación). 2007. Servicio de Información Agroalimentaria y Pesquera (SIAP). www.siap.gob.mx (Consulted: December 15, 2007).

Yúnez-Naude, A. 2002. Las remesas y el desarrollo rural. Seminario Internacional sobre la Transferencia y Uso de las Remesas: Proyectos Productivos y de Ahorro. 3-5 octubre 2001. Zacatecas, México, pp. 21-35.

8

Effect of Diesel and Biodiesel on the Growth of *Brachiaria Decumbens* Inoculated with Arbuscular Mycorrhizal Fungi

Dora Trejo[1*], Cesar Moreira[1], Isaac Bañuelos[2], Liliana Lara[1], Guillermo Alafita[1] and Araceli Reyes[3]

[1]Universidad Veracruzana, Facultad de Ciencias Agrícolas, Laboratorio de Organismos Benéficos. Circuito Gonzalo Aguirre Beltrán s.n., Zona Universitaria Xalapa, Veracruz. CP 91000, México. Email: doratrejo@gmail.com,

[2]Unidad de Servicios de Apoyo en Resolución Analítica (SARA), Dr. Luis Castelazo Ayala s/n, Col. Industrial-Animas Xalapa, Veracruz. CP 91190, México;
[3]Ciencias Básicas Dr. Luis Castelazo Ayala s/n, Col. Industrial-Animas Xalapa, Veracruz. CP 91190, México.
**Corresponding author*

SUMMARY

Arbuscular mycorrhizal fungi have been found to be associated with plants useful in soil phytoremediation. The aim of this study was to compare the effects of diesel and biodiesel in soil and sand on the growth of *Brachiaria decumbens* inoculated with mycorrhizae. Two experiments were carried out: one experiment in soil and another in sand. A two-level- factorial design with three factors was used (one on sterile and another on non-sterile soil, with and without mycorrhizae; and one with diesel and another with biodiesel). In sand, a two-factor design with two levels was used (with and without mycorrhizae and with diesel and biodiesel), both with three replications. NOVADIESEL, biodiesel and PEMEX diesel were use as contaminants, both at 7%. The fresh and dry weight of the plants and percentage of mycorrhizal colonization, were assessed 30 days after planting. In soil, biodiesel was more toxic and reduced the fresh and dry weights of plants, especially in non-sterile soil. Biodiesel yielded greater mycorrhizal colonization values that doubled those of the control. In sand, diesel was found to reduce three times the fresh and dry weights of plants, compared to the biodiesel. In sand diesel presented high values of mycorrhizal colonization in comparison with biodiesel. Plants inoculated with mycorrhizal fungi exhibited better development than non-inoculated plants, even in the presence of contaminants.

Key words: Soil contamination; fuel; symbiotic fungus; grass; phytoremediation.

INTRODUCTION

Arbuscular mycorrhizal fungi (AMF) are considered one of the most common microorganisms, constituting an important functional component of plant-soil systems in almost all habitats and climates, and they have been shown to facilitate plant growth in soils contaminated with hydrocarbons. AMF have also been found to regularly associate with the roots of plants used in soil phytoremediation (Binet *et al.*, 2001; Leyva *et al.*, 2002; Vosátka *et al.*, 2006).

AMF have been extensively studied as a potential tool in the recovery of contaminated sites by various agents; thus to date, several studies investigating the effects of diesel soil contamination have been conducted; however, no reports have assessed the effects of biodiesel. Most research work on the field of interaction of contaminant mycorrhizae has focused in heavy metals (Kirk *et al.*, 2005; Graman *et al.,* 2003; Sudova *et al.,* 2007; Alarcón *et al.,* 2008; Usman and Hashem, 2009; Hernández-Ortega *et al.*, 2012). Considering the current use of biodiesel as an energy alternative, it is critical to understand its effects on soil microflora and plants (Binet *et al.*, 2001; Leyva *et al.*, 2002; Xiaolin *et al.*, 2002).

In this paper, we consider that due to the higher chemical reactivity of the fatty alkyl acid esters (biodiesel), these will degrade faster than those diesel hydrocarbons reagents. Nevertheless, our hypothesis is that in the short time, biodiesel will have a more toxic effect on plants and AMF will help to the survival of vegetables in their early growth stages. AMF will move the necessary resources so that plants will have an acceptable development, further more they will facilitate that degrader microorganisms of pollutants proliferate in the rhizosphere.

The goal of the present study was to compare the effects of diesel and biodiesel contamination in soil and sand on the growth of *Brachiaria decumbens* Stapf., a mycorrhizal plant that grows easily in the tropical areas of southern Mexico.

MATERIALS AND METHODS

Characterization of the soil

The soil used was taken at a depth of 20 cm and it was collected from coffee farms in Xalapa, Veracruz. The physicochemical properties of the soil were determined using the Mexican Official Standard 021 RECNAT 2000. The soil had a texture sandy-loam with the following properties: 49.20% sand, 21.80% clay and 29.00% silt; 6.56% organic matter, and a $pH_{water\ (1:2\ p/v)}$ of 5.79. The concentrations of soil nutrients were as follows: 28 ppm inorganic N, 13.6 ppm P *i*, 395 ppm K, 1561 ppm Ca, 262 ppm Mg, 29.7 ppm Fe, 3 ppm Zn, 1.5 ppm Cu, and 24.8 ppm Mn.

Contamination of the substrate

The contaminant used was Diesel which was taken at random to compare with biodiesel (made from residual oils). These contaminants or similar such as petroleum, biodiesel, diesel, born oil etc. have been used many times (Lapinskiene *et al.*, 2006; Demello *et al.*, 2007; Alarcón *et al.* 2008). The substrates (soil and sand) were dried, disaggregated and sieved to particle sizes between 2.00 and 2.36 mm to achieve a more homogeneous distribution of the contaminant and to ensure the occurrence of aerobic processes.

The portion of the soil used for the control was sterilized in an autoclave operating at a temperature of 121°C and a pressure of 15 kg/m^2 for 1 hr. The remaining portion was not sterilized. Both portions were used to compare the behaviour of the microorganisms in the biodegradation in soil contaminates.

The substrates were contaminated separately for the different experimental treatments: sterile soil plus diesel (SS+D), sterile soil plus biodiesel (SS+B), sterile soil minus diesel (SS-D), sterile soil minus biodiesel (SS-D), sand plus diesel (S+D), sand plus biodiesel (S+B). An intermediate value of 7%, as reported by Lapinskiene *et al.* (2006), was used as the concentration for both contaminants. Diesel (15.2 ml) and biodiesel (16.25 ml) were dissolved in 100 ml acetone and then used to saturate 195 g of each substrate. Each substrate was subsequently mixed with an electric mixer to homogenize the contaminants and it was allowed to dry in glass trays for one week. Glass containers were filled with 32.5 g of each substrate.

Experimental design and statistical analysis

Two experiments were performed. In the first one, a randomly factorial design soil substrate was used; it conveyed three factors: sterile [(SS+) and non-sterile (SS-)]; contaminants [biodiesel (B) and diesel (D)] and mycorrhizae with mycorrhizae [(M+) and without mycorrhizae (M-)], with 12 treatments and three repetitions each. The second substrate consisted of bank sand, with a randomly factorial design: mycorrhizae [with and without (M+, M-)], two types of contaminant [biodiesel (B) and diesel (D)], six treatments and three repetitions. The experiment was conducted in a growth chamber. Conditions during the course of the trial were as follows: 25°C average temperature, 80% relative humidity and a photoperiod of 12 hours light and 12 hours dark. Data were subjected to analysis of variance (ANOVA) with the following programs: STATGRAPHICS CENTURIUN XV 15.206. Followed by Fisher's least significant difference (LSD) test at $P < 0.05$.

Inoculation

Treatment with mycorrhiza was inoculated with 5 g of MTZ1-UV mycorrhizal complex, which consists of 12 AMF species (*Acaulospora morrowiae, A. spinosa, A. scrobiculata, Funneliformis mosseae, F. geosporus, Gigaspora rosea, Gi. decipiens, Glomus macrocarpum, Gl. aggregatum, Rhizophagus intraradices, Scutellospora pellucida* and *Claroideoglomus etunicatum*) with 75% mycorrhizal colonization, provided by the laboratory investigating beneficial organisms at the Universidad Veracruzana. *B. decumbens* seeds (3 g) were sown in each of the pots and watered with 10 ml of tap water every two days.

Evaluated Variables

The weight of fresh and dry leaves was evaluated (at 60°C for 72 hours) 30 days after planting for each treatment, and we also evaluated the percentage of root colonization by mycorrhizal fungi (Giovannetti and Mosse, 1980).

RESULTS

In sand, only were found significant differences in the biomass fresh and dry weights, the contaminant-mycorrhizal interaction (Table 1) was only observed due to the effect of the pollutant. In soil, 30 days after planting, statistically significant ($P<0.01$) differences were observed for the fresh and dry weights and the percentage of mycorrhizal colonization due to the effect of the contaminants (diesel and biodiesel). For the interaction between the contaminant, sterile soil and for the three factors evaluated, differences in the fresh and dry weights were only observed when comparing the control and contaminated treatments (Table 1). The higher values of biomass fresh and dry weights were in sand the contaminant and the lower was the interaction between contaminant and mycorrhiza. In soil the higher was the interaction between contaminated soil sterilization and the lower was the interaction between the three factors.

Plants inoculated with the AMF showed a significant difference in comparison to those which were not inoculated.

The fresh weight of plants inoculated with mycorrhizal fungi was significantly higher than those non-inoculated plants (Fig. 1a). However, differences in dry weight were not observed (Fig. 1b). Thus, these differences affecting only the fresh weights were likely due to the effect of the mycorrhiza and their interaction with the contaminant.

The presence of diesel in the sand caused a significant drop (Figs. 1a y b), and compared to the uncontaminated control, the presence of diesel in the sand caused a significant reduction three times ($P < 0.001$) in the fresh and dry weight of the plants (Figs. 1a and b). Similarly, biodiesel caused significant decreases ($P < 0.001$) in the fresh and dry weights of the plants (Figs. 1a and b). When plants were grown in diesel-contaminated sand, inoculation with mycorrhizal fungi did not have any effects on the dry and fresh weights of the plants. In contrast, inoculated plants growing in biodiesel-contaminated sand showed an increase a half time in fresh and dry weights compared to non-inoculated plants grown in contaminated sand (Figs. 1a and b).

Table 1. ANOVA of mycorrhizal root colonization and dry shoot weight of *B. decumbens* in soil and sand contaminated by diesel and biodiesel.

	Sand			Soil		
	DW	FW	Col	DW	FW	Col
Contaminant (C)	***	**	NS	**	***	*
Mycorrhiza (M)	NS	NS	-	NS	*	-
C-M	NS	**	-	NS	**	-
Sterilization (S)	-	-	-	NS	NS	NS
C-S	-	-	-	***	***	NS
S- M	-	-	-	NS	NS	-
C-M-S				*	*	-

*, P < 0.05; **, P< 0.01; ***, P <0.001; DW, dry weight; FW, fresh weight; Col, colonization; NS, no significative; -, not measured.

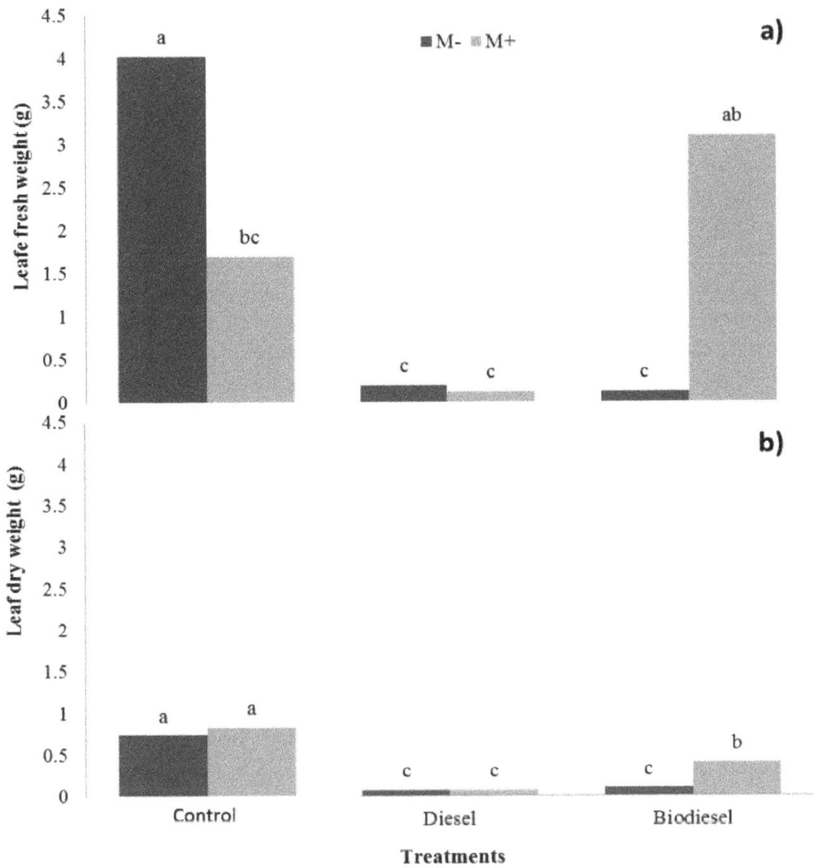

Figure 1. Leaves fresh a) and dry b) weights of *Brachiaria decumbens* inoculated and not by mycorrhizal fungi in sand with diesel, biodiesel and a control. Bars with equal letters are statistically similar (LSD, $P \leq 0.05$); n=3 M-, without mycorrhiza; M+, with mycorrhiza.

While a decreasing trend in mycorrhizal colonization was observed for sand contaminated with diesel compared to the control (Fig. 2b), an increasing trend was observed for biodiesel-contaminated sand relative to the control (Fig. 2b). The presence of diesel and biodiesel in the non-sterile soil diminished the weight of fresh and dry plants; nevertheless, biodiesel caused a higher reduction of weight in the plant leaf mass (Figs. 2 a y b).

Plants inoculated with AMF showed a reduction in fresh leaf weight; nevertheless, this was not observed in the dry weight (Figs. 2 a y b). Mycorrhizal inoculation did not have any effect on fresh or dry leaf mass, or in non-sterile contaminated soil with diesel or biodiesel (Figs. 2 a y b). However, in sterile soil mycorrhizal inoculation countered the reduction of fresh and dry leaf mass in plants only in those treatments with biodiesel. Soil sterilization did not affect mycorrhizal colonization However, only the presence of biodiesel increased colonization relative to the control and diesel (Fig. 3a).

DISCUSSION

The results of the present study demonstrate that biodiesel is more toxic to plants than diesel (Figure 1). This may be due to the fact that the soil particles stay impregnated by the biodiesel due to the facts that obstructs the pore spaces and hamper the adequate flux of water and nutriments, which at the same time may be the cause of the damage to the plants, such as has been reported by (Lapinskienė *et al.,* 2006).

On the other hand, despite the fact that biodiesel seems to be more toxic to plants, it has been reported that can be used as a biological solvent in soils contaminated with diesel, because of the transformational co-metabolic process in which microorganisms use a second substrate as a source of energy to degrade the first substrate (Pasqualino *et al.*, 2006). It had been mentioned that biodiesel is more viscous (3.5-5.0) than diesel itself (2.4-4.5) and has a very low steam pressure (Fangrui and Hanna, 1999). Nonetheless, the degrading percentage of biodiesel is 80% in 28 days, while that of diesel is

45% in the same amount of time, but disappears more easily than biodiesel (Lapinskienė *et al.*, 2006; Pasqualino *et al.*, 2006).

Moreover, the limited growth of *B. decumbens* was observed at concentrations of 7% v/v. Even in the presence of diesel, plants developed better than those in soil contaminated with these observations could be due to various factors, such as the volatilization of the contaminant (diesel is more volatile than biodiesel) (Biodisol, 2005; Martínez, 2008), in turn diesel contains compounds of intermediate and low molecular weight in addition to organic compounds of higher molecular weight, while biodiesel (a mixture of methyl or ethyl esters of fatty acids) does not (Adams and Morales-Garcia, 2008). Furthermore, these compounds are characterized by their non polar nature (hydrophobic), which confers greater volatility to diesel (Adams and Morales-Garcia, 2008), and the concentration of diesel may have therefore declined. On the other hand, the lower density and viscosity of diesel (may have allowed it to leach rather than being retained in the area of the rhizosphere (Knothe and Steidley, 2005; Baroutian *et al.*, 2010)

No effect of soil sterilization was observed possibly because the sampling time was very short (30 days). Alternatively, it may not have been possible to detect effects due to the presence of microbial complexes involved in the degradation of the pollutants, particularly for biodiesel. Several studies (Makareviciene and Janulis, 2003) have indicated that biodiesel is more readily degraded by microorganisms, tales como *Bacillus, Proteus, Pseudomonas, Citrobacter* and *Enterobacter* (Leahy and Colwell, 1990; Lutz *et al.*, 2006).

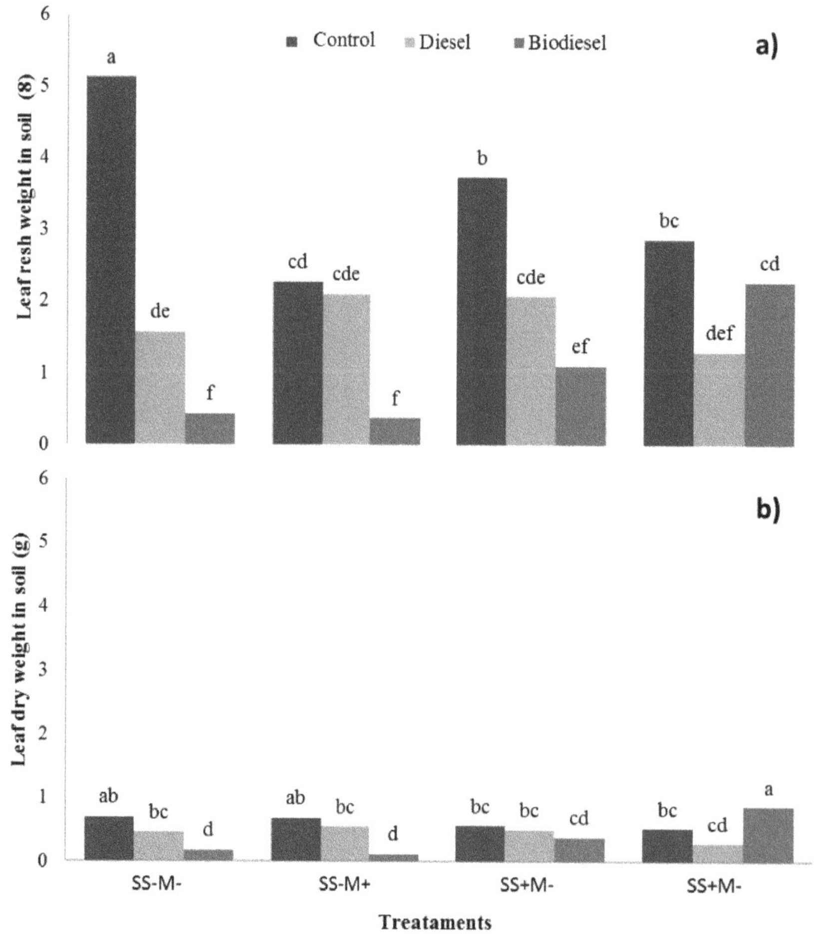

Figure 2. Leaves mass fresh (a) and dry (b) weights of *B. decumbens* with and without mycorrhizal fungi inoculation in soil with diesel, biodiesel and the control. Bars with the same letter are significantly different (LSD, $P \leq 0.05$); n = 3. SS-, non-sterile soil; SS+, sterile soil; M-, without mycorrhiza; M+, with mycorrhiza.

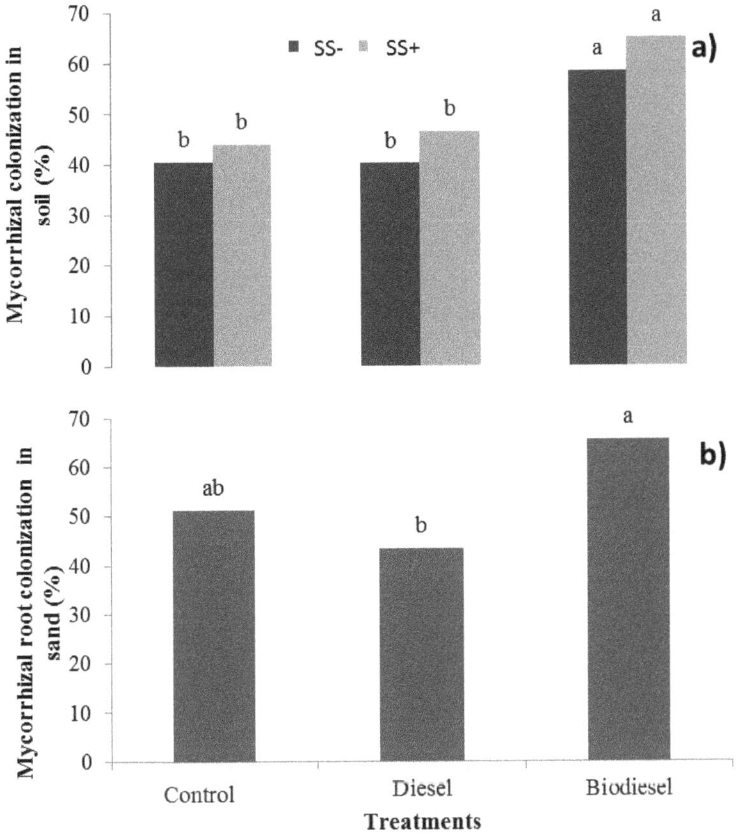

Figure 3. Mycorrhizal colonization (%) in sterile and non-sterile soil contaminated with diesel and biodiesel (a) and in sand contaminated with diesel and biodiesel (b). Bars with the same letter are significantly different (LSD, $P \leq 0.05$); n = 3. SS-, non-steril soil; SS+, steril soil.

A significant effect was observed when the interaction among the contaminant, soil sterilization and mycorrhizal inoculation was analysed. There was a clear tendency of clustering due to the effect of the pollutant. The greatest fresh and dry weights were observed with the controls, and biodiesel was found to have a more negative effect on plant growth than diesel. This result suggests that arbuscular mycorrhiza tolerates the presence of biodiesel in the ground, as shown in several studies investigating soils contaminated with hydrocarbons (Leyval *et al.*, 2002, Jeffries *et al.*, 2003 Volante *et al.*, 2005). In unsterilized soil, however, it appears that native microflora do not permitted to establish the symbiosis, and the beneficial effect of AMF in promoting growth is suppressed, which is consistent with the findings reported by Hetrick and Wilson (1990). Similar trends were observed when sand was used as the substrate, and colonization increased when the substrate contained biodiesel. When diesel was used as the contaminant in sand, there was little difference in the percentage of colonization compared to the control. Similar results for biodiesel were obtained by Alarcón *et al.* (2008), who observed that colony-forming units (cfu) of filamentous fungi increased when the inoculated soil was contaminated with oil.

CONCLUSIONS

The presence of diesel in soil affected negatively mycorrhizal colonization and plant biomass. In contrast, biodiesel promoted greater mycorrhizal colonization but led to reduced plant development. In sand, diesel contamination resulted in higher values of mycorrhizal colonization compared to contamination with biodiesel. However, plant biomass was nearly three times higher in sand contaminated with biodiesel than with diesel. In the presence of the contaminants, plants inoculated with mycorrhiza developed better than non-inoculated plants. *Brachiaria decumbens* can be used as plant shape for the study of the interactions AFM and contaminants of the type of the hydrocarbons, since they present an excellent and quick response to the settling mycorrhizae (20-30 days); they have short life cycle; and its minimal pigmentation at the roots facilitates the observation of mycorrhizal fungal structures. Likewise, these plants have a relative degree of tolerance to the presence of fossil and non-fossil fuels, which allows for the completion of the plant life cycle. It will be critical to further investigate the extent of degradation of biodiesel and diesel in the rhizosphere of contaminated soils and the potential use of mycorrhiza as an alternative for soil remediation in response to fuel spills.

ACKNOWLEDGEMENTS

We would like to acknowledge the Faculty of Agricultural Sciences for lending their facilities, especially to the Laboratory of Beneficial Organisms. With special recognition to PhD. Samuel Cruz Sanchez and MSc. Jacob Bañuelos for their support at analysing the data and doing the statistical inform; to MSc. Ingrid Karina López Guzman for the analysis of samples and to PhD. Ulises Coffen, from the Bioprocess Solutions, for donating the biodiesel used for this experimentation.

REFERENCES

Adams, R. H. And Morales García, F. 2008. Concentración residual de hidrocarburos en el suelo del trópico y consideraciones para la salud pública y protección al ganado. Interciencia. 33:476-482.

Alarcón, A., Davies, Jr. F.T., Autenrieth, R. L., Zuberer, D. A. 2008. Arbuscular mycorrhiza and Petroleum-degrading microorganisms enhance pythoremediation of petroleum-contaminated soil. International Journal of Phytorremediation. 10:251-263.

Baroutian, S., Aroua, M. K., Raman A.A.A., and Sulaiman, N. M. N. 2010. Viscosities and Densities of Binary and Ternary Blends of Palm Oil + Palm Biodiesel + Diesel Fuel at Different Temperatures. Journal of Chemestry Engineering and Data. 55:504–507.

Binet, P., Portal, M., And Leyva, C. 2001. Application of GC-MS to the study of anthracene disappearance in the rhizosphere of ryegrass. Organic Geochemestry. 32:217-222.

Biodisol. 2005. biodiesel ¿Qué es el biodiesel?, http://www.biodisol.com/biodiesel-que-es-el-biodiesel-definicion-de-biodiesel-materias-primas-mas-comunes/, recuperado en el año 2010.

Demello, J.A., Carmichael, C.A., Peacock, E.E., Nelson, R.K., Arey, J.S., Reddy, C.M. 2007. Biodegradation and environmental behaviour of biodiesel mixtures in the sea: an initial study. Marine Pollution Bulletin. 54:894–904.

Fangrui, M., and Hanna, M.A, 1999. Biodiesel production: a review. Bioresource Tech. 70:1-15.

Giovannetti, M., and Mosse, B. 1980. An evaluation of techniques for measuring vesicular–arbuscular mycorrhizal infection in roots. New Phytologist. 84:489–499.

Hernández-Ortega, H.A., Alarcón, A., Ferrera-Cerrato, R., Zavaleta-Mancera, H.A., López-Delgado, H.A., Mendoza-López, M. R. 2012. Arbuscular mycorrhizal fungi on growth, nutrient status, and total antioxidant activity of *Melilotus albus* during phytoremediation of a diesel-contaminated substrate. Journal of Environmental Management. 95:319–324.

Hetrick, B.A.D., and Wilson. G.W.T. 1990. Relationship of native and introduced mycorrhizal fungi to mycorrhizal dependence of Andropogon gerardii and Koeleria pyranidata. Mycologia. 82:779-782.

Jeffries, P., Gianinazzi, S., Perotto, S., Turnau, K., Barea, J.M. 2003. The contribution of arbuscular mycorrhizal fungi in sustainable maintenance of plant health and soil fertility. Biology and Fertility of Soils. 37:1-16.

Kavanova, M., Grimaldi, A.A., Lattanzi, F.A. and Schnyder, H. 2006. Phosphorus nutrition and mycorrhizaeffects on grass leaf growth. P status-and size-mediated effects on growth zone kinematics. Plant Cell and Environment. 29:511-520.

Kirk, J.L., Moutoglis, P., Klironomos, J., Lee, H., and Trevors, J.T. 2005. Toxicity of diesel fuel to germination, growth and colonization of Glomus intraradices in soil and in vitro transformed carrot root cultures. Plant and Soil. 270:23-30.

Knothe, G., Steidle, K.R. 2005. Kinematic viscosity of biodiesel fuel components and related compounds. Influence of compound structure and comparison to petrodiesel fuel components. Fuel. 84:1059–1065.

Lapinskiene, A., Martinkus, P. and Rébzdaite, V. 2006. Eco toxicological studies of diesel and biodiesel fuels in areated soil. Environmental pollution. 142:432-437.

Leahy, J.G., and Colwell, R.R. 1990. Microbial degradation of hydrocarbons in the environment. Microbiological Reviews. 54(3):305-315.

Leyva, C., Joner, E.J., Del Val C. and Haselwandter, K. 2002. Potential of arbuscular mycorrhizal fungi for bioremediation. In: (eds.) Gianinazzi, S., Schüepp, H., Barea, J.M. Haselwandter, K. Mycorrizal Technology in Agriculture. pp. 175-186.

Lutz, G., Chavarría, M., Arias, M.L., and Mata-Segreda, J.F. 2006. Microbial degradation of palm (Elaeis guineensis) biodiesel. Revista de Biología Tropical. 54(1) 59-63.

Makareviciene, V., and Janulis, P. 2003. Environmental effect of rapeseed oil ethyl ester. Renewable Energy. 28:2395–2403.

Malgorzata Hawrot, H.P. 2011. The influence of diesel of fuel on soil microbial biomass. Polish Journal of Environment. 20(2):497-501.

Martíncz, S. 2008, Carburantes diesel y el futuro http://www.silviamar.com/Spanish/Documentos/diesel.htm recuperado en el año del 2010.

Pasqualino, J.C., Montané, D., and Salvadó, J. 2006. Synergic effects of biodiesel in the biodegradability of fossil-derived fuels. Biomass and Bioenergy. 30(10):874–879.

Sudova, R., Pavlikova, D., Macek, T., and Vosatka, M. 2007. The effect of EDDS chelate and inoculation with the arbuscular mycorrhizal fungus Glomus intraradices on the efficacy of lead phytoextraction by two tobacco clones. Applied Soil Ecology. 35 (1):163-173.

Volante A., Lingua, G., Cesaro, P., Cresta, A., Puppo, M., Ariati L. and Berta, G. 2005. Influence of three species of arbuscular mycorrhizal fungi on the persistence of aromatic hydrocarbons in contaminated substrates. Mycorrhiza. 16:43–50.

Vosátka, M., Rydlová, J., Sudová, R., and Vohní, M. 2006. Mycorrhizal fungi as helping agents in phytoremediation of degraded and contaminated soils. Phytoremediation Rhizoremediation. 9A:237-257.

Xiaolin, L., Chen, B., Feng, G. and Christie, P. 2002. Role of arbuscular mycorrhizal fungi in alleviation of Zn phytotoxicity and mineral nutrition of host plants. Acts of the 17th World Congress of Soil Science, Thailand, Symposium No. 42, paper no. 1649.

9

Biofertilization with Rhizobacteria and a Consortium of Arbuscular Mycorrhizal Fungi in Citrus Rootstocks

Roberto Gregorio Chiquito-Contreras[1], Francisco Osorio-Acosta[2], Eliseo García-Pérez[2*], Juan Antonio Villanueva-Jiménez[2], Ramón Zulueta-Rodríguez[1] and Doris Guadalupe Castillo-Rocha[1]

[1]Universidad Veracruzana, Facultad de Ciencias Agrícolas, Campus Xalapa. Circuito Gonzalo Aguirre Beltran s/n Zona Universitaria, CP 91091, Xalapa, Veracruz, México.
[2]Colegio de Postgraduados, Campus Veracruz, Km 88.5 Carretera Federal Xalapa-Veracruz, CP 91700, Tepetates, Veracruz, México.
[*]Corresponding Author: geliseo@colpos.mx

SUMMARY

The effect of individual and combined biofertilization with the plant growth promoting rhizobacteria (PGPR) strains FCA-8, FCA-56 and FCA-60 of *Pseudomonas putida* and arbuscular mycorrhizal fungi (AMF) on the growth of *Citrus volkameriana* and Rangpur lime rootstocks grafted with Tahiti lime was evaluated under nursery conditions. Plants were inoculated individually and combined with the three rhizobacterial strains and with AMF, and received 50% fertilization. These plants were compared with a control group that received nursery management with 100% fertilization. A 2 x 10 factorial split-plot experimental design with five replications per treatment was established. Variables measured included plant height, stem-base diameter, root length and volume, fresh and dry plant biomass, and bacterial and mycorrhizal colonization. Individual and combined biofertilization with PGPR and AMF promoted the growth of *C. volkameriana* and Rangpur lime grafted with Tahiti lime in the nursery, similar to the control with 100% fertilization. The three strains of *P. putida* and AMF applied individually or combined provided similar results to the control in the nutrient content of Tahiti lime leaves. The *P. putida* strains showed good capacity for colonization of the rhizosphere of *C. volkameriana* and Rangpur lime individually or combined with AMF.

Key words: *Pseudomonas putida*; plant growth; nutrition; mycorrhizal colonization.

INTRODUCTION

Citrus cultivation is of great importance in countries with subtropical and tropical climates due to the quantity of jobs that are generated during the production process, as well as by the income obtained from the industrialization and commercialization of fresh or processed fruit (Agustí, 2003; FAO, 2009). In Mexico the production of citrus is ranked 5th in global production and is carried out in 23 states (SIAP, 2009).

Modern agriculture is characterized by the excessive use of inorganic fertilizers and agrochemicals to manage and increase production and to control pests and diseases. This production system has generated serious environmental contamination problems, public health issues and reduced soil fertility (Leach and Mumford, 2008; Berg, 2009). Thus, it is necessary to implement sustainable production strategies to reduce the damage to human health and contribute to soil and water conservation, important components for the equilibrium and optimal operation of agroecosystems (Barea *et al.*, 2005; Rives *et al.*, 2007; Ryan *et al.*, 2008). Currently, agricultural biotechnology promotes alternatives for profitable and viable production such as biofertilization, which improves crop performance without damaging the environment or human health, and at the same time contributes to the improvement of soil fertility (Van Loon, 2007; Lambers *et al.*, 2009). The growing interest in the biofertilization of agricultural production systems is based on the use of beneficial living microorganisms, such as plant growth-promoting rhizobacteria (PGPR) and the formation of arbuscular mycorrhizal fungi (AMF). These organisms are capable of stimulating the growth, development and health of the plant and improving soil fertility (Song *et al.*, 2007).

Many genera of bacteria exist within the rhizobacteria, including *Pseudomonas*, *Bacillus*, *Azospirillum*, *Azotobacter* and *Serratia*. The genus *Pseudomonas* includes free-living saprophytic microorganisms in the soil, decomposing organic matter, aquatic ecosystems and is associated with plants and animals (Berg, 2009). The most beneficial species in *Pseudomonas* are *P. putida* Trevisan and *P. fluorescens* Migula (Gravel *et al.*, 2007), which fix atmospheric nitrogen, degrade aromatic compounds, promote the synthesis of plant growth regulators and other antagonistic metabolites such as siderophores, lithic enzymes (glucanases and chitinases) and antibiotics (Richardson *et al.*, 2009; Ryan *et al.*, 2009). These species have been extensively studied for their positive interactions on the growth, development and health of crops such as corn, wheat, tomato, lettuce, rice and citrus (Rives *et al.*, 2007; Bhattacharjee *et al.*, 2008; Santoyo *et al.*, 2010).

Among the AMF are the genera *Gigaspora*, *Glomus*, *Acaulospora* and *Sclerocystis*, which establish symbiotic associations with more than 90% of plant species. This symbiosis influences the nutrition and growth of the plants, improves tolerance to water stress and resistance to soil pathogens, as well as promotes increased root volume and intervenes in the solubilization and absorption of phosphorous (Vogelsang *et al.*, 2006; Bever *et al.*, 2009). The potential of AMF has been evaluated in crops of agricultural importance such as corn, wheat, sugarcane, tomato, onion, coffee, pastures and citrus (Wang and Qiu, 2006; Lambers *et al.*, 2009; Rigamonte *et al.*, 2010).

The process of biofertilization using PGPR and AMF has viable potential for agricultural production plans that seek to improve plant growth because these microorganisms are positively synergistic (Bever *et al.*, 2009; Lambers *et al.*, 2009). Thus, the objective of the present study was to evaluate in a nursery the effect of individual and combined inoculation with three rhizobacterial strains and a consortium of AMF on the growth and nutrition of *Citrus volkameriana* and Rangpur lime grafted with Tahiti lime.

MATERIALS AND METHODS

Nursery location

The study was carried out in Martínez de la Torre, Veracruz, Mexico (20° 04' N and 97° 04' W, at an altitude of 151 m a.s.l.), in the San Manuel nursery which is certified for citrus production. Management of the plants used normal nursery operations with the difference that inoculation with rhizobacteria and/or mycorrhiza was according to the corresponding treatment.

Plants used

Considering the difference in growth that rootstocks conferred to scion, *C. volkameriana* Tan. & Pasq. and Rangpur lime (*Citrus* x *limonia* Osbeck) were selected; *C. volkameriana* is more vigorous than Rangpur lime (Curti-Díaz *et al.*, 2000). The graft used was Tahiti lime (*Citrus latifolia* Tan.). When the rootstocks presented three pairs of true green leaves in the seedbed, they were transplanted to black polyethylene bags of 3.0 kg in capacity containing fertile river sediment (clay-loam), previously disinfected with Bunema® 55 GE (Buckman) at 20 mL L^{-1}. The graft with Tahiti lime was performed 120 days after the transplantation.

Propagation and maintenance of the inoculations

Rhizobacteria

The PGPR (*Pseudomonas putida*) used were supplied by the Universidad Veracruzana. The strains were labeled as FCA-8, FCA-56 and FCA-60, and were grown in the B-King (BK) culture medium [glycerol (IDQ®) 10 mL L^{-1}, peptone (BD-Bioxon®) 15 g L^{-1}, $MgSO_4$ (IDQ®) 1 mL L^{-1}, K_2HPO_4 (J.T. Baker®) 1.5 g L^{-1}]. Each strain was cultured in a total of 450 mL of BK culture medium, divided into three flasks with 150 mL in each; the media was sterilized for 15 min at 120 °C (Garrity *et al.*, 2005). Strain incubation was at 25 °C for 96 h according to Díaz *et al.* (2001).

Mycorrhiza

The AMF were supplied by the Universidad Veracruzana. This consortium is comprised of infective propagules of *Funneliformis mosseae* (T.H. Nicolson & Gerd.) C. Walker & A. Schüßler comb. nov., *Claroideoglomus etunicatum* (W.N. Becker & Gerd.) C. Walker & A. Schüßler comb. nov., *Glomus aggregatum* N.C. Schenck & G.S. Sm., *Glomus* sp., *Gigaspora albida* N.C. Schenck & G.S. Sm., *Acaulospora morrowiae* Spain & N.C. Schenck and *Acaulospora scrobiculata* Trappe. The propagation of the consortium was carried out in a sterilized mixture 2:1 (v/v) of soil and river sand (Zulueta *et al.*, 2010).

Biofertilization of *Citrus volkameriana* and Rangpur lime

Rootstock inoculation was carried out at the moment of transplantation from the seedbed to the polyethylene bags. The PGPR were prepared to a concentration of 10^9 cells mL^{-1} of BK culture media, adjusted with sterile physiological saline solution using a digital spectrophotometer (Thermo-Spectronic, Genesys 20®) and according to the McFarland scale (Rives *et al.*, 2009). The inoculation with the PGPR was carried out by immersing the roots in 500 mL of bacterial suspension for 20 min. The inoculation of AMF consisted of completely impregnating the roots with 10 g of inoculum per plant.

In addition to the inoculation with the rhizobacteria and the AMF, the plants were fertilized at 30 days after transplantation with the formula 18-46-00 (N-P-K). The control was fertilized according to nursery management protocol, which was the application of 5 g per plant, and in those treatments with biofertilization 2.5 g per plant was applied.

Variables evaluated

Variable measurement was carried out equally between the rootstocks and grafts. For *C. volkameriana* and Rangpur lime, the variables measured were: plant height (cm), stem base diameter (mm), root length (cm) and volume (cm^3), dry and fresh plant biomass (g), bacterial colonization (CFU g^{-1} of roots) according to Bashan *et al.* (1996), as well as the percentage of total mycorrhizal colonization (hyphae, arbuscules and vesicles) using the technique of clearing and staining roots (Phillips and Hayman, 1970) and observing them under an optical compound microscope (Nikon®) (Biermann and Linderman, 1981).

Variables evaluated for the Tahiti lime graft included: height (cm), graft base diameter (mm), leaf area (cm^2), dry and fresh plant biomass (g), and leaf content of N, P, K, Ca, Mg, Fe, Cu, Zn and Mn (Maldonado *et al.*, 2008). For nutrient determinations a compound sample from five plants per treatment was used, which did not include replications. All variables were measured when the plants reached 12 months of age, a time defined by the technical-productive system of the nursery.

Experimental design and statistical analysis

A split-plot factorial design was used in which the plots represented by *C. volkameriana* and Rangpur lime were divided into 10 sub-parcels (10 treatments) having five replications, with plant as the experimental unit. The treatments evaluated were: 1) Control, 2) FCA-8, 3) FCA-56, 4) FCA-60, 5) AMF-1, 6) Mix (mixture of the three rhizobacterial strains), 7) Mix + AMF, 8) FCA-8 + AMF, 9) FCA-56 + AMF and 10) FCA-60 + AMF. The control was according to the normal management of the nursery. Analyses of variance and comparison of means (Tukey test; $P \leq 0.05$) were used to analyze the data in SAS for Windows (Rebolledo, 2002).

RESULTS AND DISCUSSION

Growth of *Citrus volkameriana* and Rangpur lime

The analysis of variance showed significant differences in root volume ($P = 0.0063$), fresh plant biomass ($P = 0.0008$) and dry plant biomass ($P = 0.0073$). Dry plant biomass in *C. volkameriana* between Mix and AMF was significantly different. For Rangpur lime, significant differences in plant height were revealed for Mix and FCA-8 in relation to the control (Table 1). According to the results obtained on the growth of *C. volkameriana* and Rangpur lime (Table 1), there are various studies that show responses to inoculations with AMF and PGPR on the growth, development, health and performance of different plant species (Barea *et al.*, 2005; Berg, 2009; Santoyo *et al.*, 2010). Rives *et al.* (2009) mention that the use of rhizobacteria significantly favors the growth and development of plants. They found that 10 local strains of *P. putida* in rice promoted growth and development, with significant differences in dry and fresh biomass and root length. These results were not detected in the present study, possibly because the control under nursery management included a doubling of chemical fertilizer compared to the other treatments, as well as being influenced by the substrate. Similar values to the control were found indicating that the biofertilization treatments substituted 50% of the missing chemical fertilization. The results obtained provide continued incentive to test potential native *Pseudomonas* rhizobacteria to promote growth in plants from varied agricultural interests. As well, they indicate the importance to test these microorganisms in the field.

Table 1. Effects of biofertilization with rhizobacteria and arbuscular mycorrhizal fungi on the growth of *Citrus volkameriana* and Rangpur lime.

Treatments	Plant height (cm)	Stem diameter (mm)	Root length (cm)	Root volume (cm^3)	Fresh plant biomass (g)	Dry plant biomass (g)
C. volkameriana						
Control	34.50 a	9.43 a	25.74 a	35.60 a	42.52 a	15.61ab[1]
FCA-8	34.04 a	9.21 a	29.12 a	36.00 a	54.95 a	16.99 ab
FCA-56	35.26 a	9.41 a	28.28 a	35.60 a	55.18 a	18.40 ab
FCA-60	34.46 a	9.55 a	24.30 a	38.80 a	54.97 a	19.12 ab
AMF	35.46 a	8.62 a	26.26 a	34.00 a	49.66 a	14.16 b
Mix	34.30 a	9.46 a	26.10 a	42.00 a	56.33 a	19.63 a
Mix + AMF	35.08 a	9.35 a	27.92 a	36.60 a	49.17 a	16.94 ab
FCA-8 + AMF	34.78 a	9.45 a	25.34 a	37.20 a	49.29 a	17.79 ab
FCA-56 + AMF	33.96 a	8.94 a	29.66	36.60 a	47.17 a	18.66 ab
FCA-60 + AMF	34.82 a	9.36 a	26.40	35.60 a	49.38 a	17.25 ab
Rangpur lime						
Control	32.22 b[1]	8.79 a	23.40 a	29.20 a	42.11 a	15.22 a
FCA-8	36.58 a	9.71 a	29.12 a	33.20 a	48.73 a	17.94 a
FCA-56	35.70 ab	9.47 a	25.70 a	33.00 a	47.17 a	17.40 a
FCA-60	35.36 ab	9.59 a	23.82 a	27.40 a	42.45 a	16.45 a
AMF	35.92 ab	9.49 a	24.10 a	32.40 a	45.41 a	17.10 a
Mix	36.26 a	9.18 a	25.72 a	29.00 a	43.29 a	15.90 a
Mix + AMF	35.34 ab	9.28 a	23.36 a	30.60 a	45.02 a	15.87 a
FCA-8 + AMF	34.62 ab	9.50 a	24.76 a	26.00 a	41.48 a	15.39 a
FCA-56 + AMF	35.74 ab	8.94 a	25.66 a	30.00 a	43.17 a	15.76 a
FCA-60 + AMF	34.68 ab	9.45 a	25.80 a	30.80 a	45.94 a	15.79 a

Mix = mixture of three rhizobacterial strains.
[1]Values with the same letter in the same column are not significantly different (Tukey, $P \leq 0.05$).

Growth of Tahiti lime promoted by biofertilization

The analysis of variance of the split-plots showed significant differences for the general model (P = 0.0004) and for the interaction of rootstock by treatment (P = 0.0316). The Tukey test separated the means of the five variables studied between the two rootstocks (Table 2).

Table 2. Means of the growth-related variables for Tahiti lime grafted to *Citrus volkameriana* and Rangpur lime and biofertilized with rhizobacteria and arbuscular mycorrhizal fungi.

Graft host	Stem height (cm)	Stem diameter (mm)	Leaf area (cm^2)	Fresh plant biomass (g)	Dry plant biomass (g)
C. volkameriana	53.76 a[1]	7.40 a	988.03 a	48.46 a	16.02 a
Rangpur lime	49.46 b	6.65 b	890.88 b	40.94 b	13.85 b

[1]Means with the same letter in the same column are not significantly different (Tukey, $P \leq 0.05$).

Evaluations of treatment effects on Tahiti lime in each one of the rootstocks showed significant differences in *C. volkameriana* for dry plant biomass (P = 0.0184). Means comparison tests showed that strain FCA-60 worked best in the production of dry matter in comparison to AMF and in FCA-56 + AMF treatments (Table 3). Tahiti lime grafted onto Rangpur lime did not show significant differences for any of the variables studied (Table 3).

Rhizobacteria and AMF can improve the growth and development of the citrus evaluated, similar to the findings of Alarcón *et al.* (2003), because even with half of the fertilization they provided similar results to the control. Santillana (2006) found similar results for beans and corn, and showed differences with the control for three native strains of *Pseudomonas* sp. in terms of aboveground dry plant biomass and dry root biomass.

Table 3. Effects of biofertilization with rhizobacteria and arbuscular mycorrhizal fungi on the growth of Tahiti lime grafts on two rootstocks.

Treatments	Stem height (cm)	Stem diameter (mm)	Leaf area (cm^2)	Fresh plant biomass (g)	Dry plant biomass (g)
Citrus volkameriana					
Control	49.56 a	6.97 a	935.38 a	46.47 a	15.97 ab[1]
FCA-8	55.64 a	7.29 a	989.38 a	46.38 a	15.67 ab
FCA-56	53.24 a	7.16 a	883.48 a	43.79 a	14.96 ab
FCA-60	61.92 a	7.93 a	1213.38 a	61.15 a	20.92 a
AMF	49.28 a	6.76 a	921.58 a	43.61 a	13.33 b
Mix	57.34 a	7.98 a	1111.34 a	52.93 a	17.72 ab
Mix + AMF	47.72 a	7.36 a	1024.83 a	52.83 a	16.06 ab
FCA-8 + AMF	51.60 a	7.74 a	955.94 a	46.86 a	15.97 ab
FCA-56 + AMF	57.66 a	7.30 a	910.43 a	44.60 a	14.44 b
FCA-60 + AMF	53.68 a	7.48 a	934.54 a	45.98 a	15.21 ab
Rangpur lime					
Control	50.02 a	6.48 a	803.57 a	37.52 a	13.35 a
FCA-8	53.80 a	6.57 a	923.49 a	40.14 a	13.52 a
FCA-56	48.72 a	6.97 a	911.28 a	43.66 a	15.13 a
FCA-60	44.12 a	6.70 a	923.70 a	42.25 a	14.81 a
AMF	52.62 a	6.76 a	952.23 a	45.13 a	15.34 a
Mix	49.48 a	6.77 a	812.82 a	40.92 a	12.84 a
Mix + AMF	52.36 a	6.51 a	930.58 a	40.92 a	13.43 a
FCA-8 + AMF	46.76 a	6.52 a	859.49 a	38.87 a	13.40 a
FCA-56 + AMF	47.32 a	6.60 a	881.30 a	39.37 a	12.94 a
FCA-60 + AMF	49.40 a	6.63 a	910.20 a	40.56 a	13.76 a

Mix = mix of three rhizobacterial strains.
[1]Means with the same letter are not significantly different (Tukey, $P \leq 0.05$).

Even though the effects from the combined biofertilization of PGPR with AMF were not detected in this study, diverse reports have indicated the importance of using inoculations where both are associated. This is so because the individual qualities that each type of microorganism confers to the plant strengthens the plants' phenological development when both types are combined and interact in the rhizosphere to provide a positive synergism that improves growth, development, yield and health of the plants (Bever *et al.*, 2009; Lambers *et al.*, 2009; Richardson *et al.*, 2009).

Nutrient content in Tahiti lime leaves

Table 4 shows the macro- and microelement contents in Tahiti lime leaves in the different treatments. In Tahiti lime grafted on *C. volkameriana*, the highest values for N (3.10%) and P (0.24%) were with AMF; however, the difference with that of smaller value (FCA-60) was only 0.33% and 0.04%, respectively. In the combined biofertilization using strain FCA-56 with AMF, the content of K was 0.19% greater than the control. In the microelement assessments, no tendency for any of the treatments with individual or combined biofertilization was observed. According to the variability of responses in Table 4, the treatments did not show a clear effect on nutrient content; bioinoculation permits plants to grow similar to the control where 100% fertilization was applied.

For the graft on Rangpur lime the results were very similar to those on *C. volkameriana* where no significant tendency was observed for any of the treatments. A very low value for Cu was observed for the treatment with the mixture of the three bacterial strains, while a high value for Mn was observed in the control and in treatment FCA-8 (Table 4). These values were likely due to external factors and not by treatment effects; however, to determine this it will be necessary to use known quantities of nutrients and to provide them to the plants in a sterile substrate.

Due to the capacities of the PGPR and AMF to fix atmospheric nitrogen, solubilize phosphorous, and to produce plant growth regulators and other metabolites that favor plant vigor and health, these microorganisms have been extensively used to improve crop nutrition and contribute to reducing fertilizer use, because their activity influences the plant rhizosphere and improves absorption of macro- and microelements (Ryan *et al.*, 2009; Rigamonte *et al.*, 2010).

Table 4. Influence of biofertilization with rhizobacteria and arbuscular mycorrhizal fungi on the nutrient composition Tahiti[1] lime leaves from two rootstocks.

Treatments	N	P	K	Ca	Mg	Fe	Cu	Zn	Mn
	%------					mg kg^{-1}			
Citrus volkameriana									
Control	3.06	0.19	1.65	3.10	0.28	175	28	19	97
FCA-8	3.08	0.19	1.47	4.56	0.30	200	18	22	102
FCA-56	2.86	0.20	1.68	3.31	0.25	212	45	24	55
FCA-60	2.77	0.20	1.74	3.21	0.27	159	28	22	44
AMF	3.10	0.24	1.71	3.06	0.29	205	30	23	48
Mix	2.83	0.21	1.78	4.89	0.30	210	32	21	43
Mix + AMF	2.95	0.21	1.59	3.18	0.25	188	15	19	44
FCA-8 + AMF	2.97	0.21	1.71	3.40	0.29	231	23	23	66
FCA-56 + AMF	2.95	0.23	1.84	4.22	0.30	331	35	17	53
FCA-60 + AMF	3.07	0.21	1.78	3.11	0.26	205	24	21	59
Rangpur lime									
Control	2.93	0.19	1.65	2.15	0.33	132	30	21	70
FCA-8	2.94	0.18	1.96	2.61	0.25	120	30	16	43
FCA-56	3.11	0.19	1.95	2.86	0.29	143	37	19	42
FCA-60	2.99	0.20	1.90	3.77	0.38	140	33	22	38
AMF	3.04	0.22	1.69	3.15	0.32	123	31	21	38
Mix	2.81	0.22	1.83	3.00	0.38	140	12	21	38
Mix + AMF	2.99	0.23	1.51	2.86	0.32	200	21	23	43
FCA-8 + AMF	2.97	0.23	1.81	2.88	0.22	186	40	23	43
FCA-56 + AMF	2.77	0.21	1.47	4.40	0.30	234	27	21	43
FCA-60 + AMF	2.83	0.22	1.86	3.34	0.26	170	21	18	44

Mix = mix of three rhizobacterial strains.
[1]Values obtained from one compound sample from five plants per treatment.

The growing interest in beneficial microorganisms in general, and particularly in the rhizobacteria of the genera *Pseudomonas*, *Bacillus*, *Azospirillum*, and *Azotobacter* for incorporation into systems of agricultural production, is due primarily to their ease of application. They can be inoculated on seeds, foliage, roots or directly in the ground, places where, once established in the rhizosphere of the plants, they can multiply in response to root exudates, leading to enhanced plant growth and/or the biological control of plant diseases. Such qualities are the result of aggressive root system colonization (Castro-Sowinski *et al.*, 2007; Fernández-Herrera *et al.*, 2007; Santoyo *et al.*, 2010).

Determination of rhizobacterial populations and mycorrhizal colonization

Populations of rhizobacteria obtained from the individual and combined biofertilizations differed significantly among treatments (P = 0.0001). In both rootstocks, the significant differences were between the control and the AMF which were similar to each other (Figure 1) in comparison with all other treatments which were statistically similar.

The rhizobacterial populations obtained from the treatments that combined them with the AMF indicate that no antagonistic relation between the microorganisms existed. Indeed, both are capable of being complemented and to be more effective when interacting in the plant rhizosphere, interaction that can be selective and depend on the number and type of bacterial and fungal species (Terry and Leyva, 2006; Rigamonte *et al.*, 2010).

The unidentified bacterial populations quantified in the control and AMF treatments were produced through repopulation from environmental contamination provided by the cultural practices performed during the production process in the nursery. This contamination indicates that the effect of the soil disinfectant (Bunema® 55 GE) used in the nursery was only temporary, permitting subsequent induced and natural microorganismal recolonization.

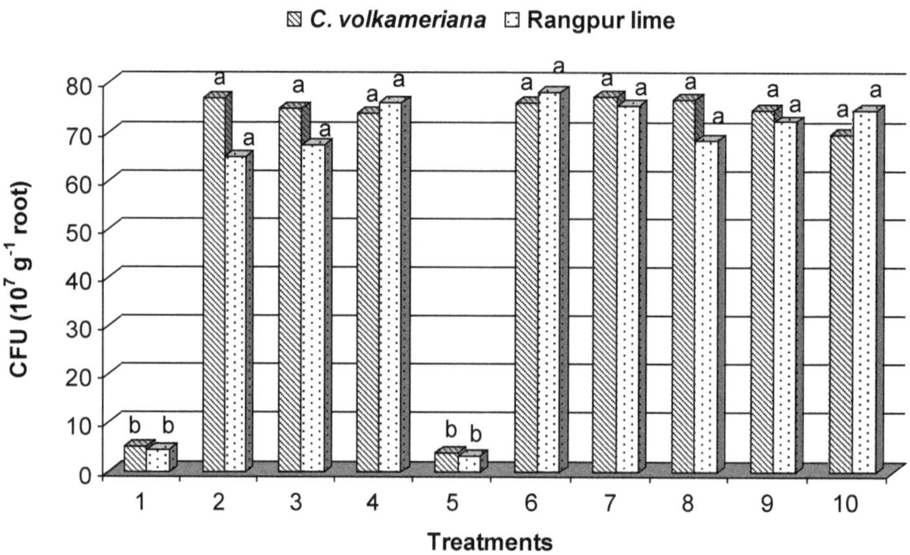

Figure 1. Abundance of PGPR in the rhizosphere of *Citrus volkameriana* and Rangpur lime: 1) Control, 2) FCA-8, 3) FCA-56, 4) FCA-60, 5) AMF, 6) Mix (mix of the three rhizobacterial strains), 7) Mix + AMF, 8) FCA-8 + AMF, 9) FCA-56 + AMF and 10) FCA-60 + AMF.

[a,b]Values with the same letter are not significantly different (Tukey, $P \leq 0.05$).

The benefit of the AMF in the plants is influenced by its capacity to colonize the root system and to establish a symbiotic relationship with its host. Given the treatments evaluated, the results of the analysis of variance showed significant differences between rootstocks ($P = 0.0013$) and among treatments ($P = 0.0001$). In *C. volkameriana*, the greatest mycorrhizal colonization (97.67%) was for individual inoculation with AMF. In contrast, for Rangpur lime 63.80% was recorded for Mix (mixture of the three rhizobacterial strains) with AMF. In the controls and treatments with individual rhizobacterial inoculation and Mix, no mycorrhizal colonization was found (Figure 2).

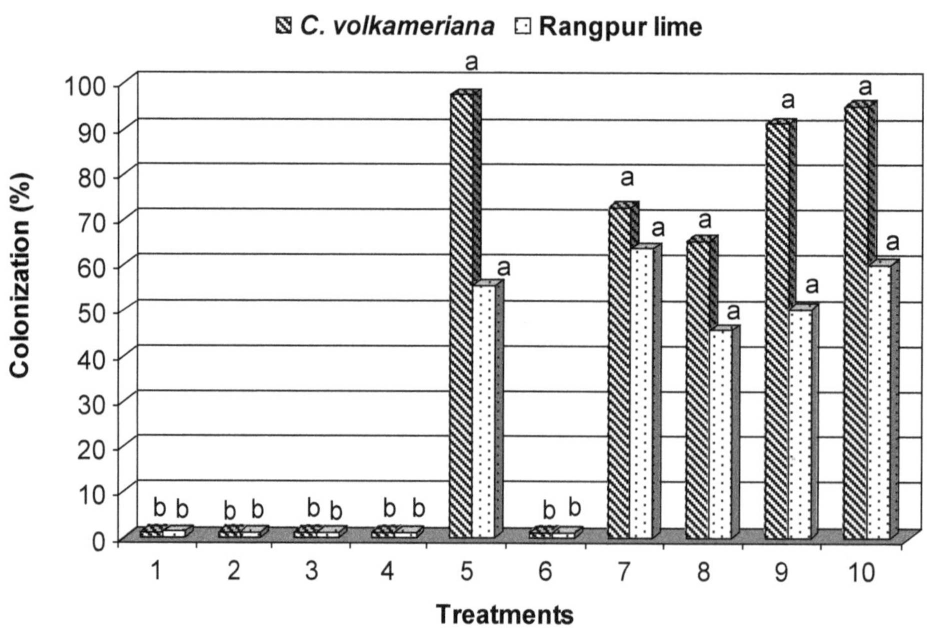

Figure 2. Rhizospheric colonization by the mycorrhizal consortium AMF in *Citrus volkameriana* and Rangpur lime: 1) Control, 2) FCA-8, 3) FCA-56, 4) FCA-60, 5) Consortium AMF, 6) Mix (mix of the three rhizobacterial strains), 7) Mix + AMF, 8) FCA-8 + AMF, 9) FCA-56 + AMF and 10) FCA-60 + AMF.

[a,b]Values with the same letter are not significantly different (Tukey, $P \leq 0.05$).

It is important to note the difference in mycorrhizal colonization percentage between the rootstocks *C. volkameriana* and Rangpur lime. This difference is related to the competition that rhizobacteria have for space and root exudates; however, coexistence of both microorganisms in the root systems of the plants was possible. Berg (2009) mentions that between communities of bacterial and AMF there is a positive synergism, where, from time to time, one of the two dominates the colonization of the rhizosphere due to its invasive capacity and effectivity for rapid establishment of the symbiotic relationship (Lambers *et al.*, 2009; Rigamonte *et al.*, 2010).

Citrus volkameriana showed greatest symbiotic capacity with the AMF. According to Tapia-Goné *et al.* (2010), this capacity depends on the infective ability and effectiveness of the fungus. The infective process refers to the capacity of the fungus to penetrate and invade the root system intensely and to explore the soil, as well as its ability to persist in the production system. The effectiveness of the fungus is demonstrated by improved development of the host indirectly or directly. The indirect benefit is protection against soil pathogens and stress, as well as increasing soil aggregation and stability, qualities that promote hyphal development. The direct benefit is the improved absorption of nutrients as phosphorous, zinc and copper. Besides the abilities of the fungus, environmental factors and the plant species play important roles in the efficient establishment of the symbiosis (Berg, 2009; Bever *et al.*, 2009).

CONCLUSION

Individual and combined biofertilization with PGPR and the AMF positively stimulated the growth of the citrus graft hosts *C. volkameriana* and Rangpur lime, as well as the Tahiti lime graft in the nursery, providing results equal to 100% chemical fertilization. The *P. putida* strains and the mycorrhizal consortium applied in individual or combined form contributed to the increased nutrient content of Tahiti lime leaves, in particular strain FCA-56 and consortium AMF, for the contents of N and P assimilated by the plants. The *P. putida* strains showed good capacity at colonizing the rhizosphere of *C. volkameriana* and Rangpur lime individually and combined with consortium AMF. Better mycorrhizal symbiosis with *C. volkameriana* was shown individually and associated with strains FCA-56 and FCA-60.

ACKNOWLEDGEMENTS

The rhizobacterial strains were provided by the Laboratorio de Química Agrícola and the mycorrhizal consortium by the Laboratorio de Organismos Benéficos, both from the Facultad de Ciencias Agrícolas, Universidad Veracruzana, Campus Xalapa. The graft host plants and the Tahiti lime grafts were provided by Ing. Julio González Contreras, Gerente Técnico of the "San Manuel" nursery. Financing for this investigation was provided by CONACYT and Colegio de Postgraduados through Campus Veracruz and the Línea Prioritaria de Investigación 2, Agroecosistemas Sustentables.

REFERENCES

Agustí, M. 2003. Citricultura. 2a ed. Mundi-Prensa. Madrid, España. 416 p.

Alarcón, A., González-Chávez, M.C., Ferrera-Cerrato, R. 2003. Crecimiento y fisiología de *Citrus volkameriana* Tan. & Pasq. en simbiosis con hongos micorrízicos arbusculares. Terra Latinoamericana. 21:503-511.

Barea, J.M., Pozo, M.J., Azcón, R., Azcón-Aguilar, C. 2005. Microbial co-operation in the rhizosphere. Journal of Experimental Botany. 56:1761-1778.

Bashan, Y., Holguín, G., Ferrera-Cerrato, R. 1996. Interacciones entre plantas y microorganismos benéficos. Terra. 14:159-192.

Berg, G. 2009. Plant-microbe interactions promoting plant growth and health: Perspectives for controlled use of microorganisms in agriculture. Applied Microbiology and Biotechnology. 84:11-18.

Bever, J.D., Richardson, S.C., Lawrence, B.M., Holmes, J., Watson, M. 2009. Preferential allocation to beneficial symbiont with spatial structure maintains mycorrhizal mutualism. Ecology Letters. 12:13-21.

Bhattacharjee, R.B., Sing, A., Mukhopadyay, S.N. 2008. Use of nitrogen-fixing bacteria as biofertilizer for non-legumes: Prospects and challenges. Applied Microbiology and Biotechnology. 80:199-209.

Biermann, B.J., Linderman, R.G. 1981. Quantifying vesicular arbuscular mycorrhizae. A proposed method towards standardization. New Phytologist. 87:63-67.

Castro-Sowinski, S., Herschkovitz, Y.,Okon, Y., Jurkevitch, E. 2007. Effects of inoculation with plant growth-promoting rhizobacteria on resident rhizosphere microorganisms. FEMS Microbiology Letters. 276:1-11.

Curti-Díaz, S.A., Loredo-Salazar, X., Díaz-Zorrilla, U., Sandoval R., J.A. ,Hernández H., J. 2000. Tecnología para producir limón Persa. Libro Técnico No. 8. INIFAP-CIRGOC SAGAR. Campo Experimental Ixtacuaco. Veracruz, México. 144 p.

Díaz V., P., Ferrera-Cerrato, R., Almaraz-Suárez, J.J., Alcántar-González, G. 2001. Inoculación de bacterias promotoras de crecimiento en lechuga. Terra Latinoamericana. 19:327-335.

FAO (Food and Agriculture Organization). 2009. Frutos cítricos frescos y elaborados. Estadísticas anuales. faostat.fao.org. (Consulted: 12/12/2009).

Fernández-Herrera, E., Acosta-Ramos, M., Ponce-Gonzélez, F., Manuel-Pinto, V. 2007. Manejo biológico de *Phytophthora capsici* Leo., *Fusarium oxysporum* Schlechtend.:Fr. y *Rhizoctonia solani* Kühn en jitomate (*Lycopersicon esculentum* Mill.). Revista Mexicana de Fitopatología. 25:35-42.

Garrity, G.M., Bell, A.J., Lilburn, T. 2005. Bergey's Manual of Systematic Bacteriology. Second edition. Vol. 2. The Proteobacteria Part B. The Gammaproteobacteria. Springer-Verlag. New York, USA. pp. 323-377.

Gravel, V., Antoun, H., Russell, J., Weddell, T. 2007. Growth stimulation and fruit yield improvement of greenhouse tomato plants by inoculation with *Pseudomonas putida* or *Trichoderma atroviride*: Possible role of indole acetic acid (IAA). Soil Biology and Biochemistry. 39:1968-1977.

Lambers, H., Mougel, C., Jaillard, B., Hinsinger, P. 2009. Plant-microbe-soil interactions in the rhizosphere: An evolutionary perspective. Plant and Soil. 321:83-115.

Leach, A.W., Mumford, J.D. 2008. Pesticide environmental accounting: A method for assessing the external costs of individual pesticide applications. Environmental Pollution. 151:139-147.

Maldonado, T.R., Almaguer, V.G. Alvaréz, S.M.E., Robledo, S.E. 2008. Diagnóstico nutrimental y validación de dosis de fertilización para limón Persa. Terra Latinoamericana. 26:341-349.

Phillips, J.M., Hayman, D.S. 1970. Improved procedures for clearing roots and staining parasitic and vesicular-arbuscular mycorrhizal fungi for rapid assessment to infection. Journal Transactions of the British Mycological Society. 55:158-161.

Rebolledo R., H.H. 2002. Manual SAS por Computadora: Análisis Estadístico de Datos Experimentales. Editorial Trillas. México. 208 p.

Richardson, A.E., Barea, J.M., McNeill, A.M., Prigent-Combaret, C. 2009. Acquisition of phosphorus and nitrogen in the rhizosphere and plant growth promotion by microorganisms. Plant and Soil. 321:305-339.

Rigamonte, T.A., Pylro, V.S., Duarte, G.F. 2010. The role of mycorrhization helper bacteria in the establishment and action of ectomycorrhizae associations. Brazilian Journal of Microbiology. 41:832-840.

Rives, N., Acebo, Y., Hernández, A. 2007. Bacterias promotoras del crecimiento vegetal en el cultivo del arroz (*Oryza sativa* L.), perspectivas de su uso en cuba. Cultivos Tropicales. 28:29-38.

Rives, N., Acebo, Y., Almaguer, M., García, J.C., Hernández, A. 2009. Actividad antagónica frente a *Pyricularia grisea* (Sacc.) y fitoestimulación en el cultivo del arroz de cepas autóctonas de *Pseudomonas putida* (Trev.). Revista de Protección Vegetal. 24:106-116.

Ryan, P.R., Dessaux, Y., Thomashow, L.S., Weller, D.M. 2009. Rhizosphere engineering and management for sustainable agriculture. Plant and Soil. 321:363-383.

Ryan, R.P., Germaine, K., Franks, A., Ryan, D.J., Dowling, D.N. 2008. Bacterial endophytes: Recent developments and applications. FEMS Microbiology Letters. 278:1-9.

Santoyo, G., Valencia-Cantero, E., Orozco-Mosqueda, M.C., Peña-Cabriales, J.J., Farías-Rodríguez, R. 2010. Papel de los sideróforos en la actividad antagónica de *Pseudomonas fluorescens* zum80 hacia hongos fitopatógenos. Terra Latinoamericana. 28:53-60.

Santillana V., N. 2006. Producción de biofertilizantes utilizando *Pseudomonas* sp. Ecología Aplicada. 5:87-91.

SIAP (Servicio de Información Agroalimentaria y Pesquera). 2009. Anuario estadístico de la producción agrícola. Limón y Naranja. www.limonmexicano.gob.mx/index.php?portal=limon.www.naranja.gob.mx/index.php?portal=naranja.(Consulted: 12/12/2009).

Song, M., Xu, X., Hu, Q., Tian, Y., Ouyang, H., Zhou, C. 2007. Interactions of plant species mediated plant competition for inorganic nitrogen with soil microorganisms in an alpine meadow. Plant and Soil. 297:127-137.

Tapia-Goné, J.J., Ferrera-Cerrato, R., Varela-Fregoso, L., Rodríguez-Ortiz, J.C., Soria-Colunga, J.C., Tiscareño-Iracheta, M.A., Loredo-Osti, C., Alcala-Jáuregui, J., Villar-Morales, C. 2010. Infectividad y efectividad de hongos micorrízicos arbusculares nativos de suelos salinos en el cultivo de lechuga (*Lactuca sativa*). Revista Mexicana de Micología. 31:69-74.

Terry A., E., Leyva G, A. 2006. Evaluación agrobiológica de la inoculación micorrizas-rizobacterias en tomate. Agronomía Costarricense. 30:65-73.

Van Loon, L.C. 2007. Plant responses to plant growth promoting bacteria. European Journal of Plant Pathology. 119:243-254.

Vogelsang, K.M., Reynolds, H.L., Bever, J.D. 2006. Mycorrhizal fungal identity and richness determine the diversity and productivity of a tallgrass prairie system. New Phytologist. 172:554-562.

Wang, B., Qiu, Y.L. 2006. Phylogenetic distribution and evolution of mycorrhizas in land plants. Mycorrhiza. 16:299-363.

Zulueta R., R., Varela, L., Aguilar E., S., Trejo A., D., Lara C., L. 2010. Estatus micorrízico de *Jacaratia mexicana* y hongos formadores de micorriza arbuscular presentes en selvas bajas caducifolias del Golfo de México. Revista Mexicana de Micología. 31:37-44.

10

Moringa Genetic Diversity from Germplasm Bank using Rapd Markers

Ana Veruska Cruz da Silva[1]*, Allívia Rouse Ferreira dos Santos[2], Ana da Silva Lédo[1], Rosana Barroso Feitosa[3], Camila Santos Almeida[3], Gilvânia Melo da Silva[3] and Maria Salete Alves Rangel[1]

[1] *The Brazilian Agricultural Research Corporation's (Embrapa)*
[2] *Universidade de Santiago de Compostela (USC)*
[3] *Universidade Federal de Sergipe (UFS)*
E-mail:anaveruska@cpatc.embrapa.br
**Corresponding author*

SUMMARY

The Moringa (*Moringa oleifera* Lam.) is a tree species with the known value of food, medicine and water treatment, and is promising for bioenergy use. Our objective was to evaluate, using RAPD markers, the genetic diversity of sixteen accessions from Germplasm Bank (BAG) of Embrapa Coastal Tablelands, Sergipe, Brazil. We estimate the diversity indices, and genetic similarity between accessions. The Shannon index and the index of genetic diversity (H) were 0.33 and 0.22, respectively. We observed by Jaccard similarity that, the accessions MO1 and MO12 are the most similar (0.27), and MO13 and MO16 are the most divergent (0.69). By UPGMA and PCoA groupings, we identified that MO1, MO2, MO12 and MO13 are genetically isolated. The results are important in designing strategies for conservation, but because of low diversity detected in this investigation, new activities of collection should be realized, with integration and characterization of new accessions, to ensure the ever increasing diversity of the collection.

Key words: *Moringa oleifera* Lam.; DNA; variability; genetic resources.

INTRODUCTION

Moringa (*Moringa oleifera* Lam., Moringaceae) is a perennial species, native of India. It is widely cultivated in tropical and subtropical countries, and it was introduced in Brazil during the 1950s. The species presents food, medicinal and cosmetic value. In addition, the seeds can be used to oil extraction and water treatment (Bezerra *et al.*, 2004).

The State of Sergipe has a potential area for Moringa production. However, more information about the species is needed, mainly concerning the characterization of genotypes of interest (Dos Santos *et al.*, 2011b). Given the importance of species and their need for conservation, the Embrapa Coastal Tablelands created in 2009, the Moringa Germplasm Bank (BAG) located in municipality of Nossa Senhora das Dores city.

The genotypes with immediate potential or future should be kept in Germplasm banks (BG's). The primary importance of BG's are the ability to provide to genetic variability for improvement programs, besides, help efforts to reconcile the conservation of agricultural biodiversity and sustainable development (Nass, 2007).

Studying the variability of such collections allows to identify duplicates, to estimate genetic linkage among accessions, to quantify genetic variability in the collection, to propose collections based on observed genetic diversity and to identify suitable breeding genotypes to improvement (Ferreira and Grattapaglia, 1995).

Molecular markers are used to detect genetic variation of genotypes of interest at the DNA level. The RAPD (Random Amplified Polymorphic DNA) is one of the tools used, especially in the characterization of genetic resources for rapid quantification of diversity (Silveira *et al.*, 2009). With the help of statistical methods, the RAPD technique has been effective in detecting the diversity of population in various types of organisms. Besides its low cost and speed, this technique has the advantage, even without prior knowledge of the genome, of requiring little amount of DNA for analysis (Goulão *et al.*, 2001).

RAPDs markers were used by Souza *et al.* (2001) successfully to evaluate the hypothesis of the existence of three races in peach palm; estimate the diversity and separation of cultivars in acerola (Salla *et al.*, 2002), peach and nectarine (Lima *et al.*, 2003), plum (Bianchi *et al.*, 2003), banana (Padmesh *et al.*, 2009), mangaba (Costa *et al.*, 2011) and moringa from Africa (Mgendi *et al.*, 2010). Muluvi *et al.* (1999) using AFLP found high level of genetic diversity in Moringa from Kenya and India, indicating the genetic potential of the species.

The aim of the study was the molecular characterization and genetic diversity of the moringa Germplasm Bank belonging to Embrapa Tabuleiros Costeiros, using RAPD markers.

MATERIAL AND METHODS

We used 16 accessions of Moringa Germplasm Bank belonging to Embrapa Tabuleiros Costeiros. These accessions are from exchanged germplasm of University of Florida, USA. It were introduced in the Campo Experimental Jorge do Prado Sobral belonging to Embrapa Tabuleiros Costeiros, located in the municipality of Nossa Senhora das Dores (Sergipe State).

The DNA extraction was based on Doyle and Doyle (1991) methodology. 300 mg young leaves were macerated in liquid nitrogen, adding 800 mL of extraction CTAB buffer (2% of CTAB, EDTA – 0.5 M pH 8.0, Tris-Cl – 1 M pH 8.0, NaCl 5M, 2% de PVP).

Each RAPD reaction was performed in 25μL volume containing 50 ng genomic DNA, 1X PCR buffer (GIBCO-BRL), 1.5 mM $MgCl_2$, 0.2 mM dNTP, 1.0 U *Taq* DNA polymerase (GIBCO), 30 ng/μL primer , 20 μL ultrapure water.

The PCR amplifications were performed using a PTC-100 thermocycler (Programmable Thermal Controller - MJ Research, Inc.) and subjected to a cycle of 96°C for 5 minutes for initial denaturation, followed by 35 cycles of denaturation at 96°C for 45 seconds, 36°C for 45 seconds for primer annealing, 72°C for 45 seconds for extension, and finally one cycle of 72°C for 5 minutes for final extension (Silva and Martins, 2006). We tested 17 primers (A3, A4, A8, A12, A15, A16, A18, IDT02, IDT3, IDT15, S01, S18, W02, W13, W19, B02, B18) with 14 primers Operon (Operon Technologies, USA) and three IDT primers (Integrated DNA Technologies, Germany) (Table 1).

For electrophoresis, 10 μL of PCR products were mixed to 1.5 μL of blue juice (0.01% bromophenol blue, 40% glycerol). We used 1.5% agarose gel (1X TEB - 89 mM TRIS, 89 mM boric acid, 2.5 mM EDTA, pH 8.3) in a horizontal electrophoresis system Sunrise (Gibco BRL), carried out at a constant voltage of 100 V for 90 minutes. Gel was stained with ethidium bromide solution (5 mg/mL) for 15 minutes, and the amplification products visualized under ultraviolet light using a Gel Doc L-Pix image system (Loccus Biotecnologia, Brazil).

The electrophoretic profile of each gel was transformed into a binary matrix of presence (1) and absence (0) and which was subsequently used for all analysis. Bootstrap procedure was applied to calculate variance of the genetic distance obtained from markers, and was obtained from 5,000 bootstrap random draws using the DBOOT software (Coelho, 2000). Polymorphic information content (PIC) is a parameter that provides an estimate of the discriminatory power of molecular marker per primer and was calculated according to Ghislain *et al.*, (1999). The marker index (MI) was determined as a product of PIC and the number of polymorphic bands per assay unit as described in Zhao *et al.*, (2007). To measure the genetic diversity, we used Genalex v.6.3 (www.anu.edu.au/BoZo/GenAlEx/) and calculated the Shannon Index (I) (Brown and Weir 1983) as well as genetic diversity (H) as described by Lynch and Milligan (1994) and Maguire *et al.*, (2002) for dominant markers.

The data matrix of the RAPD scores was generated and similarity coefficients were calculated using Jaccard's arithmetic complement index. The dendrogram was constructed using the Unweighted Pair Group Method with Arithmetic Mean (UPGMA) cluster algorithm. In order to determine the robustness of the dendrogram, the data was bootstrapped with 5,000 replications using FreeTree software (http://web.natur.cuni.cz/flegr/programs/freetree.htm). For visualization of the cluster we used the TreeView package (http://web.natur.cuni.cz/flegr/programs/freetree/TreeView.exe). Principal Coordinates Analysis (PCoA) was performed using the software XLSTAT (http://www.xlstat.com/) based on similarity matrix of Jaccard.

RESULTS AND DISCUSSION

The 17 primers generated a total of 95 fragments, 59 polymorphic (62%). Primers with the highest number of polymorphic fragments was IDT15 and S01 (seven). Primers IDT3 and B18 did not show polymorphic fragments (Table 1). The primers used resulted in a banding pattern that was distinct, as seen in Figure 1 which shows the RAPD profiles of the primer IDT 02.

The polymorphism found can be considered high (62%), and similar values were found in other studies. In a study in cultivated and non-cultivated Moringa from Tanzania (Africa), a total of 98 fragments were scored and 86 of them (89.6%) were polymorphic using 12 RAPD primers (Mgendi *et al.*, 2010). The selection stage of these markers is critical to the efficiency of the work of characterization and can be done through literature review or tests in the laboratory (Silva *et al.*, 2008).

There is a directly proportional relationship between the number of fragments analyzed and the coefficient of variation (CV) (Figure 2). The results indicate a clear of decreasing CV with increasing number of fragments. According to Moura *et al.* (2005), there is a point where the enlargement in the number of fragments does not show a significant increase in experimental accuracy and not justifying the extra effort in labor. From 55 fragments there is a stabilization of the CV, with value less than 10%, suggesting that the results obtained by the fragments used in this study (59) can be used for analysis of diversity.

To access the Moringa, the PIC value ranged from 0.00 to 0.41 and MI from 0.00 to 2.07. For the variables of diversity, the Shannon index (I) with average value of 0.33, and the genetic diversity index (H) showed a mean value of 0.22 (Table 1).

Table 1. Primers, totals (TF) and polymorphic fragments (PF%), polymorphic information content (PIC), marker index (MI), Shannon index (I) and genetic diversity (H) between 16 accessions belonging Moringa (*Moringa oleifera* Lam.) Germplasm Bank from Embrapa Coastal Tablelands (Nossa Senhora das Dores, Sergipe State, Brazil).

Primers	Sequence 5' – 3'	TF	PF%	PIC	MI	I	H
A3	AGT CAG CCA C	3	100	0.36	1.07	0.52	0.36
A4	AAT CGG GCT G	4	50	0.25	0.50	0.34	0.25
A8	GTG ACG TAG G	3	67	0.26	0.52	0.38	0.26
A12	TCG GCG ATA G	3	67	0.13	0.27	0.23	0.13
A15	TTC CGA ACC C	4	75	0.17	0.52	0.29	0.17
A16	AGC CAG CGA A	4	100	0.39	1.55	0.57	0.39
A18	AGG TGA CCG T	6	100	0.23	1.35	0.37	0.23
IDT02	TGA TCC CTG G	8	62	0.14	0.69	0.21	0.14
IDT3	TGC CGA GCT G	2	0	0.00	0.00	0.00	0.00
IDT15	GGT CGG AGA A	9	78	0.30	2.07	0.44	0.30
S01	CTA CTG GCG T	8	87	0.26	1.82	0.40	0.26
S18	CTG GCG AAC T	9	62	0.16	0.79	0.24	0.16
W02	ACC CCG CCA A	7	29	0.12	0.24	0.17	0.12
W13	CAC AGC GAC A	4	75	0.34	1.02	0.48	0.34
W19	CAA AGC GCT C	9	67	0.17	1.02	0.27	0.17
B02	AGT CAG CCA C	4	100	0.41	1.65	0.60	0.41
B18	AAT CGG GCT G	8	0	0.00	0.00	0.00	0.00
Average		6	62	0.22	0.89	0.33	0.22

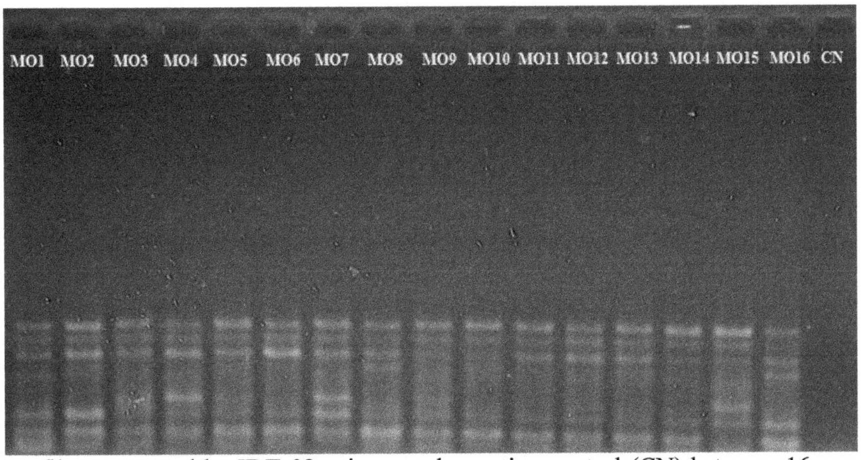

Figure 1. RAPD profiles generated by IDT 02 primer and negative control (CN) between 16 accessions belonging Moringa (*M. oleifera* Lam.) Germplasm Bank from Embrapa Coastal Tablelands (Nossa Senhora das Dores, Sergipe State, Brazil).

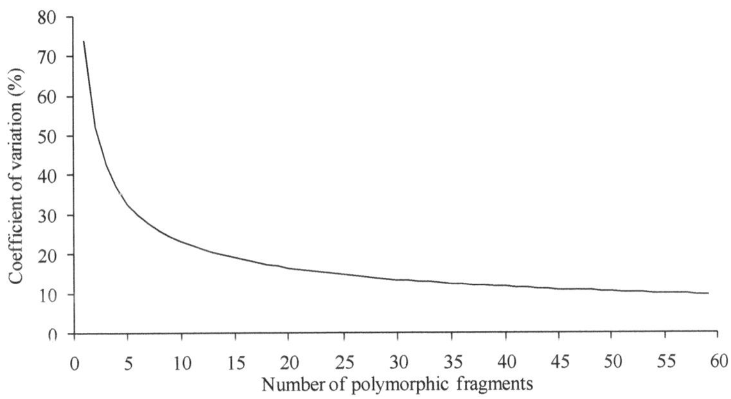

Figure 2. Coefficient of variation for the number of polymorphic fragments using RAPD markers between 16 accessions belonging Moringa (*M. oleifera* Lam.) Germplasm Bank from Embrapa Coastal Tablelands (Nossa Senhora das Dores, Sergipe).

Genetic diversity indicated by I (0.00 - 0.52) and H (from 0.00 to 0.41) in BAG Moringa can be considered low since the values were equal to or less than 0.55, or is in the BAG for a low diversity, being necessary to insert a new accessions to promote increased diversity and more likely to use these resources. The values found are similar to those obtained for other tree species *Populus tremuloides* (0.58 to 0.69, Yeh *et al*, 1995), *Fitzroya cupressoides* (0.42 to 0.56; Allnut *et al.*, 1999) and *Swietenia macrophylla* (0.41 to 0.27, Gillies *et al.*, 1999). The average genetic diversity (H) obtained is in agreement with those found for other species. To *Trichilia pallida* Swartz, ranged from 0.27 to 0.33 (Zimback *et al.*, 2004) and *Aspidosperma polyneuron*, diversity averaged 0.28 (Torezan *et al.*, 2005).

The average similarity (Table 2) between accessions was 0.48, ranging from 0.27 (MO1 and MO12) to 0.69 (MO13 and MO16). This variability expressed should be used in breeding programs and studies on conservation and multiplication of top accessions. RAPD primers were used to investigate the genetic diversity between and within cultivated and non-cultivated provenances of Moringa from Costal regions of Tanzania, and the cluster analysis showed five clusters with similarity ranging from 54% to 96% (Mgendi *et al.*, 2010).

According to UPGMA clustering, we indentified three distinct groups (Figure 3). The first group (G1) was formed by ten accessions, second (G2) by four and third (G3) that only two accessions. Among the groups, G3 stood alone, indicating that these accessions are the most genetically distant.

Table 2. Genetic similarity of 16 accessions from Moringa (*Moringa oleifera* Lam.) Germplasm Bank belonging to Embrapa Coastal Tablelands (Nossa Senhora das Dores, Sergipe state).

	MO1	MO2	MO3	MO4	MO5	MO6	MO7	MO8	MO9	MO10	MO11	MO12	MO13	MO14	MO15	MO16
MO1	-	0.45	0.30	0.33	0.35	0.42	0.41	0.46	0.49	0.45	0.47	***0.27***	0.35	0.45	0.54	0.46
MO2	0.45	-	0.33	0.38	0.46	0.41	0.40	0.42	0.38	0.45	0.42	0.29	0.38	0.44	0.50	0.46
MO3	0.30	0.33	-	0.64	0.57	0.50	0.45	0.38	0.40	0.36	0.44	0.34	0.29	0.43	0.44	0.35
MO4	0.33	0.38	0.64	-	0.65	0.55	0.50	0.45	0.47	0.41	0.46	0.36	0.34	0.47	0.51	0.45
MO5	0.35	0.46	0.57	0.65	-	0.68	0.57	0.47	0.49	0.50	0.47	0.47	0.47	0.53	0.50	0.47
MO6	0.42	0.41	0.50	0.55	0.68	-	0.62	0.50	0.45	0.55	0.61	0.43	0.51	0.53	0.54	0.46
MO7	0.41	0.40	0.45	0.50	0.57	0.62	-	0.56	0.55	0.49	0.64	0.50	0.54	0.51	0.53	0.45
MO8	0.46	0.42	0.38	0.45	0.47	0.50	0.56	-	0.63	0.50	0.44	0.47	0.44	0.60	0.61	0.57
MO9	0.49	0.38	0.40	0.47	0.49	0.45	0.55	0.63	-	0.45	0.50	0.36	0.43	0.59	0.60	0.56
MO10	0.45	0.45	0.36	0.41	0.50	0.55	0.49	0.50	0.45	-	0.56	0.38	0.52	0.45	0.54	0.58
MO11	0.47	0.42	0.44	0.46	0.47	0.61	0.64	0.44	0.50	0.56	-	0.44	0.58	0.58	0.64	0.51
MO12	***0.27***	0.29	0.34	0.36	0.47	0.43	0.50	0.47	0.36	0.38	0.44	-	0.53	0.54	0.56	0.51
MO13	0.35	0.38	0.29	0.34	0.47	0.51	0.54	0.44	0.43	0.52	0.58	0.53	-	0.58	0.55	***0.69***
MO14	0.45	0.44	0.43	0.47	0.53	0.53	0.51	0.60	0.59	0.45	0.58	0.54	0.58	-	0.64	0.64
MO15	0.54	0.50	0.44	0.51	0.50	0.54	0.53	0.61	0.60	0.54	0.64	0.56	0.55	0.64	-	0.65
MO16	0.46	0.46	0.35	0.45	0.47	0.46	0.45	0.57	0.56	0.58	0.51	0.51	***0.69***	0.64	0.65	-

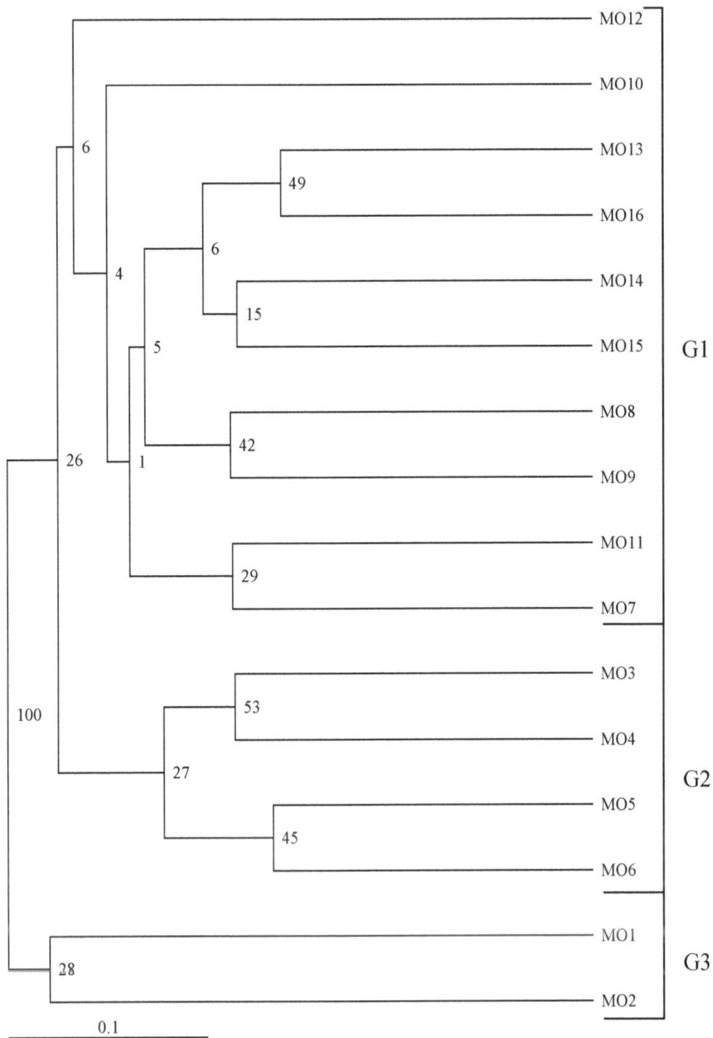

Figure 3. Cluster generated from the Jaccard similarity coefficient and UPGMA clustering using RAPD markers between 16 accessions from Moringa (*Moringa oleifera* Lam.) Germplasm Bank belonging to Embrapa Coastal Tablelands (Nossa Senhora das Dores, Sergipe state).

According by PCoA (Figure 4), the first two axes represent 38.29% of the variation. We indentified four groups, the first one (I) formed by two accessions, second (II) for seven third one (III) by two and fourth (IV) for five. The groups (I) and (III) were isolated, suggesting that these are the most different groups.

Some differences were found in contrasting two clustering methods. According to UPGMA analysis accesses MO1 and MO2 were more related, and to PCoA were MO1 and MO2, and MO12 and MO13. In addition, the group G1 of UPGMA clustering is equivalent II and III to the PCoA, indicating that the groups are subgroups.

The regrouping has been used by some authors (Costa *et al.*, 2011; Dos Santos *et al.*, 2011a) as an alternative to make the results more revealing. The use of more than one method of grouping prevents erroneous inferences because of differences in classification, optimization and ordering of groups. Thus, this option is adopted in the allocation of materials within a particular subset of genotypes (Arriel *et al.*, 2006).

Information on the genetic diversity can assist in a breeding program avoiding duplication or mixtures of genotypes in studies and germplasm conservation programs (Pinheiro *et al.*, 2003). Molecular characterization of genetic diversity from Germplasm Bank can provide useful data to assist the breeder in the identification and selection of clones or parents to establish the basic breeding program.

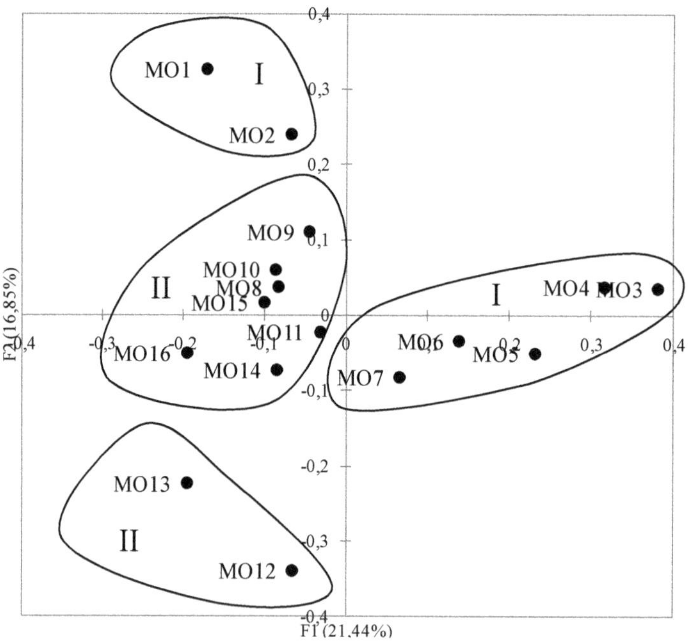

Figure 4. Principal Coordinates Analysis (PCoA) obtained from the Jaccard similarity coefficient using RAPD markers between 16 accessions from Moringa (*Moringa oleifera* Lam.) Germplasm Bank belonging to Embrapa Coastal Tablelands (Nossa Senhora das Dores, Sergipe state).

CONCLUSIONS

The accessions MO1, MO2, MO12 and MO13 are genetically isolated, and the information generated can be used to design strategies and conservation of germplasm in future breeding programs of the species. Because of low diversity detected in this investigation, to Moringa Germplasm Bank is fundamental to continue inclusion of new accessions.

REFERENCES

Allnut, T.R., Newton, A.C., Lara, A., Premoli, A., Armesto, J.J., Vergara., R., Gardner, M. 1999. Genetic variation in *Fitzroya cupressoides* (alerce), a threatened South American conifer. Molecular Ecology. 8:975-87.

Arriel, N.H., Di Mauro, A.O., Di Mauro, S.M., Bakke, O.A., Uneda-Trevisoli, S.H, Costa, M.M., Capeloto, A., Corrado, A.R. 2006.Técnicas multivariadas na determinação da diversidade genética em gergelim usando marcadores RAPD. Pesquisa Agropecuária Brasileira. 41: 801-809.

Bezerra, A.M.E., Momenté, V.G., Medeiros Filho, S. 2004. Germinação de sementes e desenvolvimento de plântulas de Moringa (*Moringa oleífera Lam.*) em função do peso da semente e do tipo de substrato. Horticultura Brasileira. 22:295-299.

Bianchi, V.J., Fachinello, J.C., Schuch, M.W. 2003. RAPDs na caracterização genético molecular e no estudo da variabilidade genética de cultivares de ameixeira. Revista Brasileira de Fruticultura. 25:272-274.

Brown, A.H.D., Weir, B.S. 1983. Measuring genetic variability in plant populations, in Isozymes in Plant Genetics and Breeding, *Part A* (Tanksley SD, Orton TJ, Editors). Elsevier Science Publ.: Amsterdam. p. 219-239.

Coelho, S.G. 2000. Software DBOOT – Avaliação dos erros associados a estimativas de distâncias/similaridades genéticas através do procedimento de bootstrap com número variável de marcadores, v. 1.1. Goiânia, Brasil, Universidade Federal de Goiás.

Costa, T.S., Silva, A.V.C., Lédo, A.S., Dos Santos, A.R.F., Silva Junior, J.F. .2011. Diversidade genética de acessos do banco de germoplasma de mangaba em Sergipe. Pesquisa Agropecuária Brasileira. 46: 499-508.

Dos Santos, A.R.F., Ramos-Cabrer, A.M., Diaz-Hernandez, M., Pereira-Lorenzo, S. 2011a. Genetic variability and diversification process in local pear cultivars from northwestern Spain using microsatellites. Tree Gen and Genome, in press.

Dos Santos, A.R.F., Silva-Mann, R., Ferrreia, R.A. 2011b. Water pre-hydration as priming for *Moringa oleifera* lam. seeds under salt stress. Tropical and Subtropical Agroecosystems. 14: 201-207.

Doyle, J.J., Doyle, J.L. 1991. Isolations of plant DNA from fresh tissue. Focus. 1:13-15.

Ferreira, M.E., Grattapaglia, D. 1995. Introdução ao uso de marcadores moleculares em análise genética. 2 ed. Embrapa Recursos Genéticos e Biotecnologia, Brasília.

Ghislain, M., Zhang, D., Fajardo, D., Huamann, Z., Hijmans, R.H. 1999. Marker assisted sampling of the cultivated Andean patato Solanumphureja collections using RAPD markers. Genetic Research Crop Evolution. 6:547–555.

Gillies, A.C.M., Navarro, C., Lowe, A.J., Newton, A.C., Hernández, M., Wilson, J., Cornelius, J.P. 1999. Genetic diversity in Mesoamerican populations of mahogany (*Swietenia macrophylla*), assessed using RAPDs. Heredity. 83:722-32.

Goulão, L., Cabrita, L., Oliveria, C.M., Leitão, M.J.M. 2001. Comparing RAPD and AFLP analyis in discrimination and estimation of genetic similarities among apples (*Malus domestica* Borkh.) cultivars. Euphytica. 119: 259-270.

Lima, M.R., Augustin, E., Choer, E., Raseira, M.C.B. 2003. Caracterização de cultivares de pessegueiro e nectarineira por marcadores moleculares. Pesquisa Agropecuária Brasileira. 38:349-355.

Lynch, M., Milligan, B.G. 1994. Analysis of population genetic structure with RAPD markers. Molecular Ecology. 3:91-99.

Maguire, T.L., Peakall, R., Saenger, P. 2002. Comparative analysis of genetic diversity in the mangrove species *Avicennia marina* (Forsk.)Vierh. (Avicenniaceae) detected by AFLPs and SSRs, Theoretical and Applied Genetics. 104: 388-398.

Mgendi, M.G., Manoko, M.K., Nyomora, A.M. 2010. Genetic diversity between cultivated and non-cultivated *Moringa oleifera* Lam. provenances assessed by RAPD markers Journal of Cell and Molecular Biology. 8: 95-102.

Moura, N.F., Chaves, L.J., Vencovisk, R., Zucchi, M.I., Pinheiro, J.B., Morais, L.K., Moura, M.F. 2005. Seleção de marcadores RAPD para o estudo da estrutura genética de *Hancornia speciosa* Gomes. Bioscience Journal. 21:119-125.

Muluvi, G. M., Sprent, J. L., Soranzo, N., Provan, J., Odee, D., Folkard. G., McNicol, J. W., Powell, W. 1999. Amplified fragment length polymorphism (AFLP) analysis of genetic variation in *Moringa oleifera* Lam. Molecular Ecology. 8: 463-470.

Nass, L.L. 2007. Recursos genéticos vegetais. Embrapa Recursos Genéticos e Biotecnologia, Brasília.

Padmesh, P., Reji, J.V., Paul, B., Mukunthakumar, S., Praveen, G., Seeni, S. 2009. Analysis of genetic variability in two diploid Musa cultivars using RAPD. Biologia Plantarum. 53: 4.

Pinheiro, J.B., Zuxxhi, M.I., Teles, F.L., Azara, N.A. 2003. Diversidade genética molecular em acessos de açafrão utilizando marcadores RAPD. Acta Scientiarum: Agronomy. 25: 195-199.

Salla, M.F.S., Ruas, C.F., Ruas, P.M., Carpentieri-Pípolo, V. 2002. Uso de marcadores moleculares na análise da variabilidade genética em acerola (*Malpighia emargunata* D.C.), Revista Brasileira de Fruticultura. 24: 15-22.

Silva, A.V.C., Martins, A.B.G. 2006. Identificação de marcas moleculares associadas à ausência de sementes em videira. Ciência Rural. 36: 801-806.

Silva, D.A.S, Antunes, L.E.C., Anthonisen, D.G., Lemões, J.S., Gonçalves, E.D. 2008. Caracterização de genótipos de mirtilo utilizando marcadores moleculares. Revista Brasileira de Fruticultura. 30: 180-184.

Silveira, D.G., Amorim, E.P., Jesus, O.N., Souza, F.V.D., Pestana, K.N., Santos, V.J., Santana, J.R.F. 2009. Variabilidade genética de populações naturais de caroá por meio de marcadores RAPD. Pesquisa Agropecuária Brasileira. 44: 283-290.

Souza, N.R., Rodrigues, D.P., Clement, C.R., Nagao, E.O., Astolfi-Filho, S. 2001. Discriminação de raças primitivas de pupunha (Bactris gasipaes) na Amazônia brasileira por meio de marcadores moleculares (RAPDS). Acta Amazonica. 31: 539-545.

Torezan, J.M.D., Souza, R.F., Ruas, P.M., Ruas, C.F., Camargo, E.H., Vanzela, A.L.L. 2005. Genetic variability of pre and post-fragmentation cohorts of *Aspidosperma polyneuron* Muell. Arg. (Apocynaceae). Brazilian Archives of Biology and Technology. 48: 171-180.

Yeh, F.C., Chong, D.K.X., Yang, R.C. 1995. RAPD variation within and among natural populations of trembling aspen (*Populus tremuloides* Michx.) from Alberta. Journal of Heredity. 86: 454-60.

Zhao, K., Zhou, M., Chen, L., Zhang, D., Robert, G.W. 2007. Genetic Diversity and Discrimination of *Chimonanthuspraecox* (L.) Link Germplasm Using ISSR and RAPD Markers. Hortscience. 42: 1144–1148.

Zimback, L., Mori, E.S., Kageyama, P.Y., Veiga, R.F.A., Mello Júnior, J.R.S. 2004 Estrutura genética de populações de *Trichilia pallida* Swartz (Meliaceae) por marcadores RAPD. Scientia Forestalis. 65: 114-119.

11

Growth Performance and Carcass Composition of Rabbits Fed on Diets of Gradual Levels of Barley Grain

M.M. El-Adawy[a], M.K. Mohsen[b], A.Z.M. Salem[a,c*], B.M.A. Mariezcurrena[c], G.S. El-Santiel[b] and M.Z. Dakron[b]

[a]*Department of Animal production, Faculty of Agriculture (El-Shatby), Alexandria University, Egypt*
[b]*Department of Animal production, Faculty of Agriculture, Kafr El-Shikh University, Egypt*
[c]*Facultad de Medicina Veterinaria y Zootecnia, Universidad Autónoma del Estado de México, México. E-mail address:* asalem70@yahoo.com
**Corresponding author*

SUMMARY

Forty-eight male New Zealand White rabbits of 6 weeks age (BW 875 ± 28.3 g) were randomly allocated among six groups in different levels of barley grains (BG); 0 (B0), 5 (B5), 10 (B10), 15 (B15), 20 (B20), 25% (B25) of the total diet on growth performance and carcass composition of rabbits . The highest live BW value ($P < 0.05$) was obtained in B20 rabbits, whereas the highest feed intake ($P < 0.05$) value was obtained in B25 rabbits followed by those fed on the other diets. The better-feed conversion ratio (FCR) was obtained in B20 rabbits followed by those of B15 and B25. Animals of B20 had better FCR than those of the other experimental diets. The relative contribution of soft feces to dry matter or crude protein intake differed ($P < 0.05$) among the experimental diets. The caecal turnover rate was increased ($P < 0.05$) in B15 and B20 rabbits than those other diets. The per-slaughter weight, hot and cold carcass weight and dressing percentage were differed ($P < 0.05$) among the experimental groups and highest values were recorded in B20 rabbits. Data suggested that a partial replacement of corn grains (CG) by 20% BG in rabbit diets was increased live body weight, feed intake and feed conversion ratio.

Key words: Growth performance; Barley; Corn; Rabbit.

INTRODUCTION

Energetic concentration is one of the main variables in the formulation of diet for growing rabbit. Dietary carbohydrates, especially starch, are the main sources of energy for mono gastric animals to maximize growth during the short rearing period (Leng, 2008). In non-ruminant animals, availability of starch for digestion depends on number of factors including source of starch in the diet, botanical origin and technological treatments. In general, cereal starches are more easily digested than legume starches as shown for pigs (Bengala-Freire, 1989) or poultry (Yutste et al., 1991). Differences exist also between cereals in relation to endosperm texture and amylose/amylopectin ratio (Perez and Aumaitre, 1979). Technological processing may also induce beneficial physical transformation of starch granules (Colonna and Champ, 1990). The high starch cereal grains such as corn and wheat are the main dietary sources used in formulation of commercial animal diets. Wheat was used primarily for human consumption, although corn grains (CG) are considered an excellent dietary source of starch and the premier feed grain for most livestock, it may not be so for rabbits because its high-energy content promotes enteritis, and it seems to be somewhat unpalatable. Moreover, it is difficult to achieve adequate crude fiber levels in diets containing high level of corn (Cheeke, 1987).

In rabbits, a low dietary amount of starch was recommended at the post-weaning stage (Maertens, 1992) in order to reduce digestive problems linked to the incomplete development of enzymatic system in young rabbits and their inability to digest starch completely (Scapinello et al., 1999). The outbreak of epizootic enterocolitis in Europe has further stimulated the reduction of dietary starch and the increase of fiber fractions as an enterocolitis prevention method during weaning and fattening, with a consequent important of diet nutritive value and feed conversion (Licois et al., 2000). Barley is the most common cereal used in rabbit diets in Europe, traditionally included 15-25% of the diet and preferred to corn because of its lower frequency of mycotoxin contamination and higher starch digestibility. This latter avoids high starch ileal escape and overflow in caecum with negative consequences on digestive physiology, caecal fermentation and health status (Blas and Gidenne, 1998). Moreover, barley has a higher protein content and better protein quality than corn. It is lower in starch and high in fiber than corn, and it contains *b*-glucans that may provoke enteritis. These characteristics make it suitable as rabbit feed ingredient (Cheeke, 1987). Studies on the effects of starch levels and sources in growing rabbits gave sometimes-controversial results (Blas et al., 1994; Xiccato et al., 1998). Indeed, few data are available on their effects on carcass, meat quality and excretion of soft and hard feces (Nizza and Moniello, 2000).

The aim of the present study was to evaluate the effect of two starch sources (corn and barley) through partial or complete substitution of corn (CG) by barley (BG) grains in the rabbit diets on growth performance, caecotrophy and carcass characteristics of growing rabbits.

MATERIALS AND METHODS

This work was carried out at Rabbit Research Laboratory and the chemical analysis was performed at the Animal Nutrition Laboratory, of Animal Production Department in the Faculty of Agriculture (El-Shatby), Alexandria University, in cooperation with Animal Production department, Faculty of Agriculture, Kaferelsheikh University.

Experimental diets

Five experimental diets were formulated to cover all essential nutrient requirements for growing rabbits according to NRC (1984) and De Blas (1986), in which 20, 40, 60, 80 and 100% of the corn grains (CG) in the rabbit diet was substituted by barley grains (BG); (0 (B0), 5 (B5), 10 (B10), 15 (B15), 20 (B20), and 25% (B25) of the total diet). Diets were formulated to be isonitrogenous and isocaloric. Commercial vitamin and mineral premix was added for all the experimental diets (Table 1).

Experimental animals and housing

Forty-eight weanlings male New Zealand White (NZW) rabbits at six weeks old, with an average initial live body weight (BW, 875 ± 28 g) were allocated at random among six comparable groups each contained eight rabbits. Average initial BW of the assigned groups was nearly similar. Each experimental group was received one of the experimental diets. The experiments were extended up to 13 weeks of age and animals were individually caged in metal galvanized cages under the same managerial conditions in well-ventilated block building.

Growth trial

Fresh water was automatically available all the time by stainless steel nipples for each cage. The experimental diets were offered to rabbits *ad libitum* in pelleted form. All rabbits were kept under the same environmental, hygienic and managerial conditions. Body weight (BW) and daily feed intake (DFI, g) were individually recorded weekly for estimating average daily gain (ADG), while feed conversion ratio (FCR) was calculated as g feed/g gain.

Performance index (PI) was calculated according to North (1981) as given below:

PI = [BW (kg) / FCR] × 100

Caecotrophy trials

Determination of soft (SF) and hard (HF) feces excretion were carried out using caecotrophy trials. Plastic neck collars were used to prevent coprophagy. SF and HF were collected according to the method described by Carabaño et al. (1989). The daily feed intake (*i.e.,* DFI) was recorded after deducting the scattered amounts. Daily SF and (HF) of each rabbit were collected and samples of both daily feces (20%) for chemical analysis. The daily samples of feces collected were spread with 10 ml of 1% boric acid solution to prevent ammonia losses during drying. Feces samples were dried at 60 - 70^0 C for 48h. Dried feces of each rabbit during the collection period were weighted, mixed, ground and kept until analysis. Relative contribution of SF and HF to dry matter (DM) and crude protein (CP) intake were calculated according to Fraga et al. (1991) as follow:

Relative contribution of SF to DM intake = (SF excretion, g DM/day) / (DFI, g DM/day + SF excretion, g DM/day) ×100.

Relative contribution of SF to CP intake = (CP excreted in SF, g/day) / (CP ingested in feed, g/day + CP excreted in SF, g/day)×100.

Caecal turnover rate was calculated according to Garcia et al. (1995) as follow:

[SF production (g DM/d) / caecal contents (g DM)] × 100.

Metabolizable energy (ME) was determined according to the equation described by Kalogen (1985) as follow:

ME (Kcal/ kg DM) = (0.588 + 0.164 X) 239

Where X is the dry matter digestion of the offered diet.

Table 1. Feed ingredients and chemical composition (g/kg) of the experimental diets.

Ingredients	Diets					
	B0	B5	B10	B15	B20	B25
Egyptian berseem hay	330	325	320	315	310	305
Yellow corn	250	200	150	100	50	-
Barley grain	-	50	100	150	200	250
Wheat bran	120	125	130	135	140	145
Soybean meal	190	188	186	184	182	180
Rice bran	47	49	51	53	55	57
Limestone	20	20	20	20	20	20
Common salt	10	10	10	10	10	10
Molasses	30	30	30	30	30	30
Vitamin and mineral premix[1]	2	2	2	2	2	2
DL-Methionin	1	1	1	1	1	1
Chemical composition						
Organic matter	895	895	896	895	895	896
Crude protein	167	167	167	166	166	165
Ether extract	32	31	32	32	32	33
Crud fiber	113	115	116	117	119	119
Nitrogen free extract	692	582	581	579	578	579
DE (Kcal/kg feed DM)[2]	2361	2367	2361	2356	2353	2348

[1]Vitamins and mineral premix per kilogram contained: Vit a 2,000,000 Iµ, Vit. K 0.33 mg, vit. B_1 0.33 g, Vit B_2 1.0 g, Vit B_6 0.33 g, Vit. B_{12} 1.7 mg, pantathonic acid 3.33 g, Biotin 33 mg, Folic acid 0.83g, Choline choride 200 mg, Zn 11.7 g, Mn 5.0 g, Fe 12.5 g, Mg 66.7 mg, Se 16.6 mg, Co 1.33 mg, Cu 0.5 g, 116.6 mg and antioxidant 10.0 g.

[2]DE (Kcal/kg feed DM) was calculated according to NRC (1994).

Chemical analysis

Ground samples of feeds and faeces were analyzed for DM by drying samples at 105°C for 24 h in forced air oven. Ash content was measured after igniting samples in a muffle furnace at 550 ◦C for 4 h. Crude protein (CP) was determined by Kjeldahl method (AOAC, 1990; ID 954.01). Ether extract was determined by Soxhlet method (AOAC, 1990; ID 920.39). Crude fiber was determined by the sample digestion in acids followed by alkali digestion.

Slaughter trial

Eighteen male rabbits (three rabbits per each group) were randomly taken at 13th weeks of age (the marketing age). Rabbits were slaughtered after fasting for 12 hours (Lukefahr et al., 1992) and body weights were recorded before slaughtering. After slaughtering and complete bleeding (within 30 minutes), carcass traits were evaluated according to Blasco et al. (1992). Hot carcass weights (liver, kidneys, heart and head besides: (lunges, esophagus, trachea, thymus and spleen) were obtained in 30 minutes after slaughter. Dressing percentage was estimated as hot carcass weights relatively to pre-slaughter body weight. Cold carcass weight was obtained after refrigerating the hot carcass between 0 and 4 °C for 24 hours. Drip loss percentage was calculated as:

[(Hot carcass weights - Cold carcass weight)/ Hot carcass weights] × 100.

Giblets weight (liver, kidneys, heart and spleen) and carcass measurements were obtained and their proportion to the live body weight was calculated.

Statistical analysis

Results were statistically analyzed for the effect of dietary treatments using the General Liner Model Program of SAS (1996) according to the statistical model: $Y_{ij} = \mu + T_i + e_{ij}$ where: Y_{ij} = An observation on jth individual, μ = Overall mean, T_i = The fixed effect of jth treatment groups, e_{ij} = Random error. Duncan's multiple-range test was performed (Duncan, 1955) to detect significant differences among means.

RESULTS AND DISCUSSION

Growth performance and feed intake

Average daily feed intake (DFI) from 6-13 weeks of age was higher ($P < 0.05$) in B25 rabbits, while the lower ($P < 0.05$) was obtained in rabbits of B5 diet. The performance index (PI) value differed ($P < 0.05$) when barley grains (*i.e.,* BG) in growing rabbit replaced corn grains (*i.e.,* CG) diets (Table 2). The improvement of the rabbit body weights obtained in groups fed diets containing BG are in agreement with results of Struklec et al. (1995) who reported that the average rabbit body weight (*i.e.,* ADG) was higher in groups fed BG diets than those fed CG diets at 31 days of age. Rajendra et al. (2000) showed that the finishing weight was higher on rabbit groups fed diets containing BG in weaned kits of rabbits. Higher ADG ($P < 0.01$) in B20 group was consistent with the results reported by Gidenne and Perez (1993) who pointed out differences ($P < 0.05$) in terms of weight gain, between animals fed on purified maize starch, maize, barley or pea at 25, 36, 39 or 41%, respectively. Likewise, replacement of barley by barley radical up to 10% (Ibrahim et al., 1999) and offering diets containing barley (Rajendra et al., 2000) resulted in a improvement ($P < 0.05$) of ADG. However, our results were in disagreement with those reported by other authors (Belenguer et al., 2002; Xiccato et al., 2002; Gidenne et al., 2005) who recorded a similar ADG of rabbits fed on diets containing BG or other cereal as mainly starch source.

The improvement ($P < 0.05$) in ADG, DFI and FCR in favor of rabbits fed diet contained 20% BG may attribute to the greater palatability of barley. In this regard, El-Gendy et al. (2000) found that feed conversion improved ($P < 0.05$) by rabbits fed barley radical diets when added at 15 and 30% of the whole diet. Similarly, replacement of barley by barley radical up to 10% (Ibrahim et al., 1999), feeding diets containing 8% malt/roots (Bagliacca and Mori, 1987) and offering 75 and 150 g barley/day in rabbit diets (Amici, 1989) resulted in an improvement ($P < 0.05$) of the FCR. However, rabbits fed on diets containing maize, barley or triticale as the only cereal/ a mixture of these cereals (Lanza et al., 1986) or barley, triticale, wheat as the only cereal/ mixture of these cereals (Sinatra et al., 1987) and complete diet containing the fresh lucerne supplemented with 40% barley, oats or ground maize (Akram et al., 1989) had similar FCR .

Improving the PI in rabbits of B15, B20 and B25 groups, may be the consequence of the increment ($P < 0.05$) of their BW and the decrease of the corresponded FCR. El-Gendy et al. (2000) and Ibrahim et al. (1999) earlier reported similar results in rabbits fed on diets containing barley and dried barley radical.

Table 2. Growth performance and viability[1] in growing NZW rabbits fed on the experimental diets during the experimental period.

	Diets						SEM	P value
	B0	B5	B10	B15	B20	B25		
No. of rabbits	8	8	8	8	8	8		
Initial body weight, g/day	875	864	882	871	891	869	-	-
Final body weight, g/day	2290^{d}	2289^{d}	2415^{c}	2483^{b}	2544^{a}	2472^{b}	62.0	0.001
Average daily gain, g/day	28.9^{c}	29.1^{c}	31.3^{b}	33.0^{a}	33.7^{a}	32.7^{a}	2.58	0.001
Daily feed intake, g/day	118.5^{bc}	117.6^{c}	120.7^{ab}	121.3^{a}	119.7^{abc}	121.6^{a}	6.15	0.001
Conversion ratio, kg feed/kg gain	4.3^{a}	4.2^{a}	4.1^{b}	3.8^{c}	3.7^{d}	3.9^{c}	0.23	0.001
Performance index	38.1^{d}	38.40^{d}	42.5^{c}	45.0^{b}	47.4^{a}	44.1^{b}	3.66	0.001
Viability, %	75	75	100	100	100	87.5	-	-

$^{a, b, c, d}$ means in the same row bearing different letters, differ significantly ($P < 0.05$).
SEM, standard error of the mean.

Table 3. Soft feces (SF) excretion and ceacal content in growing NZW rabbits at 13 weeks of age fed on the experimental diets.

	Diets						SEM	P value
	B0	B5	B10	B15	B20	B25		
No. of rabbits	8	8	8	8	8	8		
Dry matter intake (g)	120.7^{b}	121.3^{b}	127.6^{a}	129.7^{a}	128.4^{a}	126.8^{a}	13.42	0.0051
SF excretion (g DM/day)	17.2^{c}	19.5^{abc}	19.8^{ab}	21.9^{a}	21.2^{a}	18.1^{bc}	1.56	0.0039
Chemical composition of SF, g/Kg :								
Dry matter	377^{a}	378^{a}	352^{ab}	367^{ab}	349^{ab}	326^{b}	14.0	0.0124
Crude protein	312^{b}	311^{b}	335^{a}	348^{a}	349^{a}	347^{a}	17.2	0.0017
Relative contribution of SF to [1] :								
DM intake	12.5^{b}	13.8^{ab}	13.4^{ab}	14.5^{a}	14.2^{a}	12.5^{b}	2.79	0.0245
CP intake	21.0^{c}	23.0^{b}	23.7^{b}	26.2^{a}	25.8^{a}	22.9^{b}	1.45	0.0046
Cecal content (g DM)	33.5^{c}	36.9^{b}	37.6^{ab}	39.1^{a}	37.2^{ab}	34.6^{a}	2.06	0.0012
Cecal turnover rate% [2]	51.5^{b}	52.8^{b}	52.6^{b}	56.1^{a}	56.9^{a}	52.2^{b}	3.46	0.0074

$^{a, b, c}$ means in the same row bearing different letters, differ significantly ($P < 0.05$).
[1] According to Fraga et al. (1991).
[2] According to Garcia et al. (1995).
SEM, standard error of the mean.

Caecotrophy trails

Addition of BG up to 20% resulted an increase ($P < 0.05$) in soft feces (SF) excretion, whereas the highest increase was recorded in B15 rabbits followed by those fed on B20 diet. DM and CP of SF were affected ($P < 0.05$) by the BG level addition in the rabbit diets. However, DM content in SF of rabbits fed on B25 was lower ($P < 0.05$) than that of rabbits fed either on B0 or B5. In contrast, CP content of the SF was increased ($P < 0.05$) with increasing the BG level. The relative contribution of SF to DM or CP intake were differed ($P < 0.05$) among the experimental diets. The highest values ($P < 0.05$)

were in B15 rabbit while the lowest was in B0. Increasing BG level in rabbit diets to 25% (B25) decreased ($P < 0.05$) the relative contribution of SF to DM or CP intake. Caecal DM of content was higher ($P < 0.05$) in B10 and B20 rabbits than others groups. The highest ($P < 0.05$) caecal DM content was recorded in B15 rabbits, while the lowest ($P < 0.05$) was obtained in B0 (Table 3). Rabbits fed on diets containing 10 to 25% of BG (*i.e.,* B10 to B25) excrete soft feces (SF) with high CP content, as compared with B0 group. Increasing the CP in the SF of rabbits fed BG diets may be due to the increasing of caecum microbial activity for these rabbits. The intensive systems applied in rabbit production, using high concentrate diets, restrict the potential of caecal fermentation, depending on the source of cereal and level of inclusion in diet, and have been associated to higher risks of enteritis (Bellier and Gidenne, 1996). A higher arrival of starch to the caecum may be affected by environmental conditions and cellulolytic populations which may alter fiber digestion as when as it was reduced (Boulahrouf et al., 1991). Likewise, Belenguer et al. (2000) observed a higher ($P < 0.05$) concentration of total bacteria in BG diets as compared with CG diets. High inclusion of BG increased in total bacterial concentration in the ceacum, but decrease cellulolytics, leading to a shift in the type of fermentation to a less acetate-more butyrate proportions, even at similar VFA concentrations (Belenguer et al., 2000). Caecal turnover rate was increased ($P < 0.01$) in rabbits fed diets containing 15 or 20% of BG when compared with the other experimental groups suggesting a higher caecal nutrients digestibility. Carabaño et al. (1988) indicated no changes in NDF content among BG or CG diets, assuming the direct relationship between caecotrophes and cecum content composition.

Table 4. Carcass characteristics in growing NZW rabbits at 13 weeks of age fed on the experimental diets.

	Diets						SEM	P value
	B0	B5	B10	B15	B20	B25		
No. of rabbits	3	3	3	3	3	3		
Pre-slaughter weight (g)	2019.3^{b}	2068.7^{b}	2132.7^{ab}	2202.7^{a}	2214.0^{a}	2124.0^{ab}	60.17	0.0132
Hot carcass weight (g)	1201.5^{d}	1234.7^{cd}	1317.3^{bc}	1405.7^{ab}	1441.4^{a}	1380.1^{ab}	48.92	0.0003
Cold carcass (g)	1186.3^{d}	1219.6^{cd}	1300.8^{bc}	1389.8^{ab}	1424.6^{a}	1365.4^{ab}	48.66	0.0003
Carcass loss weight (g)	15.5	14.8	16.4	15.9	16.8	14.8	3.38	0.9593
Dressing %	59.5^{c}	59.7^{c}	61. 8^{b}	63.8^{a}	65.1^{a}	65.0^{a}	1.13	0.0001
Fur (g)	185.9	183. 7	196.5	185.5	195.4	196.4	20.16	0.9172
As (%) of pre-slaughter weight								
Liver	3.61	3.66	3.68	3.80	3.97	3.79	0.184	0.3097
Kidneys	0.66^{a}	0.64^{ab}	0.69^{a}	0.62^{b}	0.66^{a}	0.66^{a}	0.030	0.0452
Spleen	0.08	0.07	0.06	0.06	0.06	0.06	0.009	0.8969
Gallbladder	0.012	0.013	0.013	0.044	0.013	0.015	0.019	0.3350
Heart	0.27	0.28	0.28	0.27	0.300	0.29	0.032	0.8531
Lungs plus trachea	1.04	1.09	1.09	1.08	1.07	1.05	0.129	0.9951
Digestive tract full	21.64	21.15	22.49	21.59	22.59	21.29	1.358	0.6987
Digestive tract empty	5.47	5.59	6.24	5.90	6.44	6.23	0.617	0.3671
Caecum plus appendix full	8.41	8.78	8.85	8.35	8.40	8.11	0.680	0.7693
Caecum plus appendix empty	1.46	1.48	1.35	1.48	1.46	1.51	0.116	0.6684

a,b,c means in the same row bearing different letters, differ significantly ($P < 0.05$).
[1] Metabolizable energy (kcal/kg feed DM) .
SEM, standard error of the mean.

Carcass composition

The pre-slaughter weight of B15 and B25 rabbits was higher ($P < 0.05$) than those fed B0. Hot and cold carcass weight were increased ($P < 0.05$) in B15 and B20 rabbits, as compared with those of B0. Dressing percentage was increased ($P < 0.01$) in B10 rabbits, while the highest value ($P < 0.05$) was recorded in

B20 compared to that obtained in B0 (Table 4). Different levels of CG substitution by BG up to 25% in rabbit diets did not affect liver weight or total giblets weight changes as a percentage of pre-slaughter weight. However, kidneys percentage differed ($P < 0.05$) among the experimental groups. In addition, the relative weights of heart, spleen, gallbladder, lungs plus trachea and full were also not differed among animals of the different experimental groups. Similarly, partial or total replacement of CG by BG in rabbit diets affect ($P < 0.05$) on the relative weights of full or empty digestive tract and ceacum plus appendix full or empty of the experimental groups. Our results confirmed earlier by Belenguer et al. (2000) who reported that rabbit caecal volume and caecal content were not affected by BG or CG rabbit diets, but they were affected ($P < 0.05$) by the levels of inclusion of both materials in the diet. Similarly, Ibrahim et al. (1999) reported that replacing BG by 10% dried barley radical supplemented with enzyme resulted in a higher carcass yield, while the giblets percentage varied slightly. El Gendy et al. (2000) reported that feeding rabbits on diets with different level of barley radical did not affected dressing, carcass, giblets and alimentary tract. Our results were not consistent with those reported by Xiccato et al. (2002) who found that the dressing percentage was increased ($P < 0.05$) in rabbits offered high starch diets (corn or barley).

CONCLUSION

Results from our present study revealed that a partial replacement of corn grains (CG) by 20% barley grains (BG) in rabbit diets was reflected by an increase of live body weight, feed intake and feed conversion ratio. Further researches are required to investigate the meat quality produced of rabbits fed on 20% of BG as compared with the other experimental groups.

REFERENCES

Akram, M., Rasool, S., Aslam. M., Nawaz, S. 1989. Comparative nutritive value of barley, oats and maize in rabbits. Pakistan Veterinary Journal 9: 85-86.

Amici, A. 1989. Germinated barley in rabbit feeding. Rivista di conigcoltura. 26, 41-44.

A.O.A.C., 1990. Association of official Analytical chemists. Official Methods of Analysis. 15th ed. Washington DC., USA.

Bagliacca, M., Mori. B. 1987. Effect of introducing malt industry products in diets for rabbits, rivista di coniglicoltura. 24: 51-56.

Belenguer, A., Balclls, J., Fondevila, M., Torre, C. 2002. Caecotrophes intake in growing rabbits estimated either from urinary excretion of purine derivatives or from direct measurement using animals provided with a neck collar: effect of type and level of dietary carbohydrate. Animal Science 74: 135-144.

Belenguer, A., Fondevila, M., Balcells, J., Torre, C. 2000. Effect of the source and level of cereal in diet on the rabbit caecal environment and microbial population. In: Proc 7n World Rabbits Congress, Vol. C, 95-100.

Bellier, R., Gidenne, T. 1996. Consequence of reduced fibre intake on digestion, rate of passage and caecal microbial activity. British Journal of Nutrition 75: 353-363.

Bengala-Freire, J. 1989. La digestibilité totale et iléale des glucides chez le porcelet sevré précocement, conséquences physiologiques et métaboliques. Thése Doctorat Univérste Rennes 1, France, 150 pp.

Blas, E., Gidenne, T. 1998. Digestion of starch and sugars in: De Blas C., Wiseman J. (ed). The Nutrition of the Rabbit. CABI publishing CAB international, Walling ford Oxon, UK, 17-38.

Blas, E., Cervera, C., Fernandez Carmona, J. 1994. Effect of two diets with varied starch and fiber levels on the performances of 4-7 weeks old rabbit. World rabbit Science 2: 117-121.

Blasco, A., Ouhayoun, J., Mansoero, G. 1992. Status of rabbit meat and carcass: criteria and Terminology. Options Med: terranean's – Series Seminaries. N° 17, 105-120.

Boulahrouf, A., Fonty, G., Gouet, P. 1991. Establishment, counts and identification of the fibrolytic microflora in digestive tract of rabbit. Influence of cellulose content. Current Microbiology, 22: 21-25.

Carabaño, R., Fraga, M.J., Santoma, G., De Blas, J.C. 1988. Effect of diet on composition of cecal contents and on extraction and composition of soft and hard feces of rabbits. Journal Animal Science 66: 901-910.

Carabaño, R., Fraga, M.J., De Blas, J.C. 1989. Effect of protein source in fibrous diets on performance and digestive parameters of fattening rabbits. Journal of Applied Rabbit Research 12: 201-204.

Cheeke, P.R. 1987. Rabbit feeding and nutrition. Academic Press, I N C. Harcourt Brance Jovnovich, Publishers, Orlando San Diego New York Austin, Boston-London, Sydney, Tokyo, Toronto.

Colonna, P., Champ, M. 1990. Significance of starch damage in feeds. Sciences des Aliments 10: 877-897.

De Blas, J.C. 1986. Aimentacion del Conejo. Mund-prensa. 53-59 (In Spanish).

Duncan, D.B. (1955): Multiple range and multiple test. Biometrics 11: 1-42.

El-Gendy, K.M., Bassuny, S.M.,Shehata, A.S., Sarhan, M.A. 2000. Using radical in animal nutrition. Growth performance, digestibility and carcass traits of growing rabbits fed different levels of barley radical with or without methionine supplementation. Egyptian Journal of Rabbit Science 10: 75-88.

Fraga, M.J., Perez, P., Carabaño, R., De Blas, J.C. 1991. Effect of type of fiber on the rate of passage and on the contribution of soft feces to nutrient intake of finishing rabbits. Journal Animal Science 69: 1566-1574.

Garcia, J., De Blas, J.C., Carabaño, R., Garcia, P. 1995. Effect of type of Lucerne hay on ceacal fermentation and nitrogen contribution through caecotrophy in rabbits. Reproduction and Nutrition Development 35: 267-275.

Gidenne, T, Jehl, N., Perez, J.M., Arveux, P., Bourdillon, A., Mousset, J.L., Duperray, J.,Stephan, S., Lamboley, B. 2005. Effect of cereal source and processing in diets for the growing rabbit. 1-effects on performances and mortality by neteropathy. Animal Research 54: 65-72.

Gidenne, T., Perez, J.M. 1993. Effect of dietary starch origin on digestion in the rabbit. 1-Digestiblity measurements from weaning to slaughter. Animal Feed Science Technology 42: 237-247.

Ibrahim, R.M., El Nagmy, K.Y., Farghly, S.M. 1999. Effect of replacement of barley by dried barley radical with or without kenzyme on performance of growing male rabbits. Egyptian Journal of Rabbit Science 9: 61-71.

Kalogen, U.A. 1985. Rabbits feeding (In Russian). 112 P., Agropromozdat, Moscow.

Lanza, E., Sinatra, M.C., Fasone, V., Urso, G. 1986. Effect of using maize, barley and triticale in single cereal diets for fattening rabbits. Rivista-di-Coniglicoltura 23, 47-48.

Leng, R.A. 2008. Digestion in the rabbit –a new look at the effects of their feeding and digestive strategies. In: International Workshop Organic rabbit farming based on forages. 25-27 November , Cantho University, Cantho City, Vietnam.

Licois, D., Coudert, P., Cere, N., Vautherot, J.F. 2000. Epizootic enterocolitis of the rabbit: review of current research. Proc. 7th World Rabbit congress. Vol. B, 187-194.

Lukefahr, S.D., Van Vleek, L.D., Roberts, J.D. 1992. Estimates of components of variance and covariance of carcass traits in rabbits using animal model. Journal of Applied Rabbit Research 15: 259-273.

Maertens, L. 1992. Rabbit nutrition and feeding: a review of some recent developments. Journal of Applied Rabbit Research 15: 889-913.

Nizza, A., Moniello, G. 2000. Meat quality and caecal content characteristics of rabbit according to dietary content and botanical origin of starch. World Rabbit Science 8: 3-9.

North, M.O. 1981. Commercial chicken production. Annual. 2nd Edition. Av. Publishing company I. N. C., West-post Connecticut. U.S.A.

NRC. 1994. National Research Council. Nutrient Requirements of Poultry. 8th additional Academy Press, Washington, D.C., USA.

NRC. 1984. National Research Council Nutrient Requirements of Domestic Animals. Nutrient Requirements of Rabbits. Second Revised Edition Academy of Sciences, Washington D.C., U.S.A.

Perez, J.M., Aumaitre, A. 1979. Waxy versus regular maize: energy value for growing pigs and utilization in piglet diets. Animal Feed Science Technology 4: 109-115.

Rajendra, P., Sankhyan, S.K., Karim, S.A., Jakhmola, R.C. 2000. Growth performance and economics of broiler rabbits maintained on diets containing different energy supplements. Indian Journal of Animal Nutrition 17: 116-120.

SAS. 1996. The Statistical Analysis System for Windows, Release 6. 11. SAS Institue Inc., Cary, NC- USA.

Scapinello, C., Gidenne, T., Fortun Lamothe, L. 1999. Digestive capacity of the rabbit during the post-weaning period, according to the milk /solid feed intake pattern before weaning. Reproduction and Nutrition Development 39: 423-432.

Sinatra, M.C., Fasone, V., Barresi, S., D'urso, G. 1987. Effect of using wheat, barley and triticale in monocereal diets for rabbits. Tecnica Agricola 39: 243-248.

Struklec, M., Kermauner, A., Marinsek Logar, R. 1995. Effect of feeding with a diet based on maize, barley and oats on the digestive process and growth performance in growing rabbits. Sodobno Kmetijstvo 28: 505-508.

Xiccato, G., Trocino, A., Sartori, A., Queaque, P.I. 2002. Effect of dietary starch level and source on performance caecal fermentation and quality in growing rabbits. World Rabbit. Science 10:147-156.

Xiccato, G., Cossu, M.E., Trocino, A., Queaque, P.I. 1998. Influence du rapport amidon/fiber et de l'addition de graisse en post-sevrage sur la digestion, les performances zootechniques et la qualité bouchere du lapin In: Proc. 7 Journées de la Recherche Cunicole, Lyon, France,, pp, 154.

Yutste, P., Longstaff, M.A., Mc Nab, J.M., Mc Corquodale, C. 1991. The digestibility of semi purified starches from wheat, cassava, pea, bean and potato by adult cockerels and young chicks. Animal Feed Science Technology 35: 239-300

12

Sustainable Tilapia Production in a Backyard System for Rural Areas using Fish by-Products in Formulated Diets

Servando Rueda-López[a], María Teresa Viana[b], Daniel Badillo-Zapata[c], Carlos A. Martinez-Palacios[d] and Gabriel Correa-Reyes[b]*

[a] Programa de Licenciatura en Oceanología, Facultad de Ciencias Marinas, Universidad Autónoma de Baja California (UABC), Ensenada BC, México. Email: servando_86@hotmail.com
*[b] Instituto de Investigaciones Oceanológicas, UABC, PO Box 453, Ensenada BC 22860, México. Email: mtviana@hotmail.com, * gcorrea@uabc.edu.mx*
[c] Programa de Maestría en Oceanografía Costera, Facultad de Ciencias Marinas e Instituto de Investigaciones Oceanológicas, UABC, Ensenada BC, México. Email: danielbad00@hotmail.com
[d] Instituto de Investigaciones Agropecuarias y Forestales UMSNH, Morelia, Michoacán. Email: palacios@umich.mx
**Corresponding author*

SUMMARY

The aim of the present study was to demonstrate the feasibility of producing Tilapia (*Oreochromis niloticuss, L.*) with formulated diets containing acid fish silage made with Hancock (*Pterigoplichthys multiradiatus*) by-products in substitution of fishmeal. Thus four treatments containing increasing levels (0, 5, 10 and 15%) of acid fish silage to substitute fishmeal were used to feed tilapia juveniles in triplicate groups (average 1.5±0.7 g and 4.3±1.0 cm). Fish silage was prepared by adding citric and phosphoric acids at 2.6% each resulting in a liquefaction of the chop by-product mixture. After two weeks, the diets were prepared to contain 32-35% protein content. After 50 days of experimental period a total growth increase up to 827% was observed without significant differences among treatments. A specific growth rate from 3.70 to 4.10 and a survival rate from 96 to 99% was observed. The cost analysis shows a saving on production cost up to 23.87% using the highest incorporation level of fish silage. The use of Aquaculture Recirculation System made with components normally found in a hardware store in rural or semi urban areas is discussed. It is concluded that fish offal shows a promising activity that favors the increase in fish protein production to resolve the demand of cheap protein.

Keywords: Tilapia; Backyard System ratio; by-products diets; Recirculating Systems.

INTRODUCTION

Food production has been a worrying subject along the world history where programs or funds are oriented according to the different policies adopted for the different economies around the world like the Backyard Aquaculture. The FAO has foresee that up to year 2030 even if globally there will be enough food for a growing world population, hundreds of millions of people in developing countries will remain hungry and many of the environmental problems caused by agriculture will remain serious. Most of the poor people is located in rural areas from the third world countries, thus promoting the growth of food products in rural areas by giving rural people better access to land, water, credit, health and education, is essential to alleviate poverty and hunger. However, interestingly, FAO ("World agriculture: towards 2015/2030") states that aquaculture will be the fastest growing activity, where efforts on new species domestication and species improvements within traditional breeding, chromosome manipulation and hybridization will be of importance, a subject that somehow have already given significant contributions.

Nevertheless, aquaculture industry will grow only if sustainable practices will take place. For instance, how to optimize the use of water, either for food production or for direct human needs. The idea of reusing the water from agriculture or aquaculture is not a new idea (Nduwimana *et al.*, 2007). In such case, in order to be competitive, lowering production cost besides the intensification of fish production will be the key to success, and food or feeds in one of them.

In Mexico since the 80´s several attempts has been made to encourage fish protein consumption in rural areas by promoting the aquaculture. However this activity has been growing very slow apart from those areas where larger facilities have been developed throughout the private sector inversion. The reason can be explained due to the lack of effective rural programs concerning the necessary inversion, and the requirements to undertake a proper transfer technology. Backyard animal rearing systems on smallholdings in the rural areas in Mexico has been for centuries, where livestock have been an integral part of family production resulting in the main source of protein intake. Low cost backyard aquaculture has been already reported (Szyper, 1989), but a proper transfer technology is needed in rural areas in Mexico.

In a recirculation system, the water is optimized by reusing it several times using filters and pumps with a refill of water to compensate the evaporation. Then, the water is treated internally and the sludge from the discharge can be concentrated and eliminated or used as fertilizer. Ammonia can be also reduced by biofilters. This system has been widely described by several authors (Timmons *et al.*, 2001; Gutierrez-Wing and Malone, 2006) and named as Recirculation Aquaculture System (RAS). The equipment is operated by electricity pumps where often a high production costs are experienced and technical skills are needed to fully operate the system.

Tilapia is an excellent fish to promote backyard Aquaculture like resistant to rough handle, rapid growth under a wide range of temperatures (24 to 32°C) and intermediate water quality. Their nutrition has been thoroughly described (El-Sayed, 1999) being well adapted and fed with a variety of ingredients and their protein requirement is close to 35% (Goddard *et al.*, 2008).

Besides, the increasing demand and progressive scarcity of fishmeal in the international market has led the prices to boost and therefore the need to search for protein optimization is obliged. In rural and semi urban areas a great amount of protein is throw away throughout the fish filleting industry, waste that usually becomes an environmental problem that could be used in feeds throughout fish silage. Fish waste can efficiently be used as a feed ingredient by chopping the fish and preserve it either by fermentation or acid preserved. This last is easier to control by lowering the pH preventing the proliferation of bacteria and mold, whereas by fermentation a strict control of anaerobic conditions is needed besides adding soluble sugar and a source of fiber to promote lactic acid bacteria to hold the mixture at pH around 4-5, process that if is not appropriated controlled easily brings the fermentation to putrefaction.

The *Pterigoplichthys multiradiatus* (Hancok) a native fish from Orinoc River from tropical areas of the South America was introduced to Mexico for the aquarist as a cleaner fish for home aquariums, in Mexico known as "Pez Diablo or Armour catfifh". Today has become an invasive species accidentally introduced in lakes and rivers along the Mexico natural areas. Here it constitutes an environmental problem, and due to their aggressive reproduction behavior has led a large competition in the capturing gill nets, inside the natural reservoirs with concomitant commercial production problems. Therefore the government has made a called to researchers to investigate all possible uses of this fish to an overall production in the Michoacán, Mexico in the Adolfo López Mateos Dam (Presa infiernillo). This fish has show to have excellent meat properties but their aspect makes this product to be considered as trash. However, their hard skin and low proportion to meat has made the process to recover the meat protein cost effective and difficult to be implemented for direct consumption. However, the use of acids to promote the autolysis to produce acid fish silage could be an alternative to preserve fish protein to prepare formulated diets for fish aquaculture in substitution of fishmeal without reducing their performance. Therefore the aim of the present work is the use of fish silage from the *Pterigoplichthys multiradiatus* by products used to formulate diets with different levels to substitute fishmeal for tilapia feeds, reducing the costs of tilapia production using a backyard Recirculation Aquaculture System model.

MATERIALS AND METHODS

Diet preparation

Four experimental diets were formulated to contain 35% protein (Table 1) with four levels of fish silage incorporation in dry weight basis (0, 5, 10 and 15%) in substitution of fishmeal. Primary sources of protein were fishmeal and soybean meal (42% CP). Silage was made as described by Viana *et al.* (1993). In summary, chopped fish (P. *multiradiatus*) (heads, skin and bones) was mixed with 2.6 % phosphoric acid, 2.6 % citric acid. The mixture was blended to obtain a homogenate with a pH around 4 to 4.5 and left in buckets at 20°C. After 2 weeks the fish silage was incorporated into the diets as follows.

All ingredients were blended in a mixer (Robot Coupe, USA, model R10), to produce a homogeneous mixture. The water content was controlled to achieve moisture levels not higher to 60%. The dough-like diets were cold extruded through a meat grinder (Toro Rey, Mexico, model M32-5) into 0.5 × 0.7 mm pellets, and then dried in a convection oven for 12 hr at 60 °C. The dry pellets were placed in sealed plastic bags and stored at –40°C until use.

Analytical Methods

Proximate composition of diets was determined according to standard procedures (AOAC, 1990), in summary, moisture content of each diet was calculated from triplicate samples (4-5g) that were dried to constant weight at 60°C. Mean total nitrogen content was determined by the micro-Kjeldahl method, and percent crude protein was then calculated as % N x 6.25. Mean total dietary crude lipid was determined by extraction using chloroform-methanol (2:1 V/V) following the method of Folch *et al.* (1957). Ash content was determined by heating samples to 550^0 C for 4 h. Nitrogen free extract was calculated by difference as follow: NFE = 100 - (% crude protein + % crude lipid + % ash).

Table 1. Ingredient (g kg^{-1} dry weight) of four experimental diets at different levels of fishmeal replaced by fish silage made from *Pterigoplichthys multiradiatus* (Hancok).

Ingredients	TREATMENTS			
	T_1	T_2	T_3	T_4
Fishmeal [a]	250	200	150	100
Fish silage [b]	0	50	100	150
Soybean meal [c]	250	250	250	250
Wheat bran	150	150	150	150
Wheat flour	145	145	145	145
Cornstarch	88	88	88	88
Gelatin	50	50	50	50
Corn oil	30	30	30	30
Fish oil	20	20	20	20
Vitamin and mineral premix[d]	15	15	15	15
Stay C	2	2	2	2

[a] From Concentrados California (Ensenada, Mexico) 68% crude protein.
[b] Made from *Pterigoplichthys multiradiatus* (Hancok) according to Viana *et al.*, 1993
[c] From (42% CP).
[d] Vitamin mixture (1.3 %), Stay–C (0.4 %) and minerals mixture (3.3%), kindly donated by DSM, Guadalajara, Mexico.

Fish management

The fry of Nile Tilapias (*Oreochromis niloticus, L.*) were acquired from Sistemas Acuícolas Integrales S. de R.L. de C.V. (Ensenada, Baja California, México). The fish fry's were maintained in an economic outdoor Recirculation Aquaculture System, consisting in a one 6.5 Ft^3 "POLYGEISER" Pneumatic Drop Bead Filter (Aquaculture Systems Technologies, L.L.C., New Orleans, LA, USA); a 2,000-L water container (like sump), twelve 550-L water containers; a homemade wood structure; one centrifugal ¾ Hp water pump and pipe and connections of PVC.

Fry Nile Tilapias (660) with an average weight of 1.8 ± 0.8 g were randomly distributed is similar batch groups (P>0.5) among twelve circular plastic tanks 550-L (commercial water containers) at a stocking density of 55 fry per tank in triplicate groups. The dietary treatments were hand fed at apparent satiation three times per day (800, 1400 and 2000 h). The experiment started in June 2010 with a photoperiod of 14:10 (light/dark) and lasted after 50 days or up to a total growth higher than 400% according to the National Research Council (NRC, 2011). Every 15 days all fishes from each experimental unit were individually weighed (g) and measured (mm), and the following parameters were calculated as follows:

The total growth increment (%) was calculated as follow:

TGI = (Final body weight *100) / Initial body weight.

The growth rate (GR, mg day^{-1}) was calculated as follow:

GR = (Final body weight – Initial body weight) / Time in days.

The Specific Growth Rate (SGR) was calculated as follow:

SGR = ((ln(Final body weight) – ln(Initial body weight)) / Time in days)) X 100.

The Feed Conversion Ratio (FCA) was calculated as follow:

FCA = Dry weight of feed intake (g) / Gain in body wet weight (g).

Water temperature was recorder every hour, for the 50 experimental days, with a Tidbit temperature data logger (Onset HOBO Data Logger, USA). Dissolved oxygen from each experimental tank was measurement every day with an YSI ® D.O. Meter Y55 series, whereas ammonia, nitrite and alkalinity were measurement every week with an Aquarium colorimeter kit (API Aquarium Pharmaceutical Kit). The pH of water was recorder every week with an YSI ® EcoSense ® pH/Temp Pen.

Cost-Benefit Analysis

The costs of diets were calculated considering the ingredients, fish silage process, labor and energy expenditure. The analysis of the cost-benefit was calculated multiplying the costs of the diets by the feed conversion ratio.

The cost of production of Tilapia for each diet was calculated as follow:

Cost Kg Tilapia Tn = Feed Cost Tn * FCA Tn.

The saving for each diet (Tn) was calculated as follow:

Save Tn = Cost Kg Tilapia T1 – Cost Tn.

Statistical Analysis

All data were analyzed by one-way ANOVA to detect potential differences at the start and the end of the experiment among the experimental treatments. The normality of the samples was tested with the Kolmogorov-Smirnoff test. When significant differences were found (P<0.05), a Duncan's Multiple Test was used. Statistical analyses were made using Sigma Stat software for Windows. All data was presented as mean ± SEM.

RESULTS

All diets made with the inclusion from 0 to 15% fish waste silage (dry weight basis) resulted in a significant similarity in protein (32.96 to 37.63%) and lipid content (4.43 to 5.07%) (F=3.86; P=0.056 and P>0.05, respectively) (Table 2). However differences were found in ash content where levels from 8 to 10% ash where registered along T1 to T4 treatments.

Water quality from fish tanks was under acceptable parameters with a register values under of: 0.25 mg l^{-1} for NH_4 and NO_2; levels of dissolved oxygen ranged over up a 6.5 mg l^{-1}. Mean temperature values was recorder at 24.23±1.42 °C and pH values of 8.2±0.2. The alkalinity of the systems was adjusted every week at 140 mg l^{-1}, with Sodium Bicarbonate industrial grade to keep the balance of carbonate in the system.

Growth was significant similar among treatments (Figure 1), with a growth rate in length of 0.66 to 0.75 mm day^{-1} and 166.70 to 196.94 mg day^{-1}in weight (Table 3). The SGR and FCA resulted in significant similarities among treatments, with values between

640 and 827 % and 1.73 to 1.88, respectively. No differences were observed in survival; however from 0.6 to 3.7% organism died during the 50 days experimental procedure (Table 3).

Diet cost analysis (Table 4) shows that cost of diets decreases from 1.54 to 1.27 US$ kg-1, with fishmeal replacement with fish silage (5, 10 and 15%). The production costs decreases proportionally with the fishmeal replacing, generating from 11.76 to 23.87%.

Table 2. Proximate composition (%) of four experimental diets at different levels of fishmeal replaced by fish silage made from *Pterigoplichthys multiradiatus* (Hancok).

	TREATMENTS			
	T_1	T_2	T_3	T_4
Crude protein (%)	37.63±0.00 a	33.25±2.32 a	33.54±1.82 a	32.96±2.53 a
Total lipids (%)	4.43±0.19 a	4.65±0.29 a	4.92±0.95 a	5.07±0.11 a
Ash (%)	7.56±0.05 d	8.43±0.20 c	9.36±0.15 b	10.13±0.15 a
NFE[a]	47.79	50.87	50.10	49.79

[a] NFE = 100 - (% Crude protein + % Total lipids + % Ash).
Values in the same row with different superscripts are statistically different $p<0.05$.

Table 3. Biological indices obtained for juvenile of Tilapia (*Oreochromis niloticuss, L.*), feed four levels diets of fish silage replacement for fishmeal.

INDICES	TREATMENTS			
	T1	T2	T3	T4
Initial Length (cm)	4.23±0.77	4.20±0.75	4.26±1.01	4.31±0.72
Final Length (cm)	8.06±1.47	8.12±1.26	8.13±1.55	8.14±1.42
Length growth rate (mm/day)	0.71±0.04	0.75±0.02	0.71±0.05	0.66±0.03
Initial Weight (gr)	1.53±0.69	1.39±0.72	1.77±1.63	1.53±0.70
Final weight (gr)	11.43±1.52	11.25±1.34	11.30±1.54	11.41±1.47
Growth rate (mg day^{-1})	186.40±14.20	196.94±19.78	178.93±3.52	166.70±10.29
TGI (%)	757.73±126.05	827.30±184.91	640.46±95.61	751.37±78.04
SGR (%)	4.03±0.35	4.19±0.44	3.70±0.29	4.03±0.21
FCA	1.88±0.55	1.76±0.05	1.80±0.08	1.73±0.16
Survival (%)	99.41±1.03	98.25±1.75	96.33±4.91	98.25±1.75
Mortality (%)	0.59±1.03	1.75±1.75	3.67±4.91	1.75±1.75

Standard errors are given.
Values in the same row with different superscripts are statistically different $p<0.05$.

Table 4. Cost-benefit analysis of four experimental diets at different levels of fishmeal replaced by fish silage made from *Pterigoplichthys multiradiatus* (Hancok), based on FCA and cost per kg.

	TREATMENTS			
	T1	T2	T3	T4
Feed Cost (US$ kg-1)	1.54	1.45	1.36	1.27
FCA	1.88	1.76	1.80	1.73
Cost Kg Tilapia (US$)	2.89	2.55	2.44	2.20
Save (US$)		0.34	0.45	0.69
Save (%)		11.76	15.57	23.87

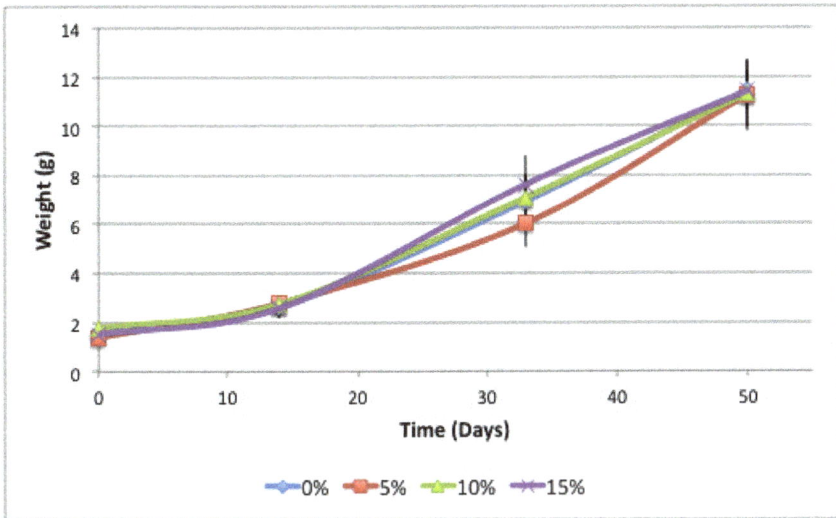

Figure 1. Growth in weight for juvenile of tilapia (*Oreochromis niloticuss*, *L.*), feed four levels diets of fish silage replacement for fishmeal. Average values and standard deviation were obtained from three independent replicates.

DISCUSSION

In the present work it was demonstrated that fish silage from *Pterigoplichthys multiradiatus waste* could efficiently replace fishmeal in tilapia feed formulations up to 60% protein replacement, a replacement that would be difficult to increase even more due to the high ash content from the by-products (up to 20%; NRC, 2011). Fish silage was made from by products from this invasive species found in large quantities in the Infiernillo Dam in Michoacán, Mexico. The silage was made out only from heads and skin with fins, whereas the flesh was used with a different purpose as food for human. This silage was prepared by adding a mixture of citric and phosphoric acids and kept at room temperature. Previously reports demonstrate that silage made in similar conditions can be store for several months without any deterioration (Viana *et al.*, 1993). The use of this acid mixture has the advantage of accessibility since both are use in the food industry, compare to sulfuric and formic acids commonly used in acid fish silage (Raa and Gildberg, 1982). Citric acid is a powder that can be measured by weight or volume, whereas the phosphoric acid in their liquid form per volume without any problems for an additional 1-2% difference than that obtained by an accurate measurement. The fish silage can be then stored in 20L plastic containers with a lock in a dry place without problems to be contaminated by flies or bugs within rural areas. After 2 weeks the silage can be used by mixing with the rest of the ingredients to prepare the formulated diets in the backyard aquaculture system. The silage was added up to 60% fishmeal replacement corresponding to a 15% from the diet considering in a dry matter content basis. This kind of preservation is cause the liquefaction of protein resulting in a liquid protein product with a content of dry matter as low as 15 to 20%, making difficult to incorporate higher amount of silage without a previous dehydration. Therefore for further inclusion of acid fish silage a previous dehydration will be needed unless the ash content is not high.

Here, all treatments resulted in an overall growth increase up to 827% during 50 days experimental procedure and even if no significant differences could be reported, a result that is considered optimal as recommended by the NRC (2011). Earlier reports claim an advantage the use of partially hydrolyze protein (Espe *et al.,* 1992) due to a more efficient use of protein, whereas a high inclusion of fish silage leads to a high amount of free amino acids and therefore a growth reduction. However, in this study we did not found significant differences at increasing amounts of fish silage on the diets, but rather the specific growth rate was hold between 3.70 to 4.19. According to Espe *et al.* (1992) the rate of peptides and amino acids uptake should be in good balance in order to promote growth. Then, when a pre-digested proteins are used in feed, if the rate of uptake exceeds the anabolic capacity of the animal then an extra amount of amino acids will rapidly be absorbed are then catabolized resulting a poor protein utilization. The inclusion of acid fish silage in substitution of fishmeal, has the advantage, not only to reduce the cost from the fishmeal (to 23%), but also for using a by-product that is scarcely used and available in most natural lakes and rivers in rural areas. According to several authors (Tacon and Metian, 2011; Tacon *et al*., 2012) fishmeal will not be sustainable due to overfishing and need to feed the increasing human population. A large amount of fishmeal comes from minor pelagic species with an increasing pressure to target the human market. Therefore the finding of any ingredient to substitute fishmeal will be positive to increase the possibilities to supply protein for fish feed artisanal manufacture. However, since the acid fish silage has a large amount of water, it will be interesting to dry it previously and then try a higher level of substitution if ash content is not higher than 20%. Moreover, these diets were composed with a mixture of fishmeal/fish silage and soybean meal in equal parts. As mentioned above the fish silage was added according to their dry matter content and it was directly replaced for fishmeal.

Today, one of the priority issues in any order of scientific knowledge relevant to food production is sustainability (FAO, 2012), so that food production goes according to the economic, social, cultural and of a healthy environment, without compromising the satisfaction of the same to future generations (Mojica-Sastoque *et al.,* 2010). Tilapia production is no stranger to this concern for sustainability, where there are even a number of studies that establish the foundations of sustainability in Mexico. Vivanco-Aranda, *et al* (2011) addresses this issue where they study the level of modernization in Mexico among several states establishing the critical factors of operation. The automation is a necessary process for a controlled and sustainable production where formulated feeds play an important role in the production cost to increase growth rate. However, the cost of food is an issue that if not studied could have a negative effect on the production of tilapia. In a traditional culture of tilapia, regardless of the level and intensity of crop you have, the cost of feed accounts for up to 60% of the production cost. Hartley-Alcocer, et al. (2006), mentioned that the 92% of the farms in Mexico employed commercial feeds (i.e. Purina, AS, El Pedregal, Malta-Clayton and Algimex). In this paper we present a cost-benefit analysis with the idea of evaluating the replacement of fishmeal by *Pterigoplichthys multiradiatus* silage,s (Hancock), finding that replacing up to 15% of fishmeal gives a saving of 23.9% of the production cost. Importantly, in this order of sustainability, that this saving involves not only the product system of Tilapia, but to reduce the consumption of fish meal as this would impact on fresh sardines could be directly feeding the population.

If such results could be applied in different product tilapia system in different states of the country, is likely to crop sustainability nationwide increase causing more competitive with the product that is imported from Asian countries and increase the sardine availability for human consumption.

Water quality was measured daily or weekly showing an optimum quality as recommended by Webster and Lim (2006). The main quality parameters of water was recorded along the experiment: ≤ 0.25 mg l^{-1} of NH_4; ≤ 0.25 mg l^{-1} of NO_2, pH = 8.2 ± 0.2; Dissolved Oxygen of ≥ 6.5 mg l^{-1} and alkalinity was controlled ≥ 140 mg l^{-1}. The Recirculation Aquaculture System used here was loaded with a fish density necessary to achieve a final biomass of 50 Kg m^{-3} per tank, an appropriated amount to support a backyard Tilapia production. The survival rate reported in this study is considered as optimal being over 90% in all cases. Under this system a production of 400.00 to 500.00 g commercial size fish would be obtained after 6 months under same growth rate. This backyard production should be promoted in rural to semi-urban areas in Mexico, where families could use the opportunity to ingest fresh protein in small quantities without the need to refrigerate the left carcass as it has been done with poultry production.

In conclusions, the use of fish by-products from the *Pterigoplichthys multiradiatus* in fish silage in substitution of fishmeal has resulted in a similar growth rate than those with fishmeal. Their use can significantly reduce the feed cost and therefore a reduction up to 23% was observed, besides the use of fishery by-products instead of minor pelagic fish species to sustain the fishery for human consumption. Therefore the introduction of backyard aquaculture should be promoted to produce high fresh quality protein in rural areas.

Acknowledgements

This work was supported by the National Council for Science and Technology Project 37147 FONDO MIXTO-CONACyT, Michoacán, México. We Thanks to DSM for donating the vitamin and mineral mixtures. This work was performed by undergraduate students in their summer practices.

REFERENCES

AOAC. 1990. 16th ed. Official Methods of Analysis of AOAC, vol. 1. Association of Official Analytical Chemists, Arlington, VA, USA.

El-Sayed A.F.M. 1999. Long-term evaluation of cotton seed meal as a protein source for Nile tilapia *Oreochromis niloticus*. Aquaculture 84, 315–320.

Espe, M., Haaland, H., Njaa, L.R. 1992. Autolysed fish silage as a feed ingredient for Atlantic salmon (*Salmo salar*). Comparative

Biochemistry and Physiology. 103A (2) 369-372.

FAO (Food and Agriculture Organization). 2012. State of World Fisheries and Aquaculture (SOFIA) 2012. FAO Fisheries Technical Paper. Rome. 209 pp.

Folch, J. Lees, M., Stanley, G. 1957. A simple method of isolation and purification of total lipids from animal tissues. Journal Biological Chemistry. 226, 497-509.

Goddard, S., Al-Shagaa, G., Ali, A. 2008. Fisheries by-catch and processing waste meals as ingredients in diets for Nile tilapia, *Oreochromis niloticus*. Aquaculture Research, 1-8.

Gutierrez-Wing, M.T., Malone, R.F. 2006. Biological filters in aquaculture: Trends and research directions for freshwater and marine applications. Aquaculture Enginnering 34(3), 163-171.

Mojica-Sastoque, F.J., Vivanco-Aranda, M., Martínez-Cordero, F.J., Trujillo-Cabezas, R. 2010. Tilapia 2020: Prospectivo del Sistema-Producto de Tilapia en México. Proyecto realizado por encargo del Comité Sistema-Producto Tilapia de México A.C. para el Comité Nacional Sistema-Producto Tilapia. Mazatlán, Sinaloa, México. pp 269.

Nduwimana, A., Yang, X.L., Wang, L. 2007. Evaluation of a cost effective technique for treating aquaculture water discharge using *Lolium perenne* Lam as a biofilter. Journal of Environmental Sciences 19(9), 1079-1085.

National Research Council. 2011. Nutrient Requirements of Fish and Shrimp. Animal Nutrition Series. National Academic Press. Washington, D.C. USA. 392.

Raa. J., Gildberg, A. 1982. Fish silage: A review. In: CRC Critical Reviews in Food Science and Nutrition, 116, 343-419.

Szyper, J. 1989. Backyard Aquaculture in Hawaii, A Practical Manual. Windward Community College, Aquaculture Development Program, Department of Land and Natural Resources, Honolulu, HI, USA. 87 p.

Tacon, A.G.J., Metian, M. 2009. Fishing for Aquaculture: Non-food use of small pelagic forage fish a global perspective. Reviews in Fisheries Science. 17(3), 305-317.

Tacon, A.G.J., Hasan, M.R., Allan, G., El-Sayed, A.F.M., Jackson, A., Kaushik, S.J., Ng, W.K., Suresh, V., Viana, M.T. 2012. Aquaculture feeds: addressing the long-term sustainability of the sector. In: Subasinghe, R.P., Arthur, J.R., Bartley, D.M., De Silva, S.S., Halwart, M., Hishamunda, N., Mohan, C.V., Sorgeloos, P. (eds.). 2012. Proceedings of the Global Conference on Aquaculture 2010: Farming the Waters for People and Food. Phuket, Thailand. 22-25 September 2010. FAO, Rome and NACA, Bangkok. 896 pp.

Timmons, M.B., Ebeling, J.M., Wheaton, F.W., Summerfelt, S.T., Vinci, B.J. 2001. Recirculating Aquaculture Systems. Northeastern Regional Aquaculture Center. New York, USA, 650 p.

Viana, M.T., Nava, C., Solana-Sansores, R. 1993. Acid fish silage. Effect of preheating and addition of phosphoric and citric acids on the biochemical quality. Ciencias Marinas. 19(4), 415-433.

Vivanco-Aranda, M., Mojica, F.J., Martínez-Cordero, F.J. 2011. Foresight analysis of Tilapia supply chains (Sistema Producto) in four states in Mexico: Scenarios and Strategies for 2018. Technological Forecasting & Social Change. 78, 481-497.

Webster, C.D., Lim, C. 2006. Tilapia: Biology, Culture and Nutrition. Food Products Press ®. An Imprint of the Haworth Press, Inc. New York, USA. 644 pp.

World agriculture: towards 2015/2030 Summary report, http://www.fao.org/docrep/004/y3557e/y3557e00.htm, November 2012.

13

Ecological Niche Modeling and Wildlife Management Units (Umas): An Application to Deer in Campeche, México

Tania Escalante*and Enrique Martínez-Meyer

Museo de Zoología "Alfonso L. Herrera", Dpto. Biología Evolutiva, Facultad de Ciencias, UNAM. Apdo. Postal 70-399, 04510 México, D. F., México. E-mail: tee@ibunam2.ibiologia.unam.mx
Laboratorio de Análisis Espaciales, Dpto. Zoología, Instituto de Biología, Universidad Nacional Autónoma de México, Apdo. Postal 70-153, 04510 México, D. F., México. E-mail: emm@ib.unam.mx
**Corresponding author*

SUMMARY

The Units for the Conservation, Management and Sustainable Use of Wildlife (UMAs) are instruments of conservation and management of specific species in Mexico. UMAs represent in southeastern Mexico an important way for deer management, but they have major problems related to the monitoring of species. In this paper, we propose a methodology based on the use of a 'niche centroid approach' for estimating ecological distances to the niche centroid in order to produce distribution maps containing information on the potential relative abundance of species to evaluate the capability of UMAs to maintain populations of deers. We modeled the abundance for *Mazama temama*, *M. pandora* and *Odocoileus virginianus* on the state of Campeche, Mexico. Our predictions of areas with most abundance of deer coincided with reports from literature. We identified the UMAs "Ik Balam" and "Ejido Carlos Cano Cruz" as areas with high proportion of suitable environment, while UMAs "Betito y Lupita", "El Huanal", "Puh", "Refugio faunístico Jalotum", "Ría Lagartos-Ría Celestun" and "Yocol Cab Balam" have not environmental conditions adequate to maintain deer populations. Although this is a preliminary study, it can be a starting point to establish institutional standards for the management of species.

Keywords: Abundance; distribution; Euclidean distance; GARP; mammals.

INTRODUCTION

The Units for the Conservation, Management and Sustainable Use of Wildlife (UMAs, from its Spanish name) are instruments of conservation and management of specific flora and fauna species in Mexico (Secretaría de Medio Ambiente, Recursos Naturales y Pesca, 1997; Diario Oficial de la Federación, 2000). UMAs are either wild or managed lands (extensive UMAs), or enclosures or greenhouses (intensive UMAs) registered to operate in accordance to an approved management plan to harvest individuals of one or more species for which there is a continuous monitoring of habitat and populations/individuals (Diario Oficial de la Federación, 2000). In the extensive UMAs, species are managed in wild conditions (*in situ*), therefore population numbers are estimated from field sampling. The UMAs system is much more developed in northern Mexico, holding more than 80% of all UMAs in the country (see March *et al.*, 2009). In southeastern Mexico, the state of Campeche holds the largest number of UMAs (Secretaría de Medio Ambiente y Recursos Naturales, 2011), where almost all of them (99%) are extensive UMAs (García-Marmolejo, 2005). However, few studies have being carried out to characterize UMAs and evaluate their effectiveness for conservation (González Marín *et al.*, 2003; García-Marmolejo *et al.*, 2008; Mandujano and González-Zamora, 2009).

A major operative problem in the extensive UMAs is to maintain reliable technical assistance and monitoring of species and to avoid illegal harvest of non-target species (March *et al.*, 2009). Estimating populations of most hunting species is especially difficult in the humid tropics (e.g., Campeche) (Weber *et al.*, 2006), and because authorization for the extraction quota depends on such estimations, it would be very useful to have a methodological approach to produce baseline data on the expected abundance of target species to serve as a reference for determining whether the numbers reported by field technicians are realistic or not.

In recent years, ecological niche modeling has developed and established as a robust approach for mapping species' geographic distributions via the characterization of suitable areas based upon occurrence records and a suite of environmental variables that are known or assumed as distributional conditionals. This approach has been broadly tested for many species of basically all groups, with generally reliable results for predicting species' distributions (see Elith and Leathwick, 2009). Although it is controversial about what is the best algorithm for modeling (Pearson *et al.*, 2006; Elith *et al.*, 2006; Guisan *et al.*, 2007; Tsoar *et al.*, 2007; see Pliscoff and Fuentes-Castillo, 2011), the Genetic Algorithm for Rule-set Production (GARP; Stockwell and Noble, 1992; Stockwell, 1999) is maintained as a good option for modeling (for example: Haverkost *et al.*, 2010; Ray *et al.*, 2011; Kumara and Suganthasakthivel, 2011; Marín-Togo *et al.*, 2012).

Although some authors have found that high quality habitat are not necessarily directly related with the density of species (Van Horne, 1983); in previous works, Díaz-Porras (2006) and Martínez-Meyer *et al.* (2013) demonstrated that abundance distributional patterns hold a strong inverse relationship with the distance to the ecological niche centroid in several species of birds, mammals and a reptile. In other words, the abundance of a species in a specific locality is determined by the niche conditions therein (Brown, 1995) and optimal conditions are around the niche centroid in ecological space, thus the closer a locality is to the niche centroid the higher its abundance (Maguire, 1973; Yañez-Arenas *et al.*, 2012).

In this paper we use the 'niche centroid approach' for estimating ecological distances to the niche centroid for three species of ungulates with the aim to produce distribution maps containing information on the potential relative abundance of species based on the ecological characteristics of the landscape, and evaluate the capability of UMAs to maintain populations of these mammals in Campeche, Mexico, as a management tool for authorities and decision makers.

MATERIAL AND METHODS

We analyzed three species of deer (Artiodactyla: Cervidae): *Mazama temama* (South American red brocket), *M. pandora* (Yucatan brown brocket), and *Odocoileus virginianus* (white-tailed deer), which have been authorized to be managed in UMAs in Campeche since 1999 to date. For these species, we obtained occurrence data points for modeling for the whole Yucatán Peninsula from Escalante *et al.* (2002), Sánchez-Cordero (2003), El Colegio de la Frontera Sur (Unidad Chetumal), and collections in Global Biodiversity Information Facility (http://www.gbif.org/), Mammal Networked Information System (MaNIS) (MaNIS, 2008), REMIB (Conabio, 2008), and Unidad de Informática para la Biodiversidad (Instituto de Biología, UNAM, 2008). The collections consulted are: Colección Mastozoológica, El Colegio de la Frontera Sur, Unidad San Cristóbal de las Casas; Colección Nacional de Mamíferos, Instituto de Biología, UNAM; Lousiana State University Museum of Natural Science; Mammal collection, The Museum of Vertebrate Zoology at Berkeley; Museo de Zoología,

El Colegio de la Frontera Sur, Unidad Chetumal; Museo de Zoología, Facultad de Ciencias, UNAM; Museum of Comparative Zoology, Harvard University; Natural History Museum, University of Kansas; and The Field Museum.

Ecological niche models (ENM) were developed in GARP (Stockwell and Noble, 1992; Stockwell, 1999) with 15 environmental layers. We used an average of monthly Normalized Difference Vegetation Index (NDVI; following Pettorelli *et al.*, 2009) obtained from MODIS (Moderate Resolution Imaging Spectroradiometer) sensor for all the Yucatán Peninsula for years 2001 and 2002, and three topographic layers: elevation data, slope and compound topographic index (USGS, 2008), all of them at cell size of 0.0028 degrees.

For *M. pandora* (less than 20 localities) we modeled with all points. For the other species, we used 75% of points for training and 25% for testing. For each species, we performed 100 individual models using all rules, and from them we selected the ten models that minimized omission and commission errors (Anderson *et al.*, 2003). Best subset models were summed to obtain a consensus map, which got values 0-10, where 0 represents the areas where all models coincided to predict the absence of the species, 1 is where 1 out of 10 models predicted the presence of the species, and so forth until 10, representing the maximum consensus areas. i.e., where all models predicted the presence of the species. Consensus maps were then clipped to the boundaries of state of Campeche (CONABIO, 2003), at scale 1:250 000.

In order to estimate the geographic distribution of relative abundance of each species, we obtained as a proxy a map representing the Euclidean distance of each pixel of the distribution to the ecological centroid. Euclidean distances were calculated as follows: (1) in a GIS, we combined the maximum consensus values of a species distribution map with the 15 environmental layers that we used to model the distribution, to obtain a matrix in which columns contained the values of each environmental variable, and rows the pixels where the species was predicted present; (2) environmental variables were *z*-standardized, thus the mean value of each variable equals zero, being this multidimensional point the niche centroid; and (3) distances to the centroid were calculated with the formula:

$$Dc = \sqrt{\sum(\bar{\mu}_i - a_{ij})^2}$$

where Dc is the distance to the ecological niche centroid, $\bar{\mu}_i$ is the mean of variable *j* and a_{ij} is the value of the variable *j* in population *i*. Euclidean distances were projected back to the landscape to generate a map which was then overlapped with the UMA polygons map (Secretaría de Medio Ambiente y Recursos Naturales, 2005). Distance values were classified in quartiles in order to identify the places of shorter distances to the centroid (i.e., higher expected abundance), which are more suitable for species management. We considered that the best conditions for species, and the higher values of abundance, will be in the first quartile, which we used for proposing some recommendations for UMA managers.

Because most UMAs have not up-to-date abundance data (needed to validate our models), we used the data of Weber (2005) for Nuevo Becal "ejido" (communal land) (polygon 21 in Fig. 1) for the three species of deer. For this polygon, we averaged the Euclidean distance values of all pixels where each species was predicted present, and compared to the abundance estimation of Weber (2005); also, we calculated the number of pixels belonging to the first quartile. Finally, we calculated the number of pixels in the first quartile of each species for all UMAs in Campeche and compared them with Nuevo Becal "ejido".

RESULTS

We obtained 59 occurrence data points for *Mazama temama*, 15 for *M. pandora* and 87 for *Odocoileus virginianus*. There are 20 UMAs in the digital map of Secretaría de Medio Ambiente y Recursos Naturales (Secretaría de Medio Ambiente y Recursos Naturales, 2005) in Campeche for mammals (Fig. 1). However, Nuevo Becal is not in that map (Fig. 1, drawn by us). Data of abundance for the three species of deer in Nuevo Becal (Weber, 2005), and results of distance are shown in Table 1. Results of validation of models are in Table 2 and the percentage of pixels for each species into each UMA is shown in Table 3. Weber (2005) found that density (deer/km^2) of *Mazama* (both species together) is higher than that of *Odocoileus* in Nuevo Becal, but for both species, the relative abundance and density in Nuevo Becal is lower than that reported elsewhere in the Neotropics. Interestingly, our results coincide with this information: although *Mazama* includes two species, we found that *Mazama* occupies a more extensive area than *O. virginianus* in Nuevo Becal, but this polygon is not the area with the highest habitat quality for the three species in Campeche. On the other hand, UMAs "Ik Balam" and "Ejido Carlos Cano Cruz" have proportionally more high-quality area within their polygons. However, there are some differences for *M. temama, M. pandora* and *O. virginianus*: in "Ejido Carlos Cano Cruz", while *O. virginianus* may be more abundant, *M. pandora* may be less abundant than the white-tailed deer. Following those two UMAs, are "Ejido Constitución" and "Macanguas".

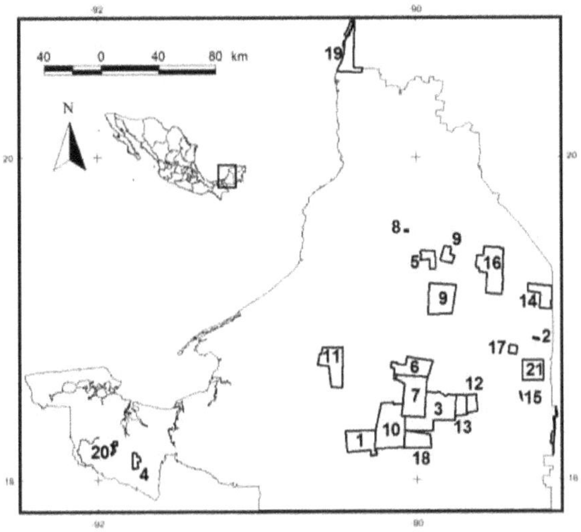

Fig. 1

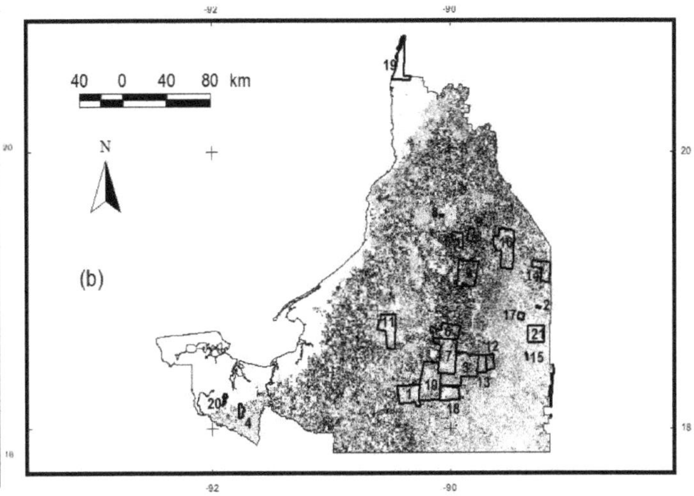

Fig. 2

Figures 1 and 2; Units for the Conservation, Management and Sustainable Use of Wildlife (UMAs) and Nuevo Becal Ejido in Campeche, Mexico (names of UMA are in Table 3; Secretaría de Medio Ambiente y Recursos Naturales 2005). Ecological niche model (ENM) for *Mazama temama* showing pixels with Euclidean distance (gray tones indicate nearness to the centroid niche). Polygons are UMAs (see Table 3).

Table 1. Abundance and Euclidean distance for three species of deer in Nuevo Becal polygon.

Deer	Abundance deer/km^2 (permanent transects of Weber, 2005)	Number of pixels with prediction* (percentage of pixels with prediction)	Euclidean distance average	First quartile value (considering all Campeche)	Number of pixels with distance in the first quartile* (percentage)
Mazama temama	0.90±0.72	743 (34.44%)	4.31	2.7393	4 (0.18%)
Mazama pandora		458 (21.23%)	4.42	2.9331	2 (0.09%)
Odocoileus virginianus	0.021	590 (27.35%)	4.47	2.7252	1 (0.04%)

*The total number of pixels in the polygon of Nuevo Becal is 2157. Each pixel size: 0.09 km^2.

Table 2. Results of validation of models by species. We show the range of the omission errors and chi-square values (x^2) for the ten best models.

Species	Omission error	x^2
Mazama temama	0-6.66	0.019-4.95
*Mazama pandora**	-	-
Odocoileus virginiana	0-4.54	0.89-3.39

*Validation was not possible for *M. pandora* because all 15 occurrence data were used for model calibration.

Habitat suitability is not equal for the three species. The largest high-quality areas in the state are for *Mazama temama*, followed by *Odocoileus virginianus*, and finally for *M. pandora*. It is interesting to observe that the two closely related *Mazama* species show contrasting results, and this has important implications for the management of the two species, since in the field the tracks of both species are hard to separate them apart (Weber, 2005) even via direct observations.

Table 3. UMA in Campeche, Mexico: area in km^2 and percentage of number of pixels with prediction in the first quartile of distance for three species of deer.

No.	UMA	*Mazama temama* km² (%)	*Mazama pandora* km² (%)	*Odocoileus virginiana* km² (%)
1	Ampliación forestal Luna	8.46 (3.16)	7.65 (2.85)	2.34 (0.87)
2	Betito y Lupita	0 (0)	0 (0)	0 (0)
3	Campamento Los Jaguares	44.73 (8.59)	45.18 (8.68)	16.70 (3.21)
4	Campo Laguna Blanca	1.17 (2.95)	0 (0)	4.14 (10.45)
5	Ejido Carlos Cano Cruz	24.84 (30.23)	14.13 (17.19)	32.76 (39.86)
6	Ejido Constitución	63.81 (23.63)	43.83 (16.23)	41.76 (15.46)
7	Ejido Xbonil	44.46 (10.30)	40.68 (9.43)	26.55 (6.15)
8	El Huanal	0 (0)	0 (0)	0 (0)
9	Ik Balam	174.33 (43.16)	119.07 (29.48)	179.55 (44.46)
10	Jaguar (*Panthera onca goldmani*) Chilam Balam	14.94 (2.81)	7.74 (1.45)	4.86 (0.91)
11	Laguna Mocu	5.58 (1.89)	6.12 (2.07)	1.8 (0.61)
12	Las Codornices	8.01 (10.11)	6.12 (7.72)	4.77 (6.02)
13	Los Pumas	5.58 (5.53)	7.47 (7.41)	1.62 (1.60)
14	Macanguas	31.5 (19.01)	21.42 (12.92)	32.31 (19.50)
15	Puh	0 (0)	0 (0)	0 (0)
16	Pool Hayin	42.84 (10.11)	23.64 (5.58)	39.06 (9.21)
17	Refugio Faunístico Jalotum	0.09 (0.32)	0.27 (0.98)	0.09 (0.32)
18	Reserva ecológica y cinegética Itzamna	4.68 (2.54)	5.4 (2.93)	1.08 (0.58)
19	Ría Lagartos-Ría Celestún	0 (0)	0 (0)	2.7 (1.59)
20	Yocol Cab Balam Ceh	0 (0)	0 (0)	0.45 (2.5)
21	Nuevo Becal	0.36 (0.18)	0.18 (0.09)	0.09 (0.04)

DISCUSSION

Many problems have been detected in the UMAs of southeastern Mexico, including poor population estimations, introduction of exotic species, overexploitation for gaming and subsistence hunting, etc. (Weber *et al.*, 2006). The first problem is very relevant to extractive activities, because extraction permits are given based on population estimations, and in many cases these are not reliable. The approach presented here aims to alleviate somehow this problem, producing spatially specific estimations of suitable areas for safe extraction without demising populations.

Mandujano and González-Zamora (2009) evaluated the UMAs and other conservation instruments to maintain viable populations of *O. virginianus* on the basis of minimum critical area for the whole country, and they found that UMAs are generally unsuitable to maintain viable populations. For *O. virginianus*, it is necessary a minimum critical area of 1,667 to 50,000 ha to support a minimum viable population of 500 deer, or 16,670 (16.67 km^2) to 500,000 ha (5,000 km^2) for long-term viability of 5,000 deer, depending on regional deer density (Mandujano and González-Zamora, 2009). Because the minimum critical area increased in a non-lineal trend as the population density diminished (Mandujano and González-Zamora, 2009), for *O. virginianus* in Nuevo Becal, with density of 0.021 deer/km^2, the minimum critical area (500 deer) will be 23,809 km^2, that means, areas extremely large. However, for species of *Mazama* with density of 0.90 deer/km^2, the minimum critical area will be 555 km^2. The total surface of Nuevo Becal is 520 km^2, just enough to maintain a minimum viable population of *Mazama*, but evidently low for *O. virginianus*.

For the other UMAs in Campeche, it should be assessed the density of deer, but even using the densities of Nuevo Becal, many of them have not the enough area to maintain minimum viable populations. For example, "Betito y Lupita", "El Huanal" and "Puh", hold 56, 27 and 39 km^2, respectively, and none of them have pixels of the first quartile. However, other bigger UMAs, like "Refugio faunístico Jalotum", "Ría Lagartos-Ría Celestun" and "Yocol Cab Balam Ceh" (305, 1,884 and 200 km^2, respectively), despite being considerably larger, do not have environmental conditions adequate to maintain populations of deer. So, the size of the UMAs is not sufficient to sustain a certain number of deer for a sustainable exploitation, but depends on the density and adequate environmental conditions.

In Yucatán (a state contiguous to Campeche), *O. virginianus*, *Cuniculus paca* and *Pecari tajacu* are the species more managed in intensive UMAs, which are a modality more similar to breeding as they raise those species in confined spaces, because they are the most demanded by consumers (González-Marín *et al.*, 2003). These authors pointed that the information about of extensive UMA is insufficient to know if they are working as a useful tool for conservation. Few efforts have focused on UMAs of Campeche. García-Marmolejo *et al.*, (2008) developed a multicriterion framework to evaluate performance and sustainability of 6 UMAs in Campeche, considering environment, economy, social development, and laws and rules. They found that the quality of technical studies to support extraction rates (including, for example, transects to estimates population densities) was generally low. Moreover, although extractive rates were generally below the allowed rates, these were derived from unreliable sampling.

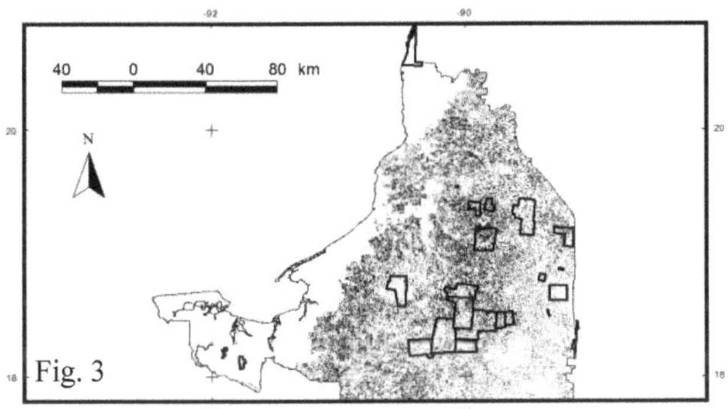

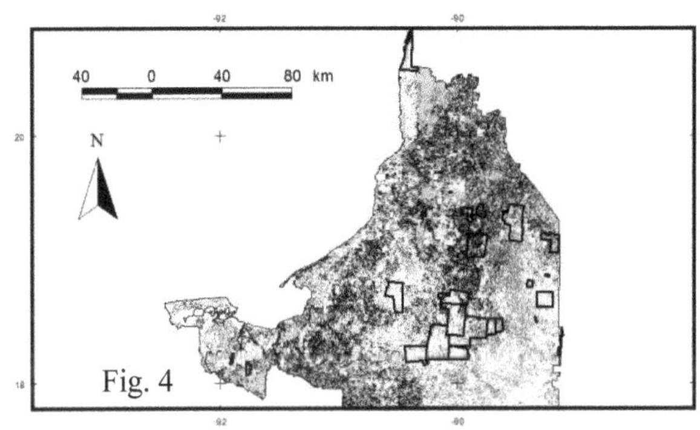

Figures 3 and 4; ENM for *M. pandora* showing pixels with Euclidean distance (gray tones indicate neareness to the centroid niche). Polygons are UMAs (see Table 3). ENM for *Odocoileus virginianus* showing pixels with Euclidean distance (gray tones indicate nearness to the centroid niche). Polygons are UMAs (see Table 3).

CONCLUSIONS

The use of this methodology will help authorities to make decisions using additional information. We propose a protocol with six steps: (1) obtain data of known presence of taxa and area of study, (2) model the ecological niche of taxa, (3) model the abundance using the Euclidean distance and map this measures, (4) test the model of abundance using field data, (5) overlap the map of abundance and UMAs and quantify areas, and (6) make decision about new UMAs, permissions to existent UMAs and evaluation of reports of UMAs for past periods.

The strategy presented here can be used to validate the technical reports of UMAs, and can be used to estimate the rates of extraction allowed by period. Following the proposal of Weber *et al.* (2006), this work can be a starting point to establish institutional standards, and a network of regional centers for management of UMAs can be the mechanism to apply the methodology proposed here.

ACKNOWLEDGEMENTS

DGVS-Semarnat provided us the digital map of UMA. E. Escobedo provided the data of Museo de Zoología, El Colegio de la Frontera Sur, Unidad Chetumal. Conabio provided data of H324, V009 and Y042 projects. The Laboratorio de Sistemas de Información Geográfica, Instituto de Biología (UNAM) provided the data of U032 project. Gabriela García-Marmolejo and M. Weber provided data of UMA and Nuevo Becal. T. E. thanks to DGAPA-PROFIP program.

REFERENCES

Anderson, R.P., Lew, D. and Peterson, A.T. 2003. Evaluating predictive models of species' distributions: criteria for selecting optimal models. Ecological Modelling. 162: 211-232.

CONABIO. 2003. Límite Nacional. Escala 1:250 000. Extraído de Conjunto de Datos Vectoriales y

Toponimias de la carta Topográfica. INEGI. Mexico, D.F.

CONABIO. 2008. Red mundial de información sobre biodiversidad. http://www.conabio.gob.mx/remib/doctos/remib_esp.html (Consulted: september, 2008).

Brown, J. H. 1995. Spatial variation in abundance. Ecology. 76: 2028-2043.

Díaz-Porras, D.F. 2006. El nicho ecológico y la abundancia de especies. Master Thesis, Instituto de Biología. Universidad Nacional Autónoma de México, Mexico.

Diario Oficial de la Federación. 2000. Ley General de Vida Silvestre. Secretaría de Medio Ambiente y Recursos Naturales. México.

Escalante, T., Espinosa, D. and Morrone, J.J. 2002. Patrones de distribución geográfica de los mamíferos terrestres de México. Acta Zoológica Mexicana (n. s.) 87: 47-65.

Elith, J., Graham, C.H., Anderson, R.P., Dudík, M., Ferrier, S., Guisan, A., Hijmans, R.J., Huettmann, F., Leathwick, J.R., Lehmann, A., Li, J., Lohmann, L.G., Loiselle, B.A., Manion, G., Moritz, C., Nakamura, M., Nakazawa, Y., Overton, J.McC., Peterson, A.T., Phillips, S.J., Richardson, K.S., Scachetti-Pereira, R., Schapire, R.E., Soberón, J., Williams, S., Wisz, M.S. and Zimmermann, N.E. 2006. Novel methods improve prediction of species' distributions from occurrence data. Ecography. 29: 129-151.

Elith, J. and Leathwick, J.R. 2009. Species distribution models: ecological explanation and prediction across space and time. Annual Review of Ecology, Evolution and Systematics. 40: 677-697.

García-Marmolejo, G. 2005. Caracterización y sustentabilidad de las Unidades para la Conservación, Manejo y Aprovechamiento sustentable de la vida silvestre en Campeche. Master Thesis. El Colegio de la Frontera Sur, Campeche, México.

García-Marmolejo, G., Escalona-Segura, G. and Van der Wal, H. 2008. Multicriteria evaluation of wildlife management Units in Campeche, México. Journal of Wildlife Management. 72: 1194-1202.

González-Marín, R.M., Montes Pérez, R. and Santos Flores, J. 2003. Caracterización de las Unidades para la conservación, manejo y aprovechamiento de fauna silvestre en Yucatán, México. Tropical and Subtropical Agroecosystems. 2: 13-21.

Guisan, A., Zimmermann, N.E., Elith, J., Graham, C.H., Phillips, S., and Peterson, A.T. 2007. What matters for predicting the occurrences of trees: Techniques, data, or species' characteristics? Ecological Monographs. 77: 615-630.

Haverkost, T.R., Gardner, S.L. and Peterson, A.T. 2010. Predicting the distribution of a parasite using the ecological niche model, GARP. Revista Mexicana de Biodiversidad. 81: 895-902.

Instituto de Biología, UNAM. 2008. Unidad de Informática para la Biodiversidad. http://unibio.ibiologia.unam.mx/ (Consulted: september, 2008).

Kumara, H.N. and Suganthasakthivel, R. 2011. Predicting the potential distribution and conservation needs of Travancore flying squirrel, *Petinomys fuscocapillus*, in Peninsular India and Sri Lanka, using GARP. Tropical Conservation Science. 4: 172-186.

Maguire, B. Jr. 1973. Niche response structure and the analytical potentials of its relationship to the habitat. American Naturalist. 107: 213-246.

MaNIS. 2008. Mammal Networked Information System. http://manisnet.org/ (Consulted: september, 2008).

Mandujano, S., and González-Zamora, A. 2009. Evaluation of natural conservation areas and wildlife management units to support minimum viable populations of white-tailed deer in Mexico. Tropical Conservation Science. 2: 237-250.

March, I.J., Carvajal, M.A., Vidal, R.M., San Román, J.E. and Ruiz, G. 2009. Planificación y desarrollo de estrategias para la conservación de la biodiversidad. In: Comisión Nacional para el Conocimiento y Uso de la Biodiversidad (ed.). Capital natural de México, Vol. II: Estado de conservación y tendencias de cambio. Comisión Nacional para el Conocimiento y Uso de la Biodiversidad, Mexico, D. F. pp. 545-573.

Martínez-Meyer, E., Díaz-Porras, D., Peterson, A.T. and Yañez-Arenas, C.A. 2013. Ecological niche structure and rangewide abundance patterns of species. Biology Letters. 9:20120637.

Marín-Togo, M.C., Monterrubio-Rico, T.C., Renton, K., Rubio-Rocha, Y., Macías-Caballero, C., Ortega-Rodríguez, J.M. and Cancino-Murillo, R. 2012. Reduced current distribution of Psittacidae on the Mexican Pacific coast: potential impacts of habitat loss and capture for trade. Biodiversity and Conservation. 21: 451-473.

Pearson, R.P., Thuiller, W., Araújo, M.B., Martínez-Meyer, E., Brotons, L., McClean, C., Miles, L., Segurado, P., Dawson, T.P. and Lees, D.C. 2006. Model-based uncertainty in species range prediction. Journal of Biogeography. 33: 1704-1711.

Pettorelli, N., Bro-Jørgensen, J., Durant, S.M., Blackburn, T. and Carbone, C. 2009. Energy Availability and Density Estimates in African Ungulates. The American Naturalist. 173: 698-704.

Pliscoff, P. and Fuentes-Castillo, T. 2011. Modelación de la distribución de especies y ecosistemas en el tiempo y en el espacio: una revisión de las nuevas herramientas y enfoques disponibles. Revista de Geografía Norte Grande. 48: 61-79.

Ray, R., Gururaja, K.V. and Ramchandra, T.V. 2011. Predictive distribution modeling for rare Himalayan medicinal plant *Berberis aristata* DC. Journal of Environmental Biology. 32: 725-730.

Sánchez-Cordero, V. 2003. Proyecto U032: Evaluación en la asignación de las categorías de riesgo de todas las especies de mamíferos terrestres de México, incluyendo las que se encuentran en la Norma Oficial Mexicana (NOM-059-ECOL-94) usando un método cuantitativo. Comisión Nacional para el Conocimiento y Uso de la Biodiversidad, Mexico, D.F. pp

Secretaría de Medio Ambiente, Recursos Naturales y Pesca. 1997. Programa de conservación de la vida silvestre y diversificación productiva del Sector rural 1997-2000. Secretaría de Medio Ambiente, Recursos Naturales y Pesca. Mexico, D. F. Pp

Secretaría de Medio Ambiente y Recursos Naturales. 2005. Unidades de Manejo para el Aprovechamiento Sustentable de la Vida Silvestre Dirección General de Vida Silvestre, digital map. Mexico, D. F.

Secretaría de Medio Ambiente y Recursos Naturales. 2011. Mapa de las UMA registradas al 15 de marzo de 2011. March, 2011. http://infoteca.semarnat.gob.mx/website/geointegrador/mviewer/viewer.htm

Stockwell, D.R. 1999. Genetic Algorithms II. In: Fielding, A.H. (ed.). Machine Learning Methods for Ecological Applications. Kluwer Academic Publishers, USA. Pp. 123-144.

Stockwell, D.R., and Noble, I.R. 1992. Induction of sets of rules from animal distribution data: A robust and informative method of data analysis. Mathematics and Computers in Simulation. 32: 249-254.

Tsoar, A., Allouche, O., Steinitz, O., Rotem, D. and Kadmon, R. 2007. A comparative evaluation of presence-only methods for modelling species distribution. Diversity and Distributions. 13: 397-405.

USGS. 2008. HYDRO1k Elevation Derivative Database.http://eros.usgs.gov/#/Find_Data/Products_and_Data_Available/gtopo30/hydro (Consulted: september, 2008).

Van Horne, B. 1983. Density as a misleading indicator of habitat quality. Journal of Wildlife Management. 47: 893-901.

Weber, M. 2005. Ecology and conservation of sympatric tropical deer populations in the Greater Calakmul Region, south-eastern Mexico. PhD. Thesis. University of Durham, Durham, United Kingdom.

Weber, M., García-Marmolejo, G., and Reyna-Hurtado, R. 2006. The tragedy of the commons: Wildlife management units in southestern Mexico. Wildlife Society Bulletin. 34:1480-1488.

Effect of the Plantation Density in the Growth and Productivity of *Cnidoscolus Chayamansa* Mcvaugh (Euphorbiaceae)

Jesús Mao Estanislao Aguilar-Luna[1*], Eduardo García-Villanueva[2], Pedro Antonio Macario-Mendoza[1], Esperanza Huerta-Lwanga[3], Ramiro de Alba-Becerra[4].

[1]El Colegio de la Frontera Sur - Unidad Chetumal. C.P. 77014. Chetumal, Quintana Roo, México.
[2]Colegio de Postgraduados Campus Montecillo. C.P. 56230. Edo. de México, México.
[3]El Colegio de la Frontera Sur - Unidad Villahermosa. C.P. 86280. Villahermosa, Tabasco, México.
[4]Fumigaciones Integrales de Quintana Roo. C.P. 77000. Chetumal, Quintana Roo, México
E-mail: laguilar@ecosur.mx
**Corresponding Author*

SUMMARY

The plantation density (PD) is a decisive factor in the agronomical resources of cultivation; as such, in this experiment, the effect of the growth and the productivity of the chaya plants were evaluated, as an associated growing in the cedar-lime system, to determine its optimum PD. The chaya plants were positioned at 1.50 x 3.00 m, utilizing cuttings without leaves; associated with young trees (less than 2 years) of *Cedrela odorata* and *Citrus latifolia* in a circular plantation design 'Nelder' of 3154 m^2. Eight PD from 2602 to 3772 plants·ha^{-1} were defined with 10 repetitions. The variables were: rhizogenic potential (RP), growth rate (GR), index of vigor (IV), index of leaf area (ILA) and production of dry biomass (PDB). The results indicated that of the 3046 to 3772 plants·ha^{-1} were found to have the best RP, the ILA (5.44) and PDB (8.48 kg·$plant^{-1}$); even though the IV (1.16) and GR (0.51 cm·day^{-1}) were the best in 2602 to 2647 plants·ha^{-1}. As the primary objective in this specie is the PDB, the PD optimal was 3046 to 3772 plants·ha^{-1}.

Key words: 'Nelder' design; response surface optimization; index of vigor; index of leaf area.

INTRODUCTION

The chaya (*Euphorbiaceae*) is a bush sub-perennial and self-seeding, that grows to a height of 5 m when it is cultivated in subtropical climates, it prospers from 0 to 1000 msnm and requires 650 to 1500 mm of precipitation annually (Stephens, 2009). In Mexico, its principal uses are as a vegetable, forage, and medicine;

being that the leaves are the part used most (Ross-Ibarra and Molina-Cruz, 2002). In spite of having a high anthropocentric value, they lack precise dates of production, due to the fact that its crossbred specie that is found associated with other vegetable components in agroforestry systems (Cuanalo de la Cerda and Guerra-Mukul, 2008).

It is a specie with a high nutritional value (Sarmiento-Franco *et al.,* 2003), the reason for which it is considered an important vegetable in the diet of the Mayan communities in the country (Cuanalo de la Cerda and Guerra-Mukul, 2008). In spite of having a high agronomical potential, its method of production has changed little from prehispanic times; since being a semidomesticated specie, in which its growth and productivity are low (Parra-Tabla *et al.,* 2004); and maintaining a negative relation in relation to the interspecific competency; it's to say, the best plantation density (PD) has less growth (Jose *et al.,* 2004; Parra-Tabla *et al.,* 2004).

A strategy that can increase the growth and productivity is the chaya plants, refers to the management of the PD. The PD is an agronomical factor in the management that influences in the development of the species and refers to the number of established plants in the determined area and has the effect in all the steps of plant development due to the competency for water, solar energy, nutriments and physical space (Mohammed-Nazeeb *et al.,* 2008). The PD can modify the environment, causing an increase or decrease in the growth and productivity (Escalante-Estrada *et al.,* 2008). Favorable management of this factor can optimize the growth space and increase productivity in individual or associated species.

The growth rate (GR), the index of vigor (IV) and the index of leaf area (ILA), can be indicators representing the growth and productivity of a plant (Hunt *et al.,* 2002; Makhabu *et al.,* 2006; Koyoma and Kikuzawa, 2009). The changes to these indicators are determined by the genetic factors, environmental and management. Of the management factors, the PD is what affects those most; its study is important because it has an effect in all the steps of the growth due to the competency effect by water, solar energy, nutriments and physical space (Mohammed-Nazeeb *et al.,* 2008).

The GR is an index that integrates the functioning of various organs of the plant and is referred to the accumulation of dry material by unit of ground area, by unit of time; and it can be expressed in $cm \cdot day^{-1}$ (Hunt *et al.,* 2002). The IV is an indicator of the capacity that the plant has to grow and the strength to develop itself (Makhabu *et al.,* 2006). The ILA refers to the proportion of the area of the leaf to the plant, with respect to the ground surface area (Koyama and Kikuzawa, 2009); serves to evaluate the intensity and assimilation by surface unit, and it is most relevant when you do the analysis of growth (Peil and Gálvez, 2005).

The growth that translates to formation of dry material or biomass can be evaluated as productivity in the distinct phases of the development of a plant. The productivity (biomass and leaf area, principally), refers to the accumulation of dry material by unit of ground area by unit of time (Peil and Gálvez, 2005). The biomass is a vegetable tissue expressed in terms of dry weight, being formed as a product of photosynthesis and the incorporation of nutriments; and can be evaluated as biological productivity in the distinct phases of development (Evert *et al.,* 2008). The work related with agronomical management of the chaya plants are scarce; because of this, the objective of this investigation was to evaluate the effect of the plantation density in the growth and productivity of the chaya plants, as a culture associated in the cedar-lime system, to find an optimum PD.

MATERIAL AND METHODS

Area of study

The experiment was realized from September 2008 to May 2010, in Noh-Bec, Quintana Roo, Mexico (19°06' N, 88°10' W and 11 msnm); the area where the experiment was established presents and annual medium temperature of 25.7 °C, 1562.4 mm of annual precipitation and 1404.3 mm of annual evaporation. The ground is calcareous "Ek-Lu´um" (following the Mayan classification) or the black rendzina soil (following the classification of the FAO-UNESCO); their principal characteristics are: black in color, shallow and fertile (WRB, 2011). During the experiment we took the soil to a depth of 0-30 and 30-60 cm, the values obtained were: 1.13 and 1.15 $g \cdot cm^{-3}$ of apparent density, 6.97 and 6.70 of pH, 30.1 and 32.4 $Cmol(+)kg^{-1}$ of CIC, 4.97 and 3.83 % of organic material, 0.14 and 0.08 % of total N, 19.0 and 24.6 $mg \cdot kg^{-1}$ de P, 0.30 and 0.33 $Cmol(+)kg^{-1}$ of K, for both of the depths respectively.

Vegetable material

Occupying 600 cuttings culinary woody of chaya plants, without leaves, at 40 cm of length and 2 cm in diameter, coming from the top sections of mature and healthy trees in the zone; and in which a indol butyric acid (IBA) was added 1500 ppm plus naphthalene acetic acid (NAA) 200 ppm to promote root growth. The cuttings planted on 3 September, 2011, in a depth of 5 cm directly in the ground, with a distancing of 1.50 m between plants and 3.00 m between lines. They were associated or made to compete inter specifically with young cedar (*Cedrela odorata* L.) and 'Persa' lime (*Citrus latifolia* Tan.), less than 2 years;

established in a 'Nelder' circular design of 3154 m^2 with a variable PD (Figure 1). The experiment developed in local ground conditions; the weed was controlled with applications in the morning with herbicide (Paraquat plus glifosato, 2 $L \cdot ha^{-1}$ every 2 months) in addition to cultural works every 2 months.

Evaluated variables

The treatments were defined with a base in the PD obtained in the area of each circle of 'Nelder' design (Nelder, 1962) to associate the chaya plants with the limes and cedars. The circles 2-9 of each species that occupied the treatments, meanwhile circles 1 and 10 was not used, to prevent the border effect in the area without cedars nor limes to the interior and exterior of the circle. In each variable utilized it showed random of 10 plants for treatment (utilizing a table of lucky numbers), to evaluate 20 months after the transplant, the following:

a) Rhizogenic Potential (RP), is the capacity of the cuttings to generate adventitious roots and form a radical functional system. Considering elements like: percentage of survival of cuttings, percentage of cuttings with roots (extending the complete plant), percentage of cuttings with callus (extending the complete plant), number of roots per cutting (extending the complete plant) and radical length (measuring the more extensive roots by the size of the base roots, utilizing a metric measuring tape).

b) Growth Rate (GR), is estimated with the equation: $GR=[(W_2W_1)/(t_2-t_1)]/S$, where W_2 and W_1: final biomass and start of the complete plant in an interval of time; t_2 and t_1: final moment start of such period; S: area of occupied terrain for the plant (Hunt *et al.*, 2002).

c) Index of Vigor (IV), is calculated by means of the formula: $IV=\beta_0(1-e^{-\beta_1 t})^{\beta_2}$, where β_0: height of plants; β_1: growth rate; β_2: diameter of the crown (the means of the calculations N-S and E-W, made with a metric measuring tape); t: time (Zhao-Gang and Feng-Ri, 2003).

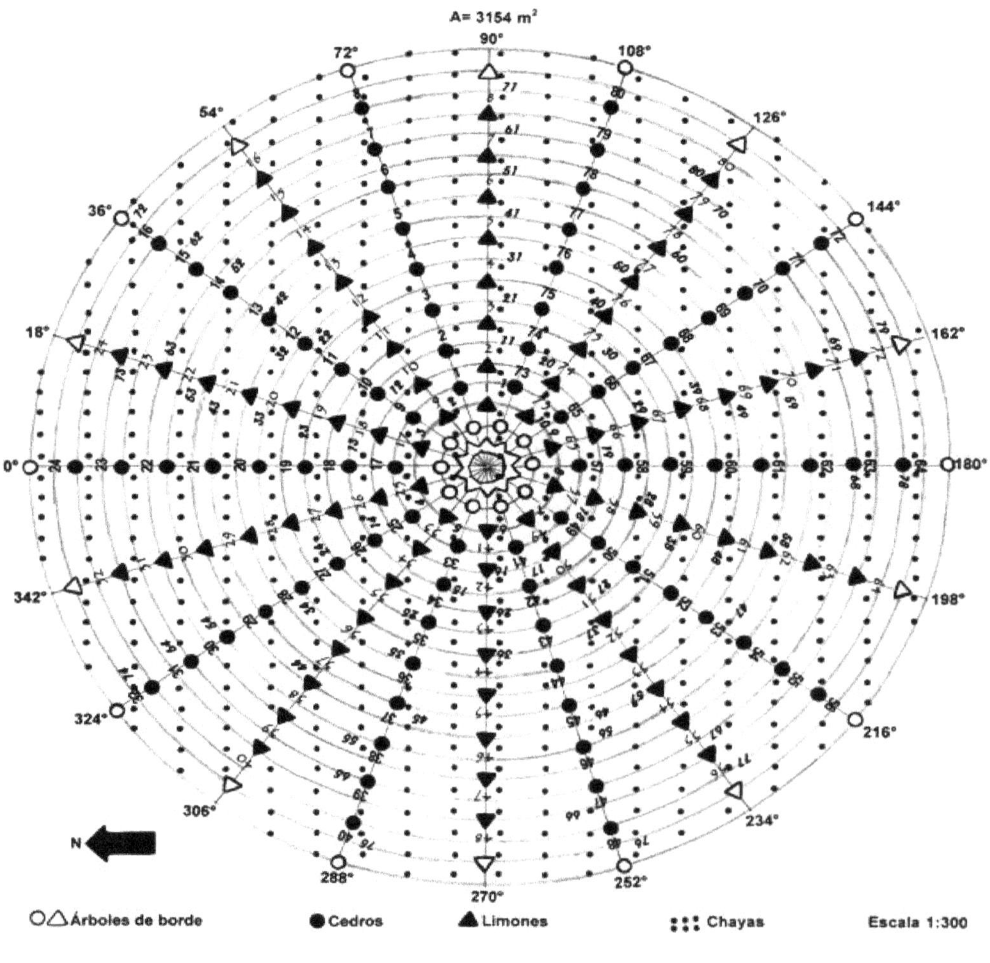

Figure 1. Plantation design in a 'Nelder' circle with interspecific competency and variable plantation density for chaya plants.

d) Index of leaf area (ILA), realizing a destructive model at the end of the experiment, utilizing 20 leaves for treatment, the leaf area (LA) measured utilizing an integrator (Model Li-3000A, Li-Cor Lincoln Nebraska, USA) and later estimated the ILA using the formula $ILA=LA(PD)/10000\ cm^2$, where LA: foliage area per plant (m^2); *PD*: plantation density (plants·m^{-2}) (Garduño-González *et al*., 2009).

e) Production of dry biomass (PDB), calculated with the formula: $PDB=W_2-W_1$, where W_2: final dry weight; W_1: initial dry weight; obtained utilizing a destructive method extending the complete plant (radical system and area portion). The fresh biomass was dried at 60 to 70 °C in a forced air circulation stove (model SL305) until obtaining a consistent weight, in order that the water did not interfere in the results.

Description of treatments and analysis of datum

The experimental unit and observation was a chaya plant, managing eight assigned treatments systemically with 10 repetitions (Figure 1 and Table 1). They found ANOVA´s independent with comparative testing means by the Tukey´s (α=0.05), for a Model of Multiple Optimization in Response Surfaces (Montgomery, 2006); using the software MINITAB 16.

RESULTS AND DISCUSSION

Rhizogenic Potential (RP)

Superimposing the plantation design in a 'Nelder' circle, established cultivation with chaya plants in a rectangular design mark, by which utilizing cuttings culinary woody. The reason for which it was possible to analyze the RP (the capacity to generate adventitious roots) in this growing. In Table 2, observe the statistical difference (α=0.05) in the distinct variables; the highest percentage of survival of cuttings was from 3297 to 3772 plants·ha^{-1} and the least percentage from 2602 to 2647 plants·ha^{-1}. The highest percentage of cuttings with roots was from 2706 to 3772 plants·ha^{-1} and the least from 2602 to 2647 plants·ha^{-1}. For the percentage of cuttings with callus, the highest values were from 2602 to 2647 plants·ha^{-1} and the lowest from 2706 to 3772 plants·ha^{-1}. The greatest number of roots per cutting was obtained from 3046 plants·ha^{-1} and the lowest in 2602, 2889 and 3772 plants·ha^{-1}. The best radical length per cutting was obtained from 3046 to 3772 plants·ha^{-1} and the least lengths from 2602 to 2647 plants·ha^{-1}. The statistical difference counted to the different physiological age of the cuttings, and the interspecific competency and a condition not homogenous to the soil; variation in depth (25 to 50 cm), MO (3.83 to 4.97 %), N (0.08 to 0.14 %) and P (19.0 to 24.6 mg·kg^{-1}).

The best percentage of survival was obtained in chaya cuttings from 3297 to 3772 plants·ha^{-1}, which is acceptable for farm production; and suggests that the PD measurement was diminishing, also affected the survival. Ross-Ibarra and Molina-Cruz (2002) and Sarmiento-Franco *et al.* (2003) indicated that the chaya cuttings easily rooted, not withstanding that they presented a radical slow growth. Statistically there was a difference (α=0.05) in the RP of the chaya cuttings (Table 2), however, it was in the best PD from 3046 to 3772 plants·ha^{-1} where they presented the highest values: up to 97 % survival, 99 % of the cuttings emitted roots and they had lengths of up to 188 cm; in comparison with a less PD from 2647 plants·ha^{-1} where the growth was low. The previous accounts for a better PD due to a better vegetable coverage, that protects the ground loss of humidity and N, contributing to the radical system of development (Petit *et al.,* 2009).

The cuttings that formed callus (up to 4 % maximum), remained latent, without developing roots nor leaves; which does not necessarily indicate they will develop a complete plant except that they will probably begin to decay using up the contents of the stem reserves (Evert *et al.,* 2008), with little to no opportunity to develop due to the fact of the interspecific competency. All of the developing roots of the chaya cuttings, formed in the nodal adjacent cut, probably in the region of the vascular cambium; this agrees with the results obtained in other cultivations such as *Malpighia emarginata* (Rivero *et al*., 2005), *Feijoa sellowiana* (Meng *et al*., 2009) and *Prosopis alba* (Oberschelp y Marcó, 2010).

In spite that cuttings without leaves were used, the auxinas contained in the buds were able to stimulate the formation of roots (Table 2); however, the physiological age of the cuttings and the successful application of IBA 1500 ppm + NAA 200 ppm, was an adequate favorable percentage for rooted cuttings with a good number of roots (up to 28 per cutting); as obtained in other cultivations such as *Malpighia emarginata* (Rivero *et al*., 2005), *Feijoa sellowiana* (Meng *et al*., 2009) and *Prosopis alba* (Oberschelp y Marcó, 2010). In all the treatments, up to 20 months after the plantation, the development the radical system was six times less in comparison with the development of the portion area. This reaffirms that the radical growth of the chaya plants are slow (Sarmiento-Franco *et al.,* 2003); and that an adequate availability of soil nutriments, is favorable to observe less extensive radical systems (De Dorlodot *et al*., 2007).

Table 1. Treatments in the 'Nelder' design for chaya plants.

N	cedar PD (plants·ha^{-1})	lime PD (plants·ha^{-1})	chaya DP (plants·ha^{-1})	Treatment (plants·ha^{-1})
1¶	1134	1666	2222	
2	691	859	2222	T_1 = 3772
3	497	578	2222	T_2 = 3297
4	388	436	2222	T_3 = 3046
5	318	349	2222	T_4 = 2889
6	270	292	2222	T_5 = 2784
7	234	250	2222	T_6 = 2706
8	206	219	2222	T_7 = 2647
9	185	195	2222	T_8 = 2602
10¶	167	176	2222	

N (number of circles). PD (plantation density in the circle).
¶ Circle used as a border.

Table 2. Rhizogenic potential in cuttings culinary woody of the chaya plants, 20 months after plantation in a black rendzina soil (Ek-Lu´um).

Treatments (plants·ha^{-1})	Survival of cuttings (%)	Cuttings with roots (%)	Cuttings with callus (%)	Number of roots for cutting	Radical length (cm)
2602	86 c	96 b	4 a	24 c	104 c
2647	88 c	96 b	4 a	26 b	112 c
2706	92 b	98 a	2 b	25 b	125 b
2784	92 b	99 a	1 b	25 b	138 b
2889	93 b	99 a	1 b	24 c	152 b
3046	94 b	99 a	1 b	28 a	164 a
3297	97 a	98 a	2 b	26 b	176 a
3772	97 a	99 a	1 b	24 c	188 a
CV (%)	12.33	10.74	11.79	16.42	12.95
MSD	1.86	0.71	0.42	0.63	3.37

CV: coefficient variation, MSD: minimum significance difference. Means with the same letter inside each column are statistically equal in accordance with the Tukey´s formula (α=0.05).

Vegetative Growth

At 20 months after plantation, the chaya plants had a maximum GR of 0.51 cm·day`1, at 2602 plants·ha^{-1}; but if to establish a better PD to this, their growth would diminish from 0.04 to 0.09 cm·day^{-1} (Figure 2). Another form to explain the patterns of growth, is measuring the dynamic of the PD base to the dates, considering how PD low (from 2602 to 2706 plants·ha^{-1}), how PD intermediate (from 2784 to 2889 plants·ha^{-1}) and how PD high (from 3046 to 3772 plants·ha^{-1}). Of the 0 to 8 months the chaya plants showed a lineal growth in all the PD; from 8 to 12 months the growth increase in PD intermediate and high, and in low PD maintained constant; from 12 to 16 months increase in PD high and low, and maintained constant in PD intermediate; from 16 to 20 months increase in PD low, intermediate and high (Figure 3).

The GR was sigmoidal, in which coincides with the other obtained cultivations such as: *Malus domestica* (Costes *et al.*, 2006) and *Solanum tuberosum* (Flores-López *et al.*, 2009); it is said, start at the origin, advance slowly augment in exponential form (in proportional relation with the size of the organism), maintain calm and augment again (Figure 3). This means that there was an alternation in the growth, in which, meanwhile the portion area found in active growth, the radical system maintains calm and viceversa (Costes *et al.,* 2006). The growth of the plants not only depends on the absorption and transformation resources in biomass, but also that the plants and resources are fixed relatively in the space (Hunt *et al.,* 2002). The GR is not limited only by the amount of water or the availability of nutriments, but also by the solar energy that the foliage was able to intercept (Jose *et al.*, 2004). From 2602 to 2647 plants·ha^{-1} the competency was more adequate for this variable, it's to say, there was a balance of water, solar energy, nutriments and physical space, that permitted the plants to express their best genetic potential.

In Figure 2, observe that from 2602 to 2647 plants·ha^{-1} the chaya plants had a maximum IV (1.16), in these PD the ILA and the PDB was less; the chaya plants, from 2706 to 3772 plants·ha^{-1} reduced its vigor from 0.14 to 0.28; similar to its response to its GR. Meanwhile higher is the IV, representing a major growth in a determined time (in this case, 20 months after plantation) (Zhao-Gang and Feng-Ri, 2003).

In Figure 4A, presents another scenario for the PD management; where the best GR (0.51 cm·day^{-1}) and IV (1.16) was found at 2602 plants·ha^{-1}; the tendency for these variables shown from 2706 to 3772 plants·ha^{-1}; the growth and vigor was less. In all, the PD and the 20 months after plantation, the values that expressed the IV in the chaya plants: plant height (from 1.70 to 2.00 m) and the volume of crown (from 0.50 to 1.70 m^3) were referred good health and appearance of the plants; because the IV was a confident indicator to measure vegetable growth. It indicated, in addition, an excellent adaptation of the chaya plants to the environment conditions of the zone (Cuanalo de la Cerda and Guerra-Mukul, 2008).

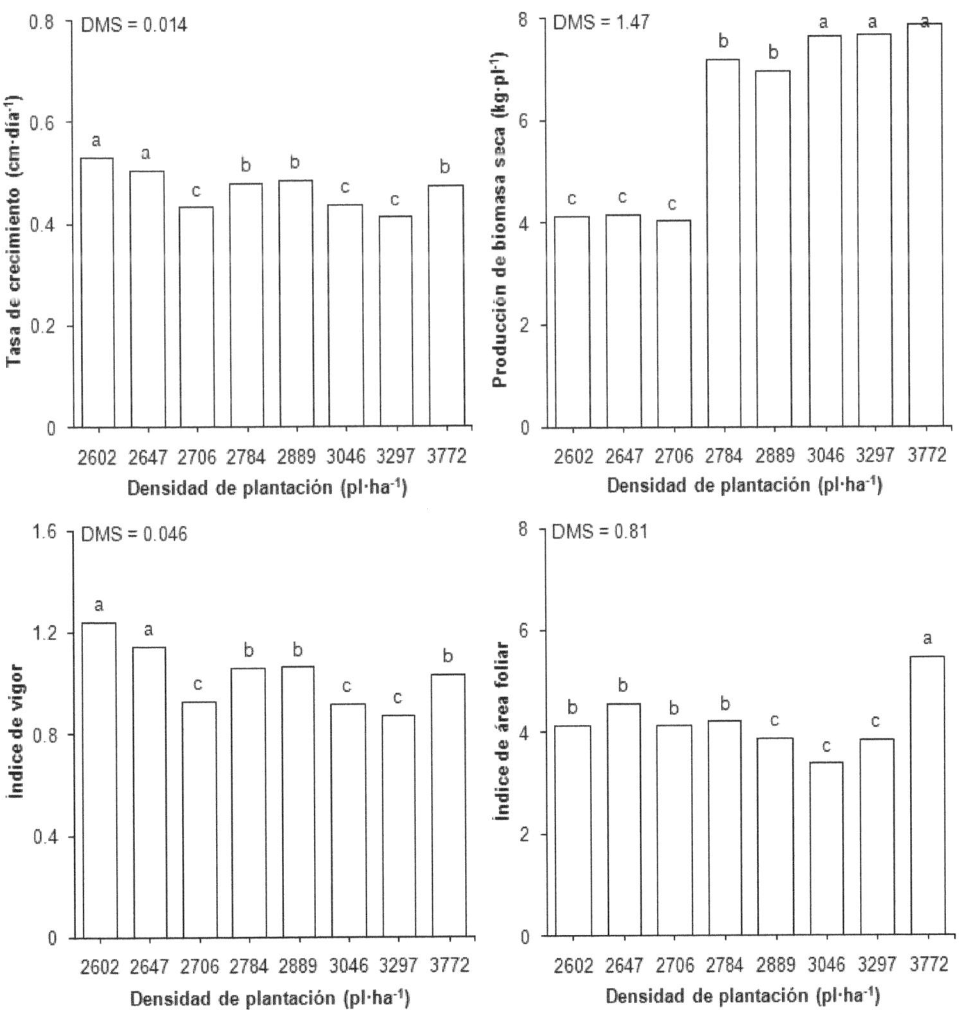

Figure 2. Growth rate, index of vigor, index of leaf area and production of dry biomass in the chaya plants, at 20 months after plantation. Bars with the same letter do not differ statistically between itself. SMD: minimum significant difference is at α=0.05.

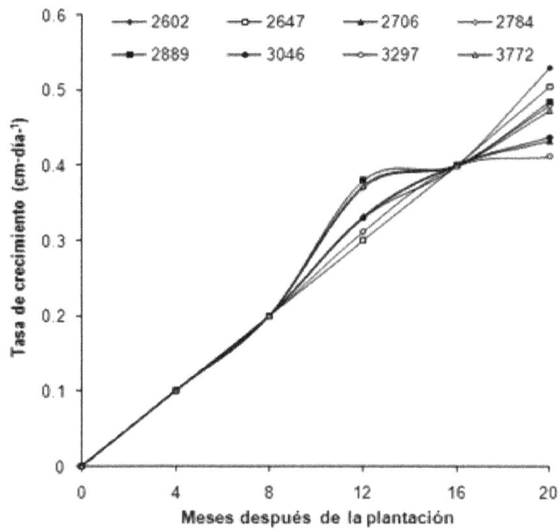

Figure 3. Sigmoidal rhythmic growth in chaya plants to different plantation density.

Biological Productivity

In Figure 2, observe a maximum ILA of 5.44 to 3772 plants·ha^{-1} in which we can see a better proportioned PDB. As such, a better ILA, better PDB, as documented by Garduño-González *et al.* (2009). The productivity was best from 3046 to 3772 plants·ha^{-1}, because the competitive interactions for water, solar energy, nutriments and physical space were minimized by the complementary (Forrester *et al.*, 2006; Petit *et al.*, 2009). In this sense, Forrester *et al.* (2006) and Garduño-González *et al.* (2009) affirmed that the total productivity by surface unit increases when the species is found associated, diminishing its efficiency in mono-cultivation systems; the reason by which, to associate the chaya plants with the cedars and limes, the biological productivity was high.

In Figure 4C, observe that the chaya plants, demonstrated a maximum ILA (5.44) at 3772 plants·ha^{-1}; with a positive lineal tendency toward a better PD, the ILA values diminished in proportion to making the PD, meanwhile, the productivity can decrease in function to the difficulty to obtain resources. From 2602 to 2442 plants·ha^{-1} the productivity for hectare was less, because it didn't occupy all of the available area, as such, the productivity for unit of area was inferior to the maximum productivity possible.

The fact to obtain a better ILA at 3772 plants·ha^{-1}, also indicate competition (remember that the chaya plants are associated with cedars and limes), which is not necessarily conductive to a less PDB (Figure 2) (Jose *et al*., 2004; Koyama and Kikuzawa, 2009). The fact is that the best productivity also is to obtain a measurement of interspecific ecological balance, such as in the below ground as in the above ground (García-Barrios and Ong, 2004).

In Figure 2, observe a maximum PDB (8.48 kg·plant^{-1}) from 3046 to 3772 plants·ha^{-1} which indicates that the foliage in these PD's were more efficient in the photoassimilation, demonstrating a better photosynthesis capacity, given that the PDB is functioning with the photosynthesis, the respiration and conversion efficiency of carbohydrates to dry material (Peil and Gálvez, 2005; Koyama and Kikuzawa, 2009). The prior results coincide with those obtained in *tuberosum* (Flores-López *et al.*, 2009), the same as in *Phaseolus vulgaris* associated with *Helianthus annuus* (Garduño-González *et al.*, 2009). Given that from 2602 to 2706 plants·ha^{-1} there presented a point of photosaturation in which no more biomass was produced even though the most solar energy exist.

The fact that the PDB for superficial unit in major PD at 3422 plants·ha^{-1} has not been incremented in significant form, indicates that the competency between plants was not able to be improved in the supervised capacity and efficiency in the use of resources from the soil (water and nutriments, principally) (Forrester *et al.,* 2006). Without considering that the stems, leaves and dead roots, are incorporated consistency to the soil, to maintain and increment its fertility (Petit *et al.,* 2009). In the cuttings, the initial growth in the portion area was rapid and the roots were slow, because of this the leaves must grow until the second year without exceeding the defoliation more than 50 %, to guarantee an adequate vegetable growth (Sarmiento-Franco *et al.,* 2003).

In Figure 4B, presents another scenario for the PD management; at 20 months after plantation the best PDB (8.48 kg·plant^{-1}) observed at 3422 plants·ha^{-1}, the tendency indicates that in the measurement the PD diminished, this value also lowered, which indicates that a competency was less than the availability of resources from the soil. In relation to this, Geyer (2006) encountered that the growth space affects the PDB in species such as: *Acer saccharinum, Robinia pseudoacacia, Populus deltoides, Gleditsia triacanthos* and *Ulmus pumila;* where the PBD by superficial unit was better in the first two years. However, after nine years the plants were losing vigor, questioning the accumulated economic cost for plantations in high PD.

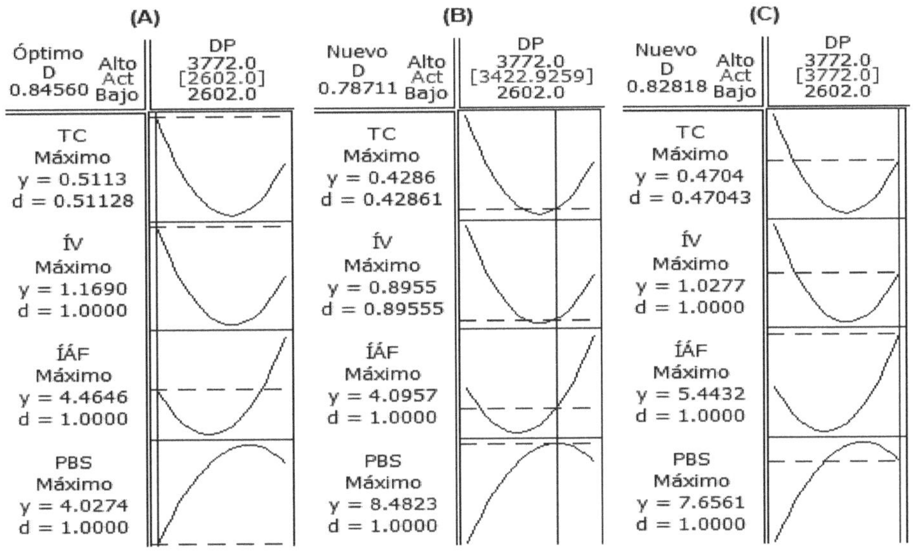

Figure 4. Scenarios with maximum values for the growth and productivity in chaya plants, in (A) growth rate (GR) and the index of vigor (IV), in (B) production of dry biomass (PDB), in (C) index of leaf area (ILA); at 20 months after plantation.

The ILA and the PDB are the two fundamental structural parameters to describe the crown of a tree in relation to the limiting factors (principally water, solar energy and nutriments) (Geyer, 2006). The PDB must be considered as the captured product for the resource most limited, and the efficiency in which the captured resource converts in biomass (García-Barrios and Ong, 2004). The competency for solar energy and its impact in the PDB, can be measured managing the PD, and the understanding of the processes of competency and the completion of the crowns (García-Barrios and Ong, 2004; Jose *et al.,* 2004).

In the search of an optimum PD, and considering such growth objectives and the productivity, a better GR and IV is obtained at 2602 plants·ha^{-1} is not conducive to obtain the same PD a better ILA and PDB (Figure 4). However, being the PBD (particularly the foliage of the portion area) primary objective in the profiting of the chaya plants (Ross-Ibarra and Molina-Cruz, 2002; Stephens, 2009), the optimum PD is defined in function from this variable, as such the optimum interval was from 3046 to 3772 plants·ha^{-1}.

CONCLUSION

The vegetative macropropagation of the chaya plants is measured factually the rooting of cuttings in the apical and middle section of mature bush; being the number of roots and the radical length, the components more important in the rooting. The growth and productivity of the chaya plants are influenced strongly by the plantation density, because to surpass from 2602 to 3772 plants·ha^{-1} the intraspecific value changes in response to the interspecific competency; as such, from 2602 to 2647 plants·ha^{-1}, was where the best growth rate was observed and index of vigor; meanwhile from 3046 to 3772 plants·ha^{-1} was where the best rhizogenic potential expressed, index of leaf area and production of dry biomass. As such, these final densities are the optimum to obtain the best productivity in the chaya plants.

ACKNOWLEDGMENTS

This work has been possible thanks to the support of the Regional Encouragement Institutional Fund for Scientific Development, Technology and Innovation (FORDECYT) for the National Counsel and Technology through the Convention 116306: social-environmental innovation for the development in areas of high poverty and biodiversity in the southern frontier of Mexico.

REFERENCES

Costes, E., García-Villanueva, E., Jourdan, C., Regnard, J.L., Guédon, Y. 2006. Co-ordinated growth between aerial and root systems in young apple plants issued from *in vitro* culture. Annals of Botany. 97:85-96.

Cuanalo de la Cerda, H.E., Guerra-Mukul, R.R. 2008. Homegarden production and productivity in a Mayan community of Yucatan. Human Ecology. 36:423-433.

De Dorlodot, S., Forster, B., Pagès, L., Price, A., Tuberosa, R., Draye, X. 2007. Root system architecture: opportunities and constraints for genetic improvement of crops. Trends in Plant Science. 12:474-481.

Escalante-Estrada, L.E., Escalante-Estrada, Y.I., Linzaga-Elizalde, C. 2008. Densidad de siembra del girasol forrajero. Agronomía Costarricense. 32:177-182.

Evert, R.F., Esau, K., Eichhorn, S.E. 2008. Anatomía Vegetal: Meristemas, Células y Tejidos de las Plantas: su Estructura, Función y Desarrollo. Vol. 1, 3ra. Edición, Ed. Omega. Barcelona, España. 614 pp.

Flores-López, R., Sánchez-Del Castillo, F., Rodríguez-Pérez, J.E., Colinas-León, M.T., Mora-Aguilar, R., Lozoya-Saldaña, H. 2009. Densidad de población en cultivo hidropónico para la producción de tubérculo-semilla de papa (*Solanum tuberosum* L.). Revista Chapingo Serie Horticultura. 15:251-258.

Forrester, D., Bauhus, J., Cowie, A.L., Vanclay, J.K. 2006. Mixed-species plantations of *Eucalyptus* with nitrogen fixing trees: a review. Forest Ecology and Management. 233:211-230.

García-Barrios, L., Ong, C.K. 2004. Ecological interactions, management lessons and design tools in tropical agroforestry systems. Agroforestry Systems. 61:221-236.

Garduño-González, J., Morales-Rosales, E.J., Guadarrama-Valentín, S., Escalante-Estrada, J.A. 2009. Biomasa y rendimiento de frijol con potencial ejotero en unicultivo y asociado con girasol. Revista Chapingo Serie Horticultura. 15:33-39.

Geyer, W.A. 2006. Biomass production in the Central Great Plains USA under various coppice regimes. Biomass and Bioenergy. 30:778-783.

Hunt, R., Causton, D.R., Shipley, B., Askew, A.P. 2002. A modern tool for classical plant growth analysis. Annals of Botany. 90:485-488.

Jose, S., Gillespie, A.R., Pallardy, S.G. 2004. Interspecific interactions in temperate agroforestry. Agroforestry Systems. 61:237-255.

Koyama, K., Kikuzawa, K. 2009. Is whole-plant photosynthetic rate proportional to leaf area? A test of scalings and a logistic equation by leaf demography census. The American Naturalist. 173:640-649.

Makhabu, S.W., Skarpe, C., Hytteborn, H., Mpofu, Z.D. 2006. The plant vigour hypothesis revisited - how is browsing by ungulates and elephant related to woody species growth rate? Plant Ecology. 184:163-172.

Meng, Z., Tang, H., Wang, D., Shao-Xiong, R., Ren-Dao, L. 2009. A study of rooting characteristics and anatomical structure of *Feijoa* cuttings. Agricultural Journal. 4:86-90.

Mohamed-Nazeeb, A.T., Tang, M.K., Loong, S.G., Syed-Shahar, S.A.B. 2008. Variable density plantings for oil palms (*Elaeis guineensis*) in Peninsular Malaysia. Journal of Oil Palm Research. 1:61-90.

Montgomery, D. 2006. Diseño y Análisis de Experimentos. 2da. Edición, Limusa Wiley. México. 686 pp.

Nelder, J.A. 1962. New kinds of systematic designs for spacing experiments. Biometrics. 18:283-307.

Oberschelp, G.P.J., Marcó, M.A. 2010. Efecto del ácido 3-indolbutírico sobre el enraizamiento adventicio y la altura de plantines clonales de *Prosopis alba* Grisebach. Quebracho. 18:112-119.

Parra-Tabla, V., Rico-Gray, V., Carbajal, M. 2004. Effect on defoliation on leaf growth, sexual expression and reproductive success of *Cnidoscolus aconitifolius* (Euphorbiacea). Plant Ecology. 173:153-160.

Peil, R., Gálvez, J.R. 2005. Reparto de materia seca como factor determinante de la producción de las hortalizas de fruto cultivadas en invernadero. Revista Brasileira de Agrociência. 11:5-11.

Petit, A.J., Casanova, L.F., Solorio, S.F. 2009. Asociación de especies arbóreas forrajeras para mejorar la productividad y el reciclaje de nutrimentos. Agricultura Técnica en México. 35:107-116.

Rivero, M.G., Ramírez, M., Caraballo, B., Guerrero, R. 2005. Enraizamiento de estacas de semeruco (*Malpighia emarginata* Sessé & Moc. ex DC). Revista de la Facultad de Agronomía de la Universidad del Zulia. 22:129-141.

Ross-Ibarra, J., Molina-Cruz, A. 2002. The ethnobotany of chaya (*Cnidoscolus aconitifolius ssp. aconitifolius* Breckon): a nutritious maya vegetable. Economic Botany. 56:350-365.

Sarmiento-Franco, L., Sandoval-Castro, C., Mcnab, J., Quijano-Cervera, R., Reyes-Ramírez, R. 2003. Effect of age of regrowth on chemical composition of chaya (*Cnidoscolus aconitifolius*) leaves. Journal of the Science of Food and Agriculture. 83:609-612.

Stephens, J.M. 2009. Chaya - *Cnidoscolus chayamansa* McVaugh. University of Florida, IFAS Extension HS578, EE. UU. 2 pp.

WRB-World Reference Base for Soil Resources. 2011. Classification key. FAO AGL. Disponible en: http://www.fao.org/ag/agl/agll/wrbnewkey [consultada: marzo 2011].

Zhao-Gang, L., Feng-Ri, L. 2003. The generalized Chapman-Richards function and applications to tree and stand growth. Journal of Forestry Research. 14:19-26.

15

Bioremediation Perspectives using Autochthonous Species of *Trichoderma* Sp. for Degradation of Atrazine in Agricultural Soil from the Tulancingo Valley, Hidalgo, Mexico

Margarita Islas Pelcastre[1,2], José Roberto Villagómez Ibarra.[1*], Alfredo Madariaga Navarrete.[3], Javier Castro Rosas[1], César Abelardo González Ramirez.[1] and Otilio Arturo Acevedo Sandoval[2]

[1] *Academic Area of Chemistry, Institute of Basic Science and Engineering, Hidalgo State University. City of Knowledge, Pachuca - Tulancingo highway Km 4.5. C. P. 42184. Mineral de la Reforma, Hidalgo, Mexico.*
*. * E-mail: jrvi@uaeh.edu.mx*
[2] *Engineering Academic Area of Agribusiness and Food Engineering, Institute of Agricultural Sciences, Hidalgo State University. Km 1 University Avenue, Ex-Hacienda of Aquetzalpa, Tulancingo, Hidalgo*
[3] *Division of Professional studies. Monterrey Tech Institute, Hidalgo Campus. Felipe Angeles Boulevard No. 2003, Pachuca, Hidalgo*
E-Mail: mislas@uaeh.edu.mx[1,2], jrvi@uaeh.edu.mx[1], jcastro@uaeh.edu.mx[1], amadaria@itesm.mx[3], ccr_gonzalez@yahoo.com[1], acevedo@uaeh.edu.mx[2].
** Corresponding author*

SUMMARY

The objective of this study was to show an *in vitro* bioremediation methodology for atrazine-contaminated soils through the use of local strains of native fungi isolated from the *Phaseolus vulgaris* L rhizosphere present in cultivable soils as well as to evaluate its resistance and capacity for atrazine degradation. A *Trichoderma* sp. species was identified in three cultivable soils from the Tulancingo Hidalgo, México region (contaminated with and without atrazine), which resisted atrazine concentrations of 10,000 mg L^{-1}. Tests showed that the strain grows exponentially in atrazine-contaminated soil over a range of 10^5-10^6 CFU g^{-1} in 15 days using atrazine as the only carbon and nitrogen source, while the control and witress showed a decrease of 10^0-10^3 UFC g^{-1} in the same period of time. For the atrazine degradation experiments, a treatment of the application of *Trichoderma* (10^4 - 10^5 CFU mL^{-1}) was applied to sterilized and non-sterilized soil contaminated with 500 mg Kg^{-1} of atrazine, evaluated at four time intervals (5, 10, 20 and 40 days). Statistical differences were found (α=0.050, Tukey) among treatments with the fungi and the test days. The native *Trichoderma* strain degraded 89% of the atrazine in 40 days. It showed that it is viable and cultivable in soil bioremediation.

Keywords: *Trichoderma*; native; biodegradation; herbicide.

INTRODUCTION

Pesticides can be a source of contamination for their use in agricultural practices. Improper handling and the persistence of some of them have caused negative impacts to soil, water, groundwater and even food, which can potentially cause damage to the environment and human health since these substances can bioaccumulate and biomagnify in living organisms or even arrive by diffusion and / or advection at different trophic levels (Madigan *et al.,* 2006; Cooper *et al.,* 2007).

In 2006, the atrazine herbicide was the most widely applied herbicide worldwide (29-34 million kg of active ingredient per year) in agricultural soils (Joo *et al.,* 2010). In Mexico, 1-2 kg of the active ingredient per hectare of this pesticide was applied in 1995, occupying the third place nationally (12.8%) of herbicides, with a clear tendency to increase its use in a rate of 10% annually (INE, 2007). Although in the country there is no official record of atrazine use, it is estimated that in only the central part of the country 376, 561 hectares of maize and sorghum the were treated (Bayer, 2009).

Atrazine (C_8 H_{14} ClN_5) is a pre-emergent herbicide with the systemic action of inhibiting photosynthesis considered toxic in the short and medium term, and it is not easily degraded. Some research shows that this product causes serious environmental changes, can act as endocrine breaker (DE), affecting reproductive function in vertebrates, and cause teratogenic effects in humans (Cooper *et al.,* 2007; Raymundo *et al.,* 2009). It has been classified by the Environmental Protection Agency (EPA) as a restricted-use pesticide and classified in group C as a possible carcinogen (EPA, 2006).

For the persistency and negative effects that it may cause, atrazine (ATZ) is on the list of toxic substances as a candidate to be prohibited or have its use minimized worldwide (UNEP, 2001). However, in Mexico, to date no regulations exist that limit or restrict its application. That is why it is necessary to find alternatives to eliminate its effects on the ecosystems. One alternative is bioremediation. This process incorporates biological organisms such as microorganisms (mainly bacteria and fungi) to the contaminated environment to induce and accelerate the biodegradation process (Gopinath and Sims, 2011).

In the case of the use of microorganisms for remediation processes, they must be capable of degrading the herbicide and, moreover, must be harmless to the environment variation. Microorganisms must be isolated from native populations present in the same soil where they are intended to act. In this regard, Robinson (1996), reports that in systems with wild populations of microorganisms, degradation processes are ecologically more adaptable, there is more population diversity, genetic flexibility, metabolic and rhizosphere interactions and horizontal resistance. Thus, specific groups of microorganisms (single strain or consortia) with herbicide degradation are isolated, identified, their potential evaluated, produced and returned to the soil in specific crops so that the bioremediation could represent the best mechanism for the removal of atrazine in soil (Barriuso *et al.,* 2008).

There are fungi and bacteria that exhibit the ability to transform the atrazine molecule through alkylation and chlorination reactions, leading to various degradation products such as Hydroxyatrazine (HA), Desethylatrazine (DEA) and Desisopropylatrazine DIA (Govantes *et al.,* 2009). However, microorganisms have also been isolated with the ability to completely mineralize this molecule (Lima *et al.,* 2009).

Atrazine degradation may be carried out by isolated species or microbial consortia (Gopinath and Sims, 2011; Ma *et al.,* 2011; Sene *et al.,* 2010; Govantes *et al.,* 2009). For this, the microorganisms are used in processes such as: bioaugmentation (inoculation of microorganisms), phytoremediation (using plants) and rizodegradation (remediation plant-microbe association at root level). It has been reported that atrazine can be degraded by various genera of bacteria such as *Pseudomonas* sp. (Monard *et al.,* 2008), *Bacillus sp.* (Getenga *et al.,* 2009), *Serrantia sp.* (So and Kyung, 2009), *Rhizobium* sp. (Chaudhary *et al.,* 2011; Abou-Shanab *et al.,* 2006), *Flavobacterium* and *Variovorax oryzihabitans paradoxus* (Belimov *et al.,* 2005), as well as fungal genera, among which are: *Aspergillus* (Gopi *et al.* 2012; Bajwa *et al.,* 2010), *Rhizopus* (Gopi *et al.,* 2012), *Fusarium moniliforme* (Bajwa *et al.,* 2010), *Penicillium* (Gopi *et al.,* 2012; Bajwa *et al.,* 2010) and *Trichoderma* (Smith *et al.,* 2005; Davidchik *et al.,* 2008; Quinto *et al.,* 2008; Sene *et al.,* 2010).

Trichoderma is an antagonist genera that can be found in the rhizosphere, where it competes with other microorganisms for nutrients and space. Some species of this genus produce enzymes that break down toxic substances and allow the cleaning of xenobiotic contaminants in soil (Perez *et al.,* 2009; Sene *et al.,* 2010), such as pesticides, organochlorine and organophosphorus compounds (Argumedo *et al.,* 2009). Organochlorine pesticides such as DDT, dieldrin and endosulfan and classified as highly toxic persistent, have been degraded by *Trichoderma* species (Llado *et al.,* 2009). This ability suggests that this as fungal genera could be used as a potential biorremediador of soils contaminated with agrochemicals.

Studies reported by Harman *et al.* (2004), mention that the *Trichoderma* spp strain T22, with genetic modification, was able to withstand concentrations up to 2,000 mg of cyanide per gram of soil exceeding the limits set by the EPA. This tolerance is quite remarkable and its activity is associated with the cellular system of intoxication based on the production of the enzyme permeases. Another study by Tang *et al.* (2009), showed that throught molecular breeding of the *Trichoderma atroviride* strain T23 in liquid medium at laboratory level, they achieved degradation levels between 81% and 96% of 600 g ml $^{-1}$ of pesticides dichlorides, compared with 72% achieved by the original strain without improvement.

There are few reports on pesticide bioremediation involving the genus *Trichoderma* sp., as a native rhizosphere strain (regularly they are exogenous GM strains) and even less research on their behavior and effects in terrestrial environments as an atrazine biodegrader. However, this genus has shown its biological potential for bioremediation activity in soil with a high level of development in agricultural applications (King *et al.,* 2004; Shmoll and Schuster, 2010). On the other hand, alternatives of bioaugmentation with genetically modified organisms are viable, but expensive and inaccessible for most farmers in the country and, in some cases, being exogenous strains, could limit or modify the native microenvironment which is to be improved.

The aim of this study was to evaluate the ability of a native rhizosphere strain of *Trichoderma* sp., of *Phaseolus vulgaris* L. to degrade atrazine under *in vitro* conditions for bioremediation purposes.

MATERIALS AND METHODS

Area of Study

To estimate the native and rhizospheric conditions of the study area, a selection criteria was set: a minimum of five continuous cycles of planting beans *(Phaseolus vulgaris),* soils with different water regimes (rained and irrigated) and soils with and without herbicides applications. It was determined that the landrace bean known as "moro" is a common crop in the region and meets the criteria for this experiment.

Three agricultural sites in the Valley of Tulancingo, Hidalgo, were selected. Two were agricultural soils of rained regime with no record of pesticide application, (El Capulin, located in the town of Cuautepec de Hinojosa, latitude: 20° 0' 23.27" N, longitude: 98° 14' 59.09" W, and the second in the town of Tepantitla: latitude: 20° 3' 2.90" N, longitude: 98° 19' 16.54" W). The third, representative of agricultural area with continued use of agrochemicals, was Santana Hueytlalpan, in the municipality of Tulancingo, Hidalgo: latitude 20° 10' 29.82" N, longitude: 98° 16' 52.72" W.

Three soil samplings were extracted during the annual cycle of the crop: soil without plant inflorescence, plant soil (rhizosphere) and dry soil. Sampling was performed according to the procedures described in the Official Mexican Norm (NOM-021-RECNAT-2000) for transport, storage and methods of microbiological, physical and chemical soil analysis.

Sampling depth was 0-25 cm using a tubular auger (7 cm diameter), and because no variability in the soil composition was observed it was not necessary a compound soil sampling, due to the uniformity of selected soils.

In the valley of Tulancingo (polygon: 20° 10' 29" N, 98° 16' 52" W, 19° 57' 20" N and 98° 15' 59" W) a subhumid temperate climate predominates (Cw), with an average annual temperature of 14.5 ° C, with maximum temperatures of 30 ° C in the months of March to July, and from August to February reaching the minimum of 3 ° C. The rainfall regime may occur from March to September with a minimum of 33.3 mm h $^{-1}$ and a maximum of 190.7 mm h $^{-1}$, with an average of between 500-553 mm per year (National Meteorological System, 2011; University Weather Station, 2009).

Physical and chemical analytical methods in soil.

Soil samples were dried at room temperature (22-25 ° C), cracked and sieved with a 32 mesh sieve of 0.5 mm aperture (AS-01 method of NOM-021-RECNAT-2000). In order to characterize these soils, the following physical properties were determined: texture (Bouyoucos, 1962); real density (Buckman and Brady, 1966), and moisture by the gravimetric method (AS-05-021-RECNAT NOM-2000).

The chemical characterization covered cation exchange capacity (Mehlich, 1948), pH: measured in water (AS-02-021-NOM RECNAT-2000); organic carbon and organic material by Walkley and Black (1934), and total nitrogen (Kjendahl, 1983).

Isolation and characterization of rhizospheric *Trichoderma* spp.

Three samples of the sites under experiment were taken during the annual crop cycle (landrace beans "Moro") and the total count of fungi was determined by the Waksma method (1922), in 6 working dilutions (10^{-1} - 10^{-6}), surface planting on Potato Dextrose Agar (PDA) without acidification. The plates were incubated at room temperature in the dark for seven days. At the end of the incubation time, and for the condition of the soil (with and without chemicals) a viable cell count was performed, recording the morphologic similarities and macroscopic characteristics (shape, texture, growth, structure, hyphae, conidia, etc.). The colonies of fungi with similar characteristics among the three sites were purified for their isolation by the hyphal puncture technique and stored in PDA in a slant tube for later identification.

The fungal identification at genus level, was related to the macro and microscopic description of each isolate. Subsequently, the strains were examined by microscopy micro culture (Riddell, 1950) using photomicrographs obtained with a Motic BA300 microscope, objectives brightfield, phase contrast that are coupled to a digital camera Moticam 480 (Motic Images Plus 2.0 software). Images obtained with lactophenol blue staining were compared, and morphological criteria (Barnett and Hunter, 2006) identified *Trichoderma* as the main genera.

Resistance trials

The fungal strains, resistance to atrazine was determined by applying Minimum Inhibitory Concentration (MIC) tests to 23 different strains of fungi isolated by the qualitative disk diffusion method (Bauer *et al.,* 1966), with three repetitions. PDA dishes were inoculated with the different fungus genera identified by the surface extension. Sterile filter paper discs (0.5 cm in diameter) impregnated with 10 µL of atrazine solutions at concentrations of 0, 500, 1000, 1500, 2000, 2500, 5000, 7500 and 10,000 mg L^{-1} were used. The inhibition zone (no growth area) present on each disk, in relation to the mycelial growth, was determined.

Trichoderma sp. biomass production

Massive biomass production of *Trichoderma* sp. was carried out in petri dishes containing PDA (propagated by the hyphal tip technique) at room temperature and with five days in darkness. Mycelium was removed by scraping the dish with three consecutive washes of 10 mL sterile water (Guigon *et al.,* 2010). The suspension of conidia obtained from each dish was kept in a 1 L sterile glass jar with screw cap.

Trichoderma sp. growth kinetics

Samples of 100 g of sterile soil, saturated at field capacity with a concentration of 100 ppm of atrazine (Gesaprim 50FW ®), were placed in 250 mL jars with screw cap. The samples were inoculated with a suspension of strains of *Trichoderma* sp. (Concentration of about 10^4 - 10^5 CFU g^{-1} of mycelium). The jars were stored at room temperature for 15 days. The total count of the fungus (Waksma, 1922) was monitored at the first day of storage (time 0) and at 5, 10 and 15 days as described above. A Blank (sterile soil) and a Control (non-sterile soil), under the same conditions was set.

Atrazine degradation in soil

The soil for this experiment came from the Tempantitla experimental site without pesticide application. Samples were air-dried and sieved with a No. 16 mesh 1 mm opening. A homogeneous mixture 80/20 v/v Agrolita ® - perlite was prepared. 50 g of soil mixture were placed in jars and sterilized at 110 ° C for 45 minutes for three consecutive days. The jars were saturated at field capacity (Rodriguez and Rodriguez, 2002) with a solution of 500 ppm of technical grade atrazine (Gesaprim 50FW ®).

Two soil conditions were used: sterile and non-sterile. Flasks were inoculated with 10 mL of the suspension of *Trichoderma* (10^4 – 10^5 CFU * mL^{-1}) and stored for 40 days under greenhouse conditions. To evaluate the treatments, the soil was analyzed at 5, 10, 20 and 40 days after inoculation. A blank (sterile soil inoculated with (*Trichoderma*) and a control (non-sterile soil inoculated with *Trichoderma*) was set. In each time period, the soil was air dried and stored in plastic bags for herbicide removal.

Atrazine extraction and quantification

5 g of dry soil were placed in 50 mL conical tubes and homogenized with 10 mL of methanol and stirred orbitally at 300 revolutions per minute (rpm) for 24 hours in darkness. The tubes were centrifuged 10 min at 5000 rpm and 14 ° C. The supernatant was filtered and collected in 10 mL glass jars. Methanol was evaporated to dryness at 25 ° C, the residue was re-suspended in 5 mL of Liquid Chromatography High Resolution (HPLC) grade acetonitrile. Finally, the samples were filtered with a 0.45 microns Millipore Nylon® membrane and the liquid was collected in jars for HPLC analysis (Fragoeiro and Magan, 2005).

The liquid filtrate was analyzed with a HPLC Agilent 1200®, Infinity Series, diode array detector with a Poroshell® column 120 EC-C18 4.6 x 50mm 2.7 micron. The elution in the chromatography mobile phase began in 70/30 water/acetonitrile (v/ v) at a flow rate of 0.5 mL min $^{-1}$ and sample injection of 10 microliters. The retention time was 2.84 minutes for atrazine with a detection limit of 1 mL hg $^{-1}$. The working solutions for the calibration curve of atrazine (0, 0.2, 0.4, 0.6, 0.8 mg mL $^{-1)}$ were prepared from the stock solution of 1 mg mL $^{-1}$ (atrazine 98% analytical reagent) by dilution with HPLC grade acetonitrile.

The HPLC values obtained were fitted to Equation 1, of the standard curve:

Y = 8694X +684.7 [1], R^2 = 0.987 (Equation 1)

This equation gives the retained mass of atrazine (X). The overall mass balance for atrazine in the test was:
M-input = M + M-retained-downgraded (Equation 2)
Where:
M = mass of atrazine,
From equation (2) the degraded atrazine milligrams were obtained:
M-degraded = M-M-input-hold (Equation 3)

Experimental design

For the characterization of physical and chemical soils a simple variance of means statistical analysis with a completely randomized design (Tukey: α = 0.05) model was used.

The biodegradation of atrazine was a statistical model of the analysis of the variance of one factor in a completely randomized design with fixed effects under repeated measurements over time. The Tukey test (α = 0.05) was applied for means comparison between treatments.

In both cases, the data obtained were analyzed with the statistical program Number Cruncher Statistical System (NCSS), version 2007.

RESULTS AND DISCUSSION

Physical and chemical parameters of the soil samples

The values of the physical and chemicals parameters found in the soil samples collected at the three study sites are given in Table 1. The Capulín and Santana sites showed a moderately acid pH levels, and Tepantitla (NOM-021-RENACT-2000) was neutral. In these intervals, nutrients are more available, as are contaminants in the soil solution (Rinnan *et al.*, 2007).

The total soil nitrogen results were variable for each study site and within the Official Mexican Norm (NOM-021-RENACT-2000) they are classified in intervals of high (0.15 to 0.25%) and very high (> 0.25). Regarding the content of organic matter (OM) and organic carbon (OC), the norm states that values between 4.1-6.0% MO, indicate a low fertility content. On the contrary for Rodriguez and Rodriguez (2002), the amount of organic matter and organic carbon in agricultural classification is high and very high (> 3.5%). Finally, regarding texture, soils are classified as sandy loam and loam (NOM-021-RENACT-2000).

Table 1. Mean comparison of the chemical and physical characteristics of the soils studied.

	Agricultural soils of the Tulancingo Valley, Hidalgo, Mexico.			
	Capulín	Tepantitla	Santana	Standard error
% Humidity	14.89_a	15.86_a	15.20_b	0.20
pH (1:2 in water)	6.35_a	6.69_a	5.78_b	0.12
% Total Nitrogen	0.18_a	0.24_b	0.30_c	$9.48*10^{-3}$
% Organic carbon	2.86_a	3.36_b	4.08_c	0.012
% Organic material	4.94_a	5.79_b	4.04_c	$8.60*10^{-3}$
CIC (Cmol(+)kg^{-1})	29.34_a	59.63_b	25.46_c	3.63
Density (g cm^{-3})	2.38_a	2.43_a	2.29_a	0.11
Texture	Sandy loam	loam	loam	

For each line different letters indicate significant difference (with α = 0.05, Tukey).

The physical and chemical results showed that there are significant differences in the Santana soil (pesticide application site) compared to the soil without pesticide application. Furthermore, one must consider that these properties are not only important in the genesis and fertility of the soil, but also affect retention capacity, adsorption and atrazine degradation in soil (Bridges *et al.,* 2008; Raymundo *et al.,* 2009). Comparing the results with bibliographical observations and references, it was found that the three sites are at intervals of agricultural fertility, but the soil of Santana, because of its record of intensive agricultural use of pesticides, could be vulnerable to deterioration and negative environmental impacts in the short term.

Isolation and characterization of the rhizospheric *Trichoderma* sp.

It was observed that there is a higher total count of fungi in the rhizosphere (rhizospheric soil) in a ratio rhizosphere / soil (R / S) of 2:1, at the three sites Capulín (C), Tepantitla (T) and Santana (S). (Figure 1).

The proportion of Colony Forming Units per gram of soil (CFU g^{-1}) is within the range estimated by Maier and Pepper (2009), of 10^5 to 10^6 CFU g^{-1} in rhizospheric soils. Mainly, because two of the agricultural soils (C and T) are temporary regime without fertilization. These results indicate that the diversity and number of fungal genera present are in an appropriate ranges.

From macro and microscopic laboratory descriptive tests and experimental data, it was possible to identify 7 kinds of fungi (23 species posibles), as follows: *Alternaria* (one), *Mucor* (two), *Clandosporium* (one), *Fusarium* (three), *Aspergillus* (one), *Penicillium* (seven) of *Trichoderma* (five) and three unidentified fungal strains.

The growth of *Trichoderma* sp. in culture media began with the observation of colonies with small green specks inconidium cushions, and with visible airborne spores. The microculture microscopic observations showed hyaline conidiophores, branched, not verticillated, in groups of phialides producing conidia (fialosphoros) hyaline in ovoid shape and in terminal clusters (Barnett and Hunter, 2006; Sources and Ferrera, 2007).

Resistance tests and biomass production of *Trichoderma* sp.

The identified 23 strains of fungi showed no growth inhibition halos of mycelium for atrazine concentrations of 0 to 10.000 mg $L^{-1,}$ by the CMI method. This behavior shows that the genera isolated resist high concentrations of agrochemical (Gopi *et al.,* 2012; Gopinath and Sims, 2011; Sene *et al.,* 2010).

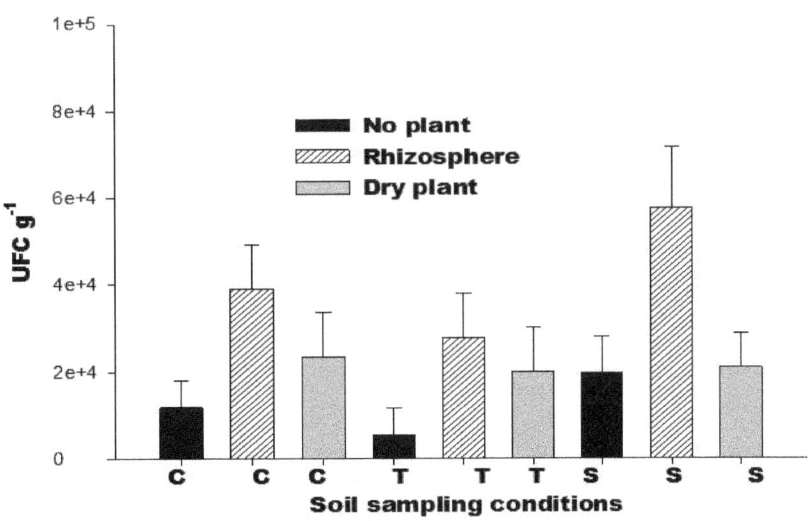

Figure1. Total count of fungi (CFU g $^{-1)}$) of the three agricultural soils form Tulancingo Valley, Hidalgo, Mexico. C (Capulín) and T (Tepantitla) are temporary regime (no herbicide application) and S (Santana) is with irrigation and atrazine application. The soil sampling conditions were: no plant and dry plant rhizosphere.

The biomass production test was performed with a strain identified as *Trichoderma genus,* since there are references that point out that this genera represents major advantages in relation to other identified genera. This fungus is characterized as a symbiotic rhizospheric in plants and is used in the bioremediation of soils for extraction and the biodegradation of contaminants (Harman *et al.,* 2004; Argumedo *et al.,* 2009; Ghopita *et al.,* 2011).

Of the five strains of *Trichoderma* sp. (T_1, T_2, T_3, T_4 and T_5), it was identified as T_5 by its uniform growth, pH ranges (4.5 to 7.5) and because it was the only strain that developed mycelium both on agar and on liquid media. 0.12 g of dry mycelium biomass of *Trichoderma* sp. grown on 78.53 / cm ^{2} of each solid PDA plate was obtained.

Trichoderma sp. growth kinetics

Experimental data indicate that *Trichoderma* sp. (Strain T_5) grew steadily for 15 days at room temperature. On the other hand, the control (non-sterile soil) increased its growth to a lesser extent and the control showed a progressive decrease in CFU g $^{-1}$ soil (Figure 2).

The concentration of *Trichoderma* sp. increased from $1x10^4$ to $1.8x10^6$ CFU g^{-1} soil in 15 days at room temperature using atrazine as the sole carbon and nitrogen source. It can be estimated that bioaugmentation with the fungus was favorable, compared to the control (unidentified microbial consortium).

Atrazine degradation in soil

Significant differences ($\alpha = 0.05$) between treatments from five to 20 days were found. At the end of the experiment (40 days), the amount of atrazine biodegraded by the *Trichoderma* sp. (T) was higher compared to that achieved by the Blank (B) and the Control (Ct) at the same time (Table 2).

Another point to emphasize is the behavior of the treatment with *Trichoderma* sp. in relation to time, where the treatment biodegraded 447.1 mg L^{-1} from an initial concentration of 500 mg kg^{-1} of atrazine (8.9 mg atrazine / 50 g soil), representing 89.4% of the herbicide. It is observed that from day 20 there is a difference in the fungal genera (T) respecting the Control and Blank (Table 2 and Figure 3).

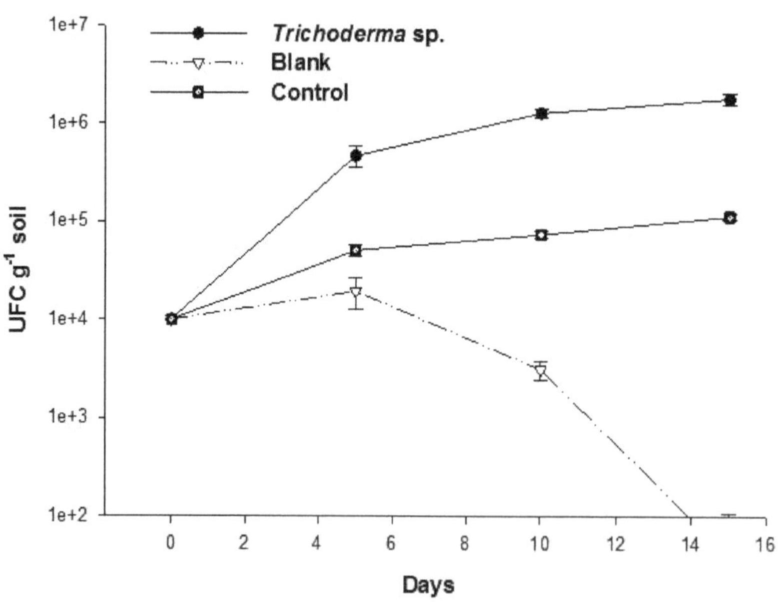

Figure 2. Growth kinetics in CFU g $^{-1}$ of *Trichoderma* sp. in sterile soil in relation to Blank and Control at 0, 5, 10 and 15 days of incubation with an initial 100 ppm atrazine concentration.

Table 2. Biodegradation of atrazine to different treatments.

Treatment	Atrazine (mg L $^{-1}$)			
	5 days	10 days	20 days	40 days
Trichoderma sp. (T)	114.9452_a	168.5607_a	442.3395_b	447.1820_b
Blank (B)	29.6909_a	67.0079_a	59.9570_a	207.0662_b
Control (Ct)	119.8873_a	174.3655_a	112.2000_a	147.4810_a
Standard error	±13.5272	±13.5272	±13.5272	±13.5272

Column means with different letters differ statistically (α = 0.050, Tukey)

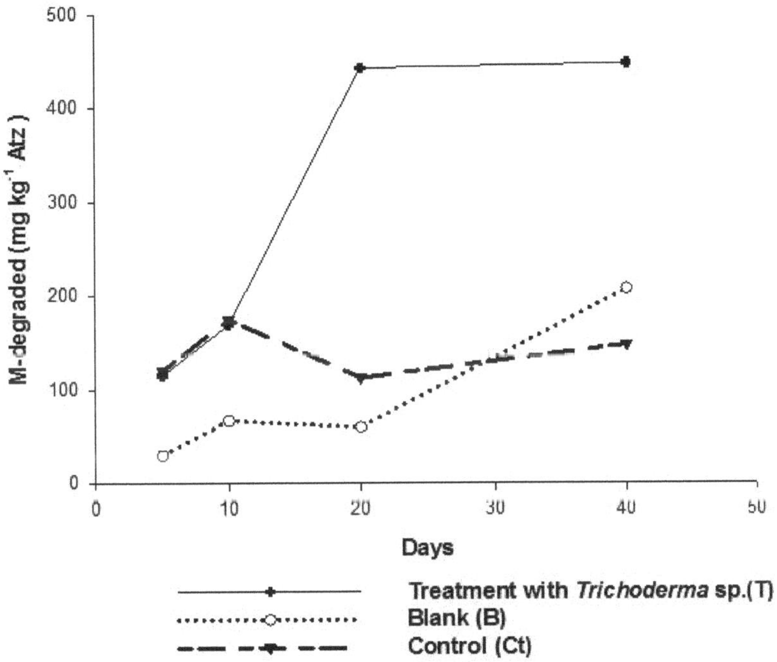

Figure 3. Atrazine mass (mg kg $^{-1}$) degraded with *Trichoderma* sp. (T) inoculation compared to Blank (B) and Control (Ct), recorded at 5, 10, 20 and 40 days after inoculation.

This behavior is different from that reported by Przybulewska and Sienicka (2008), where the same genera mineralized (up to NH_3 and CO_2) 2.5 mg/50 g of soil in 112 days (10^6 -10^7 CFU g $^{-1}$), indicating that the results are 3.5 times higher than those reported, 2.8 times less in time, and with the same initial amount of inoculum.

CONCLUSION

The genus *Trichoderma* sp. was isolated, identified and characterized as a native strain of the rhizosphere of *Phaseolus vulgaris* L., which belongs to the agro ecosystem of the valley of Tulancingo, Hidalgo, Mexico, and showed resistance to atrazine herbicide up to 10,000 mg L $^{-1.}$ It is able to degrade up to 89% of the atrazine in 500 mg kg $^{-1}$ of soil under laboratory conditions.

This study also shows that it is feasible to cultivate the native *Trichoderma* sp strain under laboratory conditions with degradation results for this herbicide superior to those already reported.

The next step in this research should include trials with the native rhizospheric strain of *Trichoderma* sp at greenhouse and field level to evaluate is effectiveness for atrazine bioremediation in cultivated soils contaminated with atrazine.

Acknowledgments.

Margarita Islas Pelcastre wishes to thank the National Council of Science and Technology of Mexico (CONACyT) for providing funding for her Phd studies through the number 33565 scholarship granted.

REFERENCES

Abou-Shanab, R.A., Angle, J.S., Chaney, R.L. 2006. Bacterial inoculants affecting nickel uptake by Alyssum murale from low, moderate and high Ni soils. Soil Biology & Biochemistry. 38: 2882-2889.

Adebusoye, S.A., Picardal, F.W., Ilori, M.O., Amund, O.O., Fuqua, C. 2008. Characterization of Multiple Novel Aerobic Polychlorinated Biphenyl (PCB)-Utilizing Bacterial Strains Indigenous to Contaminated Tropical African Soils. Biodegradation. 19:145-159. Agencia de Protección Ambiental. 2012. http://www.epa.gov/oppsrrd1/reregistration/atrazine.

Argumedo, D.R., Alarcón, A., Ferrera, C.R., Peña, C.J. 2009. El género fúngico *Trichoderma* y su relación con contaminantes orgánicos e inorgánicos. Revista Internacional de Contaminación Ambiental. 25:257-269.

Bajwa, R., Khokhar, I., Mukhtar, I., Mushtaq, S. 2010.Isolation and Identification of Filamentous Fungi from Different Industrial Effluents. Plant Product Research Journal. 14: 32–35.

Barnet, H.L., Hunter, B.B. 2006. Illustratrated general of imperfect fungi, Fourth edition, Apss Press, Minnesota, USA.

Barriuso, J., Ramos, S.B., Lucas, A.L., Probanza, L.A. 2008. Ecology, Genetic Diversity and Screening Strategies of Plant Growth Promoting Rhizobacteria (PGPR). Ahmad, I, Pichtel, J. and Hayat, S.(Eds), Plant-Bacteria Interactions. Strategies and Techniques to Promote Plant Growth, (pag. 1.13), WILEY-VCH Verlag GmbH & Co. KGaA, Weinheim, USA.

Bastos, A.C., Magan, N. 2009. *Trametes versicolor* : Potential for atrazine bioremediation in calcareous clay soil, under low water availability condition. International Biodeterioration and Biodegradation. 63:389–394.

Bauer, A.W., Kirby, W.M.M., Sherris, J.C., Turck, M.1966. Antibiotic susceptibility testing by a standardized single disk method. American Journal of Clinical Pathology. 45:493-496.

Bayer Cropscience: http://www.annualreport2009.bayer.com/en/homepage.aspx [Última consulta 13 Julio 2012].

Belimov, A.A., Hontzeas, V.I., Safronova, S.V., Demchinskaya, G., Piluzza, S., Glick, B.R. 2005. Cadmium-tolerant plant growth-promoting bacteria associated with the roots of Indian mustard (*Brassica juncea* L. Czern.). Soil Biology & Biochemistry. 37: 241-250.

Bridges, M., Brien, H.W., Shaner, D.L., Khosla, R., Westra, P., Reich, R. 2008. Spatial Variability of Atrazine and Metolachlor Dissipation on Dryland No-tillage Crop Fields in Colorado. Journal of Environmental Quality. 37:2212–2220.

Bouyoucos, G.J. 1962. Hydrometer method improved for making particle size analysis of soils. Agronomy Journal. 54:464-465.

Buckman, H. and Brady, N.C. 1966. The Nature and Properties of Soils. The Macmillan Company. United Kingdom.

Chaudhary, K.S., Inouhe, M., Rai, N.M.U., Mishra, K., Gupta, K.D. 2011. Inoculation of *Rhizobium* (VR-1 and VA-1) induces an increasing growth and metal accumulation potential in *Vigna radiata* and *Vigna angularis* L . growing under fly - ash. Ecological Engineering. 37:8:1254-1257.

Cheyns, K., Martin, L.F., Bru, D., Aamand, J., Vanhaecke, L., Diels, J., Merckx, R., Smolders, E., Springael, D. 2011. Long-term dynamics of the atrazine mineralization potential in surface and subsurface soil in an agricultural field as a response to atrazine applications. Chemosphere. 10:450-465.

Cooper, R.L., Laws, S.C., Goldman, J.M., Narotsky, M.G. 2007. Atrazine and reproductive function: mode and mechanism of action studies. Birth Defects Research. 80:98-112.

Davidchik, V.N., Kulikova, N.A., Golubeva, L.I., Stepanova, E.V., Koroleva, O.V. 2008. *Coriolus hirsutus* Laccase Effect on Atrazine Adsorption and Desorption by Different

Types of Soil. Applied Biochemistry and Microbiology. 44:440-445.

Fragoeiro, S., Magan, N. 2005. Enzymatic activity, osmotic stress and degradation of pesticide mixtures in soil extract liquid broth inoculated with *Phanerochaete chrysosporium* and *Trametes versicolor*. Environmental Microbiology 7:348–355.

Fuentes, D.G., Ferrera, C.R. (Eds.). 2007. Ecología de la raíz (segunda edición). Cd. Obregón, Sonora, México.: Sociedad Mexicana de Fitopatología, A.C.

GESAPRIM 50 FW®. Syngenta 2009. Hoja de Seguridad Resolución 350/99. http://www.jedys.com.ar/data/HojaDeSeguridad_359.pdf . [Última consulta 13 Julio 2012].

Getenga, Z., Dörfler, U., Iwobi, A., Schmid, M., Schroll, R. 2009. Atrazine and terbuthylazine mineralization by an Arthrobacter sp. isolated from a sugarcane-cultivated soil in Kenya. Chemosphere. 77:534–539.

Gopi, V., Upgade, A., Soundararajan, N. 2012. Bioremediation potential of individual and consortium Non-adapted fungal strains on *Azo dye* containing textile effluent.Advances in Applied Science Research. 3:303-311

Gopinath, K.R. and Sims, K.G. 2011. Biostimulation for the Enhanced Degradation of Herbicides in Soil. Applied and Environmental Soil Science [Versión electronica] Disponible en: http://www.hindawi.com/journals/aess/2011/843450 .[Última consulta 13 Julio 2012].

Govantes, F., Porrúa, O., García, G.V., Santero, E. 2009. Atrazine biodegradation in the lab and in the field: enzymatic activities and gene regulation Microbial Biotechnology. 2:2. 178–185.

Guigón, L.C., Guerrero, P.V., Vargas, A.F., Carvajal, M.E., Ávila, Q.G.D., Bravo, L.L., Ruocco, M., Lanzuise, S., Woo, S., Lorito, M. 2010. Identificación Molecular de Cepas Nativas de *Trichoderma* spp. su Tasa de Crecimiento *in vitro* y Antagonismo contra Hongos Fitopatogenos. Revista Mexicana de Fitopatología. 28:87-96.

Harman, G.E., Lorito, M., Lynch, J.M. 2004. Uses of *Trichoderma* spp. to Alleviate or Remediate Soil and Water Pollution. Advances in Applied Microbiology. 56:330-331.

Instituto Nacional de Ecología 2007. Estudio: precisión del Inventario de plaguicidas obsoletos y sitios contaminados con éstos http://siscop.ine.gob.mx/descargas/gt/informe_final/informe_final_plaguicidas.pdf.

Joo, H., Choi, K., Hodgson, H. 2010. Human metabolism of atrazine. Pesticide. Biochemistry and Physiology. 98:73–79.

Kjeldahl, J. 1983. Sur une nouvelle méthode de dosage de l'azote dans les substances organiques, Résumé du compte-rendu des travaux du Laboratoire de Carlsberg, Copenhague, en commission chez H. Hagerup, Imprimerie de Thiele.

Kolombet, V.L., Zhigletsova, K.S., Kosareva, I.N., Bystrova V.E., Derbyshev, V.V., Krasnova, P.S., Schisler, D. 2008. Development of an extended shelf-life, liquid formulation of the biofungicide Trichoderma asperellum. World Journal of Microbiology and Biotechnology. 24:123–131.

Llado, S., Jiménez, N., Viñas, M., Solanas A.M. 2009. Microbial populations related to PAH biodegradation in an aged biostimulated creosote-contaminated soil. Biodegradation. 20:593–601.

Lima, P.D., Viana, P. André S., Chelinho, S., Costa, C., Ribeiro, R., Sousa, J.P., Fialho, A.M., Viegas, C.A. 2009. Evaluating a bioremediation tool for atrazine contaminated soils in open soil microcosms: The effectiveness of bioaugmentation and biostimulation Approaches. Chemosphere. 74:187–192.

Liu, X., Parales, E.R. 2009. Bacterial Chemotaxis to Atrazine and Related *s* -Triazines. Applied and Environmental Microbiology. 75:5481-5482.

Ma, Y., Prasad, M.N., Raikumar, M., Freitas, H. 2011. Plant growth promoting rhizobacteria and endophytes accelerate phytoremediation of metalliferous soils. Biotechnology Advances. 29:248-258.

Madigan, M.T., Martinko, J.M., Parker, J. 2006. Brock Biología de los Microorganismos. Décima Edición, Pearson Prentice Hall, Madrid, España.

Mehlich, A. 1948. Determination of catión and anion-exchange properties of soils. Soil Science. 66: 429-445.

Monard, C., Martin, L.F., Vecchiato, C., Francez, A.J., Vandenkoornhuyse, P., Binet, F. 2008. Combined effect of bioaugmentation and bioturbation on atrazine degradation in soil. Soil Biology Biochemistry. 40:2253–2259.

NOM-021-RECNAT-2000. Norma Oficial Mexicana, que establece las especificaciones de fertilidad, salinidad y clasificación de los suelos. Estudios, muestreo y análisis. México. http://www.profepa.gob.mx/innovaportal/file/3335/1/nom-021-semarnat-2000.pdf [Última consulta 13 Julio 2012].

NOM-112-SSA1-1994. Norma Oficial Mexicana, que establece bienes y servicios.Determinación de bacterias coliformes. Técnica del número más probable. México. http://bibliotecas.salud.gob.mx/gsdl/collect/nomssa/index/assoc/hash8ceb.dir/doc.pdf [Última consulta 13 Julio 2012].

Pérez, B.Y., Cantillo, P.T., Ramos, R.E., González, G.M., López, M.M. 2009. Prospección de hongos de suelo con potencialidades para el control biológico en suelos de agroecosistemas cubanos. Fitosanidad. 13:3-5.

Pochon, J., Tardieux, Y. 1962. Techniques d' Analyse en Microbiologie du sol. Editurial de Tourelle, París, France.

Prasad, M.N.V. and Katiyar, S.C. 2010. Drill cuttings and fluids of fossil fuel exploration in North-Eastern India: environmental concern and mitigation options. Current Science. 98:1566-156.

Programa de las Naciones Unidas para el Medio Ambiente (PNUD). 2001. Convenio de Estocolmo sobre contaminantes orgánicos persistentes. http://redalyc.uaemex.mx/pdf/539/53906901.pdf.

Przybulewska, K. and Sienicka, K. 2008. Decomposition of atrazine by microorganism isolated from long-term herbicide experimet soil. Ecological Chemistry and Endengineerings. 15: 501-511.

Quinto, C., Cárdenas, L. 2007.Diálogo para ganar: interacción simbiótica entre una bacteria del suelo y el frijol. Biotecnología. 14:273-280.

Raymundo, E., Nikolskii, I., Duwig, C., Prado, B., Hidalgo, C., Gavi, F. 2009. Transporte de atrazina en un andosol y un versitol de México, Interciencia. 34:330-337.

Razak, A.A., Bachman, G., and Farrag, R. 1999. Activites of microflora in soils of upper and lower Egypt. The African Journal of Mycology and Biotechnology. 98:2557-2551.

Reporte meteorológico universitario de la estación meteorológica del Instituto de Ciencias Agropecuarias de la Universidad Autónoma del Estado de Hidalgo. 2009. http://www.uaeh.edu.mx/principal/clima/index.html . [Última consulta 13 julio 2012].

Rey, M., Llobell, A., Monte, E., Scala, F., Lorito, M. 2004. Genomics of *Trichoderma.* Applied Mycology and Biotechology . 4: 225-248.

Riddell, R.W. 1950. Permanent strained mycologycal preparation obtained by slide cultures. Micología. 82: 265-270.

Rinnan, R., Michelsen, A., Bååth, E., Jonasson, S. 2007. Mineralization and carbon turnover in subarctic heath soil as affected by warming and additional litter. Soil Biology and Biochemistry. 39:3014-3023.

Robinson, R.A. 1996. Return to Resistance: Breeding Crops to Reduce Pesticide Dependence. Davis, CA, agAccess. United States of America.

Rodríguez, H., Rodríguez, J. 2002. Métodos de análisis de suelos y plantas, Criterios de interpretación, Trillas, DF México.

Scott, C., Lewis, E.S., Milla, R., Taylor, M.C., Rodgers, J.W.A., Dumsday, G., Brodie, E.J., Oakeshott G.J., Russell, J.R. 2010. A free-enzyme catalyst for the bioremediation of environmental atrazine contamination. Journal of Environmental Management. 91:2075-2078.

Sene, L. Converti, A., Ribeiro, G.A.S., Cássia, R. 2010. New Aspects on Atrazine Biodegradation. Brazilian Archives of Biology and Technology an International Journal. 53:487-496.

Schuster, A., Schmoll, M. 2010. Biology and biotechnology of Trichoderma. Applied Microbiology & Biotechnology. 87:3. 787.

Sistema Meteorológico Nacional. 2011. http://smn.cna.gob.mx/index.php?option=com_content&view=article&id=103&Itemid=68.

Smith, D., Alvey, S., Crowley, E.D. 2005. Cooperative catabolic pathways within an atrazine-degrading enrichment culture isolated from soil. Microbiology Ecology. 53:265–273.

So, Y.K., and Kyung, S.C. 2009. Isolation and Characterization of a Plant Growth-Promoting Rhizobacterium, Serratia sp. SY5. Journal Microbiology. Biotechnology. 19:1431-1438.

Tang, J.,Liu, L., Hu, S., Chen, Y., Chen, J. 2009. Improved degradation of organophosphate dichoruos by Trichoderma atroviride transformats generated by restriction enzyme- medianted interation (REMI). Bioresurce Technology. 100:480-483.

Tappin, I.A., Loughnanc, J.P., McCarthy, J.A., Fitzsimons, F.M. 2012. Removal of atrazine from river waters by indigenous micoorganismos. Environmental Chemistry Letters. 10: 89-96.

Waksman, S.A. 1922. A method of counting the number of fungi in soil. Journal of Bacteriology. 7(3):341-399.

Walkley, A., and Black, I.A. 1934. An examination of the Degtjareff method for determining organic carbon in soils: Effect of variations in digestion conditions and of inorganic soil constituents. Soil Science. 63:251-263.

Yang, C., Yang, L., Kun, Z., Wang, X., Cuiqing, M., Hongzhi, T., Ping, X. 2010. Atrazine Degradatión by a simple consortium of *Klebsiella* sp. A1 and *Comaonas* sp. A2 in nitrogen enriched medium. Biodegradation. 21:97-105.

Zafar, S., Aqil, F., Ahmad, I. 2007. Metal tolerance and biosurption potential of filamentous fungi isolated from metal contaminated agricultural Soil. Bioresource Technology. 98 (13):2557-2561.

16

How can Rabbit Research in the Laboratory Contribute to Cuniculture on the Farm?

Gabriela González-Mariscal

Centro de Investigación en Reproducción Animal, CINVESTAV-Universidad Autónoma de Tlaxcala, Apdo. Postal 62, Tlaxcala Tlax 90000, México.

E-mail: gabygmm@gmail.com
** Corresponding author*

SUMMARY

Rabbit research coming from the laboratory can have a profound impact on Cuniculture, performed on the farm, and vice versa. This bi-directional communication is scarce at present but, by finding issues of common interest, an effective interaction between these two niches can be promoted. I will present five examples from Reproductive Neuroendocrinology where I have identified evidence that research in one niche has had (or can have) an impact on the other one, specifically: 1) distinguishing between pregnancy and pseudopregnancy; 2) preventing death of kits due to deterioration of the maternal nest and/or loss of maternal behavior; 3) facilitating the management of groups of does mated according to the "biostimulation" method; 4) increasing the success of the "biostimulation" method; improving the welfare of rabbits housed in the laboratory and on the farm. Promoting communication between "the 5) lab and the farm" will lead to new ways of exploring key scientific questions and to better management practices on the farm.

Keywords: rabbit; pregnancy; lactation; pseudopregnancy; mating; cuniculture; animal welfare.

INTRODUCTION

Rabbits have been studied for many years in the laboratory and on the farm. Their physiology (e.g., digestion, reproduction), disease susceptibility (to pathogens or toxic substances), behavior (aggressive, social, reproductive), or use (of their meat and fur) have been the subject of numerous investigations published in journals specialized in the corresponding field or devoted specifically to rabbit research. Sadly, this rich information has been separated into two areas that constitute the niches of scientists working in the laboratory ("basic" research) or on the farm (agronomic research). Usually, scientists from one niche read the publications, attend the meetings, and interact with the colleagues from their own area and are unaware that similar research lines are being pursued by investigators from "the other" niche. This division is, of course, impoverishing to science as the strategies used in the laboratory and on the farm are complementary to each other; the models used in each field can reveal new ways of approaching the same scientific question. As emphasized in earlier publications (González-Mariscal and Roselli, 2007; González-Mariscal et al., 2007) it is necessary to promote a larger interaction between "the lab and the farm". Thus, it is the objective of this article to present specific examples that illustrate how these two fields can converge and interact effectively to improve rabbit production and to generate new scientific knowledge.

WHAT ISSUES OF COMMON INTEREST TO REPRODUCTIVE NEUROENDOCRINOLOGY AND CUNICULTURE HAVE WE IDENTIFIED?

Before attempting to bring together two fields that are related but operate independently one needs to identify common interests. It is necessary to find out if a given scientific question is relevant to both areas and if it is (or has been) investigated by researchers working in the laboratory or on the farm. In the following sections five specific topics will be presented that meet these criteria; the main findings and missing information identified in each one will be described.

Distinguishing between pregnancy and pseudopregnancy.

Rabbit does are induced ovulators, i.e., they ovulate only in response to copulation (Ramírez and Beyer, 1988). Although, theoretically, this would "ensure" that all mated does became pregnant and gave birth to young in reality a variable proportion of such rabbits enter a peculiar endocrine state known as pseudopregnancy. This is characterized by an endocrine profile similar to the one of rabbits in early pregnancy, i.e., high levels of progesterone (P) in blood, produced by the corpus luteum (Erskine, 1999). This ovarian structure, however, has a shorter life span in pseudopregnant (14 days) than in pregnant (around 30 days) does. As a consequence P levels drop in the former about two weeks earlier than in the latter (Niswender and Nett, 1994) and pseudopregnant does -of course- do not produce a litter. The mechanisms underlying the occurrence of pseudopregnancy –rather than pregnancy- following mating in rabbits have been little explored. However, it is known that vaginocervical stimulation received during mating reaches the central nervous system to promote surges of prolactin that maintain the corpus luteum (Erskine, 1999). For the "lab scientist" the occurrence of pseudopregnancy is a fascinating research area in Reproductive Neuroendocrinology; for the professional trying to breed tens to hundreds of rabbits in the farm pseudopregnancy entails an important financial loss. Experienced rabbit producers can distinguish between pregnant and pseudopregnant does by palpating their ventrum at around 14-20 days post-copula. However, this method can give false negatives if pregnant does are carrying few embryos and the effectiveness of the "palpation" method depends on the experience of the person performing the procedure. We found in our laboratory that the frequency of a particular form of scent-marking, consisting of the rubbing of the rabbit's chin on any solid object in the environment ("chinning"), is directly related with the degree of sexual receptivity of a doe (González-Mariscal et al., 1990). Thus, unmated females kept under a long photoperiod and given adequate nutrition are sexually receptive (Edey and Casida, 1972) and show a high frequency of chinning (González-Mariscal et al., 1990). Pregnant does, by contrast, are not sexually receptive and show low chinning frequencies. What about pseudo-pregnant rabbits? Interestingly, their chin-marking activity is indistinguishable from that of truly pregnant does but *only* during the first 14 days post-mating. From then onwards chinning frequency increases steadily in pseudopregnant animals, to reach levels characteristic of estrous does (Figure 1; González-Mariscal et al, 1990). Thus, monitoring chinning frequency for about ten days (starting at around 8-10 days post-copula) can be a reliable, inexpensive, and simple method for determining if a mated female is truly pregnant.

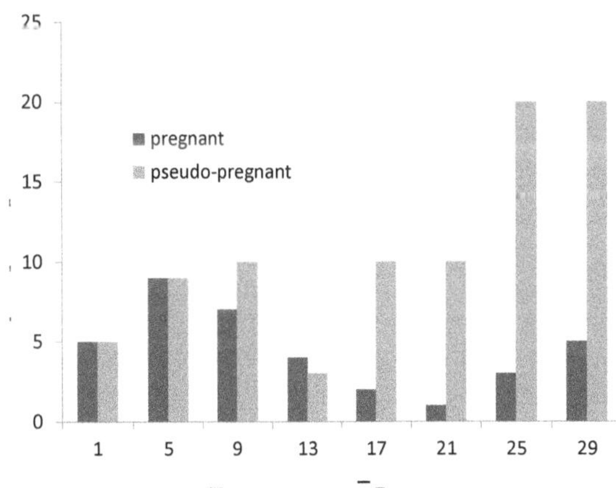

Figure 1. Chin-marking (chinning) frequency does not significantly differ between pregnant and pseudopregnant rabbits for the first 14 days post-mating. From then onwards, chinning frequency increases steadily *only* in pseudopregnant females (modified from: González-Mariscal et al, 1990).

Preventing death of kits due to deterioration of maternal nest and/or loss of maternal behavior.

Some mammals that give birth to a single offspring (like sheep and goats) nurse only their own young - which they identify largely by smell- and refuse to allow suckling from other lambs or kids, respectively. This is called exclusive nursing (for review see: González-Mariscal and Poindron, 2002). By contrast, mammals that deliver a litter (such as sows, rats, and bitches) can nurse even young that are not their own.

To determine whether something similar occurs in rabbits we explored: a) if does accepted suckling from another doe's kits and b) whether placing either the mother's own or an alien litter inside a nest different from their own modified their willingness to nurse (González-Mariscal and Gallegos, 2007). As shown in Fig. 2 placing the mother's litter inside a nest different from her own (i.e., made with synthetic or male hair) significantly increased her latency to enter the nest box for nursing. This effect was also evident when alien kits were provided, although the magnitude of the difference was smaller.

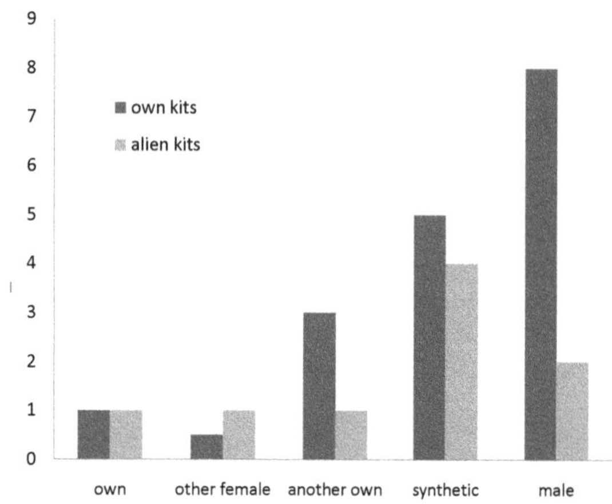

Figure 2. Latency to enter nest box of mothers provided with their own or foster kits in various types of nests (modified from: González-Mariscal and Gallegos, 2007).

Yet, once the does entered the nest box provided, they all nursed the litter placed inside it (i.e., their own or an alien one) for the usual amount of time (around 3 min; Table 1).

Table 1. Time (min; mean±s.e.) does spent inside different types of nest boxes nursing their own or foster kits (modified from: González-Mariscal and Gallegos, 2007).

Type of kits	Type of nest box where kits were placed				
	own	another female	another own	synthetic	male
own	3±1.0	3±1.0	4±1.0	4±1.0	3±1.0
foster	3±0.1	4±0.3	3±0.3	4±0.4	4±0.3

These results confirm that rabbits *do not* show exclusive nursing and that their willingness to nurse is not dependent on the type of nest in which the kits are placed. These findings, obtained in the laboratory, can be used as a managing strategy on the farm in cases where: a) the mother's original nest deteriorates or b) she loses her maternal behavior (especially in the case of primiparous does). Under such conditions a recommendation can be made -based on scientific evidence- that new nests (made from fresh materials, including synthetic hair) be provided and/or kits be relocated to other lactating does. The latter may be particularly important in cases where –for a variety of reasons- some kits from a litter die within the first days of life. As shown in Fig. 3 the maintenance of maternal behavior across lactation relies heavily on the number of suckling kits, particularly for primiparous does. Nursing only one or two young leads to a gradual loss of the mother's willingness to nurse. This observation, combined with the fact that lactating rabbits consume more feed than unmated or pregnant ones (González-Mariscal et al., 1994, 2009a), emphasizes the need to recommend the relocation of kits to "alien" mothers for optimizing rabbit production on the farm.

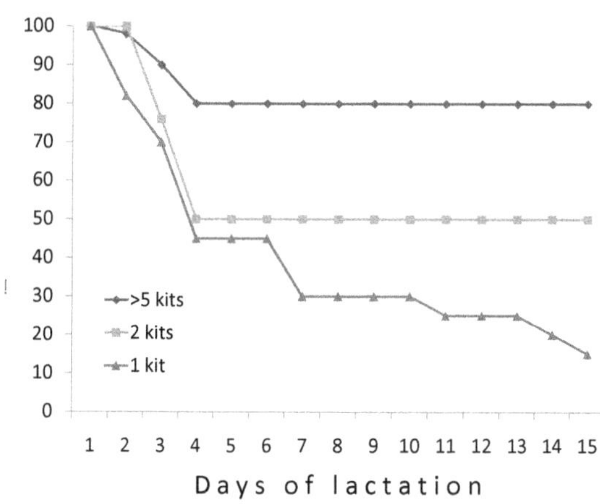

Figure 3. Number of suckled young determines maintenance of maternal behavior across lactation days 1-15.

Facilitating the management of groups of does mated according to the "biostimulation" method.

To maximize the number of kits produced/doe/year rabbit breeders have adopted a mating strategy commonly known as "biostimulation". This method takes advantage of the fact that rabbits can sustain lactation concurrently with pregnancy. Following parturition does have a so-called post-partum estrus, lasting several hours, during which they are sexually receptive. If mated at such time they can become pregnant without compromising the nursing of the newborn litter. This capacity for concurrent pregnancy and lactation has allowed that, in nature, rabbits leave a large progeny in their rather short lives. In the laboratory the endocrine profile and milk output have been compared between does mated or not at post-partum estrus. Such studies have revealed that: a) progesterone concentrations in blood are smaller in pregnant-lactating does than in pregnant-only rabbits across gestation days 7-17 and b) milk output declines precipitously from lactation day 20 onwards *only* in does that are concurrently pregnant-lactating (Fig. 4; Fortun et al, 1993; González-Mariscal et al, 2009a; Lebas, 1972). This difference in milk production has the unwanted consequence that kits suckling from pregnant-lactating does gain less weight than litters nursed by lactating-only rabbits. However, the fact that milk output declines in concurrently pregnant-lactating does only from day 20 onwards implies that if such females are mated on lactation day 10 (rather than at parturition), the litter can obtain a normal amount of milk for a full 30 days.

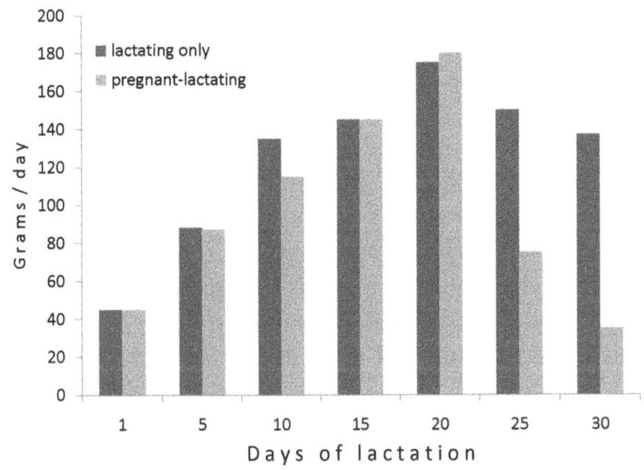

Figure 4. Milk production in lactating only vs concurrently pregnant-lactating rabbits (modified from: González-Mariscal et al, 2009).

This possibility, however, has the obstacle that lactating does are *not* sexually receptive (Beyer and Rivaud, 1969). Yet, a number of studies have revealed that cancelling a single nursing bout in early lactation (usually between days 8 to 10) leads to the restoration of estrus 24 h later and the consequent pregnancy following natural or artificial insemination (Alvariño et al, 1998; Castellini et al, 1998; Theau-Clément and Mercier, 1999). This "biostimulation" mating strategy has been adopted worldwide to maximize productivity on rabbit farms. Nonetheless, for this method to be effective it is essential that: a) nursing is prevented on a specific day; b) nursing has occurred before the litter is removed or the door leading to the maternal nest is closed (otherwise the kits risk 48 h of milk deprivation and can die). Verifying individual kits for the presence of milk in their stomachs is, of course, a time-consuming, labor-demanding process. In small farms lacking automatic systems -that open and close the doors to the nest at a pre-programmed time- it is common that litters stay in the maternal nest throughout lactation and does have round-the-clock access to the kits. Under such conditions breeders cannot be sure whether a nursing bout occurred or not on a given day (unless they inspect the kits individually). This is a major problem for breeders wanting to use the biostimulation strategy to mate lactating does because, if nursing did not occur, kits will not survive a 48-h fast. We recently found that, under a photoperiod of 14 h light:10 h darkness (lights on at 0700 h), rabbits spontaneously nurse the litters, kept inside the nest box and placed within their home cages, at around 0215 h. The rigorous statistical analysis used to determine this and the large number of suckling episodes recorded (20 lactating does across 15 days of lactation) allowed us to obtain a level of confidence of $p<0.001$ (González-Mariscal et al, 2009b). This finding can allow breeders to safely assume that nursing has, indeed, occurred before the litters are removed in the morning for "biostimulation" without having to verify individual kits.

Increasing the success of the "biostimulation" method.

Although cancelling a suckling episode in early lactation indeed restores estrus in lactating rabbits the proportion of females that do become pregnant following natural mounting or artificial insemination can vary greatly (Alvariño et al, 1998; Bonanno et al, 2002; Theau-Clément and Mercier, 1999). The sources of this variability are hardly known and, in fact, the factors underlying the effectiveness of the "biostimulation" method have been little explored. To our knowledge only the concentrations of prolactin and estradiol in blood have been compared between lactating and "biostimulated" does in early lactation. In general, the latter show higher prolactin levels before insemination (as a nursing bout was cancelled) and higher estradiol concentrations after insemination relative to control does in which nursing was not interrupted (Fig. 5; Ubilla et al, 2000).

However, the relevance of these differences for determining the likelihood of a successful pregnancy after insemination are unknown. In addition, external factors, such as: photoperiod duration, feed composition, and temperature have been reported to impinge on the relative success of the "biostimulation" method through as yet unidentified ways (Gómez et al, 2004; Marai et al, 2002; Szendrö et al, 2004; Theau-Clément et al, 1998). For the neuroendocrinologist working in the laboratory exploring the mechanisms by which steroid hormones, peptides, metabolic products, and environmental factors act on specific brain regions and peripheral organs to regulate sexual receptivity, fertility, and pregnancy is an intriguing research challenge. For rabbit breeders, employing a variety of management strategies and housing conditions, information coming from the laboratory may provide a solid basis for adjusting specific parameters to improve the relative success of the "biostimulation" method.

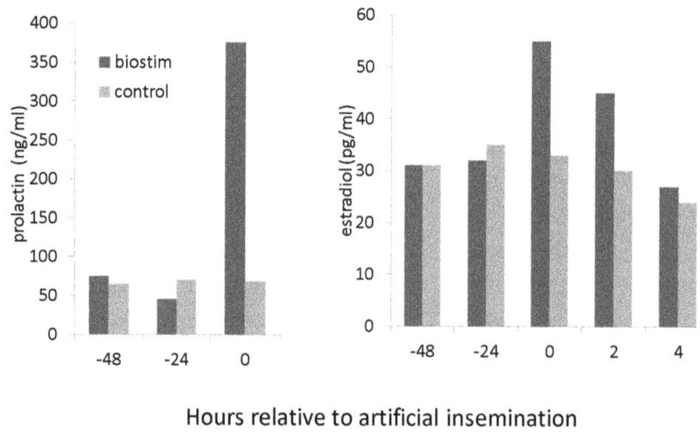

Figure 5. Effect of biostimulation on prolactin and estradiol concentrations in blood (modified from: Ubilla et al, 2000).

Improving the welfare of rabbits housed in the laboratory and on the farm.

The physical and psychological well-being of animals kept under laboratory or farm conditions is a major concern worldwide. Needless to say the requirements of food, space, social interaction, tranquility, duration of photoperiod, etc., vary greatly across mammals and, of course, among vertebrate classes (for review see: Moberg and Mench, 2000). Moreover, the ways by which information can be obtained regarding the presence of stress, the malaise provoked by sickness, the perception of pain, or the occurrence of complex mood states (such as anxiety and boredom) is largely dependent on the species under study. That is, animals are variable not only in the ways they react to the environment; they also differ in the types of responses (e.g., motor, endocrine) they can show under a specific setting. Thus, to design methods that effectively measure welfare/stress it is essential to know what types of responses a "normal" animal -of a particular species (age and sex)- can provide. Consequently, we should ask: what behavioral, physiological, and endocrine responses can we measure in rabbits that vary as a consequence of their physical and psychological well-being? Moreover, we need to inquire whether the regular management practices used in the laboratory and on the farm (regarding type of cages, feed, temperature, photoperiod, transportation, etc.) have an impact on the different categories of rabbits (i.e., bucks, juveniles, pregnant/lactating does) that usually occupy breeding facilities. As the main endocrine indicator of stress is the responsiveness of the hypothalamus-pituitary-adrenal axis to a variety of conditions, some works have measured the concentration of corticosterone or cortisol in blood in relation to: exposure to heat, cold, noise or a social mix (De la Fuente et al, 2007), group vs single housing (Whary et al, 1993), and rearing conditions in early life (Brecchia et al, 2009). The main findings from these works indicate that heat, followed by loud noise, is the main stressor for *adults of both sexes* and that group-housing *does* has no impact on their blood corticosterone levels, especially after several weeks of lodging under this condition. By contrast, a major difference in the responsiveness to a specific stressor in *adult females* was noted between those raised "normally" (i.e., that received a single daily nursing bout throughout a 30-day lactation) and the ones that missed *a single* nursing episode on lactation day 10 (as a consequence of the "biostimulation" procedure their mothers were subjected to). As shown in Figure 6 injecting saline in the thigh provoked a large increase of corticosterone in blood, evident from 30 to 90 min post-injection, but *only* in the control group: rabbits that were separated from their mothers for 48 h in early lactation showed a blunted response to stress (Brecchia et al, 2009).

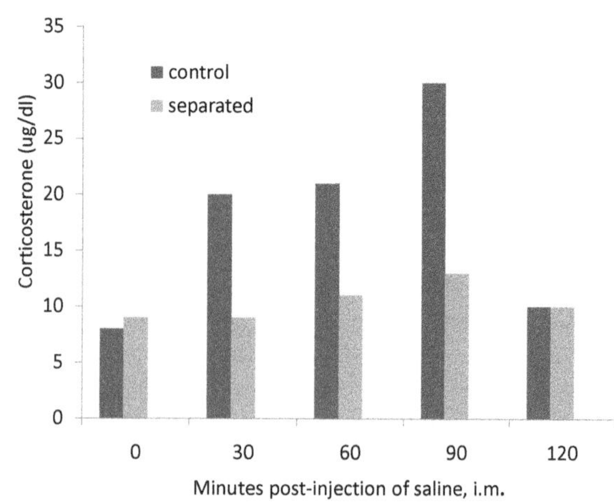

Figure 6. Impact of mother-young separation (days 9-11) on stress reactivity as adults (modified from: Brecchia et al, 2009).

Additional evidence for the impact of manipulations during the early neonatal period on the permanent modification of the stress response comes from a study in which kits were handled by humans at different times around nursing during the first week of lactation. When tested at weaning (day 28) rabbits that were handled at 15 min before or 30 after nursing showed a smaller latency to approximate a human and a larger number of approaches in the 5-min test than control (i.e., non-handled) kits (Fig. 7; Pongrácz and Altbäcker, 1999).

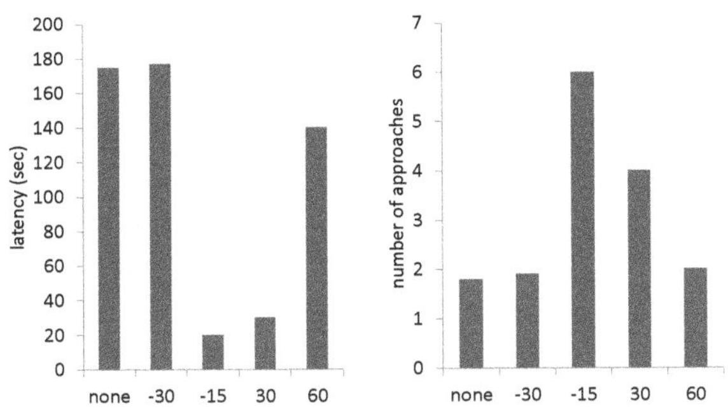

Figure 7. Effect of handling kits at different times before or after nursing on their reactions to humans at weaning (modified from: Pongrácz and Altbäcker, 1999).

Other works have studied pregnant rabbits, subjected to vibrations that simulate transportation or to noise: neither the duration of pregnancy nor the number of liveborn kits was affected by such environmental factors (only a slight increase in the number of dead young was noted; Stephens and Adams, 1982). By contrast, pregnant or lactating does housed collectively showed fewer "stereotyped" behaviors (like biting the cage bars), more sniffing of the environment, and fewer rearings than did females housed individually. However, reproductive parameters (e.g., sexual receptivity, fertility, milk production) were practically unaffected by the type of lodging condition (Dal Bosco et al, 2004). Taken together, the above findings show that important contributions are being made in the laboratory that reveal the sensitivity of rabbits to management practices and specific characteristics of the environment. They also indicate that more research is needed in this regard to explore other indicators of welfare/stress and to determine the long-term effects of particular procedures on the responses of rabbits of different breeds, ages, sex, and reproductive state.

CONCLUSIONS

Contact between the researchers devoted to studying Reproductive Neuroendocrinology and those working on Cuniculture can lead to: 1) a richer understanding of complex phenomena of rabbit physiology and behavior; 2) the generation of new scientific questions amenable to investigation; 3) better management practices on the farm. Such consequences are desirable because: a) the rabbit is a unique model for exploring key issues in several fields of "basic" science (e.g., chronobiology, behavioral neuroendocrinology, developmental psychobiology, energetic metabolism); b) as the world faces new challenges to feed the human population (Paillard et al, 2011) rabbit production has many advantages over that of other domestic animals. New ways to promote a constant exchange of ideas between the scientists of the laboratory and the researchers working on the farm are needed. The education of students needs to be modified according to the vision that the scientific study of rabbits is richer as it is broader. It is hoped that this article contributes to achieve such goals.

REFERENCES

Alvariño, J.M.R., Del Arco, J.A., Bueno, A. 1998. Effect of mother-litter separation on reproductive performance of lactating rabbit females inseminated on day 4 or 11 post partum. World Rabbit Science 6:191-194.

Beyer, C., Rivaud, N. 1969. Sexual behavior in pregnant and lactating domestic rabbits. Physiology and Behavior 4:753-757.

Bonanno, A., Di Grigoli, A., Alabiso, M., Boiti, C. 2002. Parity and number of repeated doe litter separation treatments affect differently the reproductive performances of lactating does. World Rabbit Science 10:63-70.

Brecchia, G., Bonanno, A., Dall'Aglio, C., Mercati, F., Zerani, M., Di Grigoli, A., Boiti, C. 2009. Neuroendocrine responses in neonatal mother-deprived rabbits. Brain Research 1304:105-112.

Castellini, C., Canali, C., Boiti, C. 1998. Effect of mother-litter separation for 24 hours, by closing the nestbox or change of cage, on rabbit doe reproduction performance. World Rabbit Science 6:199-203.

Dal Bosco, A., Mugnai, C., Castellini, C., Laudazi, S. 2004. A prototype of colony cage for improving the welfare of rabbit does: preliminary results. 8 World Rabbit Congress. Puebla, México. pp 1229-1234.

De la Fuente, J., Díaz, M.T., Ibáñez, M., González de Chavarri, E. 2007. Physiological response of rabbits to heat, cold, noise and mixing in the context of transport. Animal Welfare 16:41-47.

Edey, T.N., Casida, L.E. 1972. Effects of fasting on oestrus and ovulation in the oestrous rabbit. Australian Journal of Biological Sciences 25:397-403.

Erskine, M. 1999. Pseudopregnancy. In: Knobil, E., Neill J.D. (eds.). Encyclopedia of Reproduction. Academic Press, NY. pp. 120-124.

Fortun, L., Prunier, A., Lebas, F. 1993. Effects of lactation on fetal survival and development in rabbit does mated shortly after parturition. Journal of Animal Science 71:1882-1886.

Gómez, R.B., Becerril, P.C.M., Torres, H.G., Pro, M.A., Rodríguez de Lara, R. 2004. Relationship among feeding level, change of cage and fasting with vulva color and sexual receptivity in New Zealand white and Californian nulliparous does. World Rabbit Congress Proceedings 8:270-275.

González-Mariscal, G., Díaz-Sánchez, V., Melo, A.I., Beyer, C., Rosenblatt, J.S. 1994. Maternal behavior in New Zealand white rabbits: quantification of somatic events, motor patterns and steroid plasma levels. Physiology and Behavior 55:1081-1089.

González-Mariscal, G., Gallegos, J.A. 2007. New Zealand white rabbits show non-selective nursing in various types of nests. World Rabbit Science 15:167-172.

González-Mariscal, G., Gallegos, J.A., Sierra-Ramírez, A., Garza Flores, J. 2009a. Impact of concurrent pregnancy and lactation on maternal nest-building, estradiol and progesterone concentrations in rabbits. World Rabbit Science 17:145-152.

González-Mariscal, G., Lemus, A.C., Aguilar-Roblero, R. 2009b. The number of suckling young modulates circadian periodicity of nursing in rabbits. Developmental Psychobiology 57:586.

González-Mariscal, G., Melo, A.I., Zavala, A., Beyer, C. 1990. Variations in chin-marking behavior of New Zealand female rabbits throughout the whole reproductive cycle. Physiology and Behavior 48:361-365.

González-Mariscal, G., Poindron, P. 2002. Parental care in mammals: immediate, internal and sensory factors of control. In: Pfaff, D., Arnold, A., Etgen, A., Fahrbach, S. and Rubin, R. (eds.). Hormones, Brain, and Behavior. Academic Press, San Diego. pp. 215-298.

González-Mariscal, G., Roselli, C.E. 2007. Foreword. Hormones and Behavior 52:1.

González-Mariscal, G., McNitt, J.I., Lukefahr, S.D. 2007. Maternal care of rabbits in the lab and on the farm: endocrine regulation of behavior and productivity. Hormones and Behavior 52:86-91.

Lebas, F. 1972. Effet de la simultanéité de la lactation et de la gestation sur les performances laitières chez la lapine. Annales de Zootechnie 21:129-131.

Marai, I.F.M., Habeeb, A.A.M., Gad, A.E. 2002. Rabbits' productive, reproductive and physiological performance traits as affected by heat stress: a review. Livestock Production Science 78:71-90.

Moberg, G.P., Mench, J.A., eds. 2000. The Biology of Animal Stress. CABI Publishing, United Kingdom. 377 pp

Niswender, G.D., Nett, T.M. 1994. Corpus luteum and its control in infraprimate species. In: Knobil, E., Neill J.D. (eds.). The Physiology of Reproduction. Raven Press, NY. pp. 781-816.

Paillard, S., Treyer, S., Dorin, B., eds. 2011. Agrimonde: scenarios and challenges for feeding the world in 2050. Éditions Quae, France.

Pongrácz, P., Altbäcker, V. 1999. The effect of early handling is dependent upon the state of the rabbit (*Oryctolagus cuniculus*) pups around nursing. Developmental Psychobiology 35:241-251.

Ramírez, V.D., Beyer, C. 1988. The ovarian cycle of the rabbit: its neuroendocrine control. In: Knobil, E., Neill J.D. (eds.). The Physiology of Reproduction. Raven Press, NY. pp. 1873-1892.

Stephens, D.B., Adams, C.E. 1982. Observations on the effects of vibration stress and sound on pregnancy, parturition and respiration in the rabbit. Laboratory Animals 16:341-347.

Szendro, Zs., Gerencsér, Zs., Gyovai, M., Metzger, Sz., Radnai, I., Biró-Németh, E. 2004. Effect of photoperiod on the reproductive traits of rabbit does. World Rabbit Congress Proceedings 8:354-357.

Theau-Clément, M., Castellini, C., Maertens, L., Boiti, C. 1998. Biostimulations applied to rabbit reproduction: theory and practice. World Rabbit Science 6:179-184.

Theau-Clément, M., Mercier, P. 1999. Effect of a doe-litter separation on rabbit doe reproductive performance and growth of the young. World Rabbit Science 7:177-179.

Ubilla, E., Rebollar, P.G., Pazo, D., Esquifino, A., Alvariño, J.M.R. 2000. Effects of doe-litter separation on endocrinological and productivity variables in lactating rabbits. Livestock Production Science 67:67-74.

Whary, M., Peper, R., Borkowski, G., Lawrence, W., Ferguson, F. 1993. The effects of group housing on the research use of the laboratory rabbit. Laboratory Animals 27:330-341.

17

Impact of Agricultural Policy on Relative Price Variability of Food Crops and Inflation in Nigeria

Ifeoluwa Akin Babalola* and Kabir Kayode Salman

Department of Agricultural Economics, University of Ibadan, Ibadan, Nigeria
Email: Ifebab212@yahoo.com, k_salimonu@yahoo.com
**Corresponding author*

SUMMARY

Prices of food crops in Nigeria tend to exhibit similar trend with inflation. The study therefore established quantitatively relationships among agricultural policy, relative price variability (RPV) of food crops and inflation in Nigeria. Data for the study includes annual producer prices (nominal) and output of food crops and annual inflation rate obtained from the publications of the Central Bank of Nigeria, Nigerian Bureau of Statistics, Food and Agricultural Organisation and Nigerian Institute of Social and Economic Research covering the period of 1970- 2009. Analytical tools used were RPV index and Error Correction Method (ECM). The results showed that the variables are stationary at their levels. As inflation increases, RPV of food crops also increases both in short run (0.0002) and the long run (0.0310). Civilian Post-Structural Adjustment Period Policies (CPSAP) caused a significant reduction in inflation and consequently reduced the RPV of food crops in the long run. There is a need for policies that will buffer the food crop sub-sector from the effects of inflation. Policies that reduce the rate of inflation and minimise RPV among food crops are needed. Effective management of inefficiencies and misallocation of resources in the sub-sector should be explored.

Key words: Relative Price Variability; Food Crops; Inflation; Agricultural Policy.

INTRODUCTION

Agriculture plays an important role in the rapid growth and development of Nigeria. Agriculture in the context of the economy is tied with the various sectors and is essential for generating broad based growth necessary for development. Agriculture is fundamental to the sustenance of life. It is the bedrock of economic development, especially in the provision of adequate and nutritious food vital for human development, and raw materials for Industries. It provides food for the growing population, employment for over 65% of the population, and raw materials and foreign exchange earnings for the development of the industrial sector (Ajibefun, 2004).CBN (2010), reports that the agricultural sector accounts for an average of 33.5% of the Gross Domestic Product (GDP) between 2005-2009. Hence, a rapid development in agriculture would serve as a catalyst for the improvement of the standard of living of the majority of the population.

Agricultural price is a key variable that affects demand and supply of agricultural products and motivation of producers. Agricultural prices are structurally prone to fluctuations because of the short-run inelasticity of supply and demand for agricultural products (Cohen, 1999). These price changes create price risk against which those engaged in agriculture seek protection. A commodity's relative price is defined as its nominal price divided by the average price of all commodities in the group. Changes in relative prices is called relative price variability or volatility and is used as an indicator of the real costs of inflation in relation to its effect on commodity price changes (Loy and Weaver, 1998). Relative price variability (RPV) is therefore defined as the variance across a set of commodities of the rates of change of individual nominal prices (Lapp and Smith, 1992). It is measured as the standard deviation of the individual price changes around the average inflation rate.

When relative prices within agriculture vary because of inflation, such movements may decrease economic welfare for society as a whole and the agricultural sector in particular, efficiency of resource allocation decreases because decision makers have less useful information on prices to guide their decisions (Ukoha, 2007). Variation in producer price of food crops is more reflected in the relative price of food crops, therefore if inflation should relate to relative prices of food crops in any policy period, the extent and direction of this relationship may reduce the economic welfare of both producers and consumers of food crops during the policy regimes. Many Nigerians depends on food crops for their daily dietary need but it is observed that inflation affects the prices of food crops.

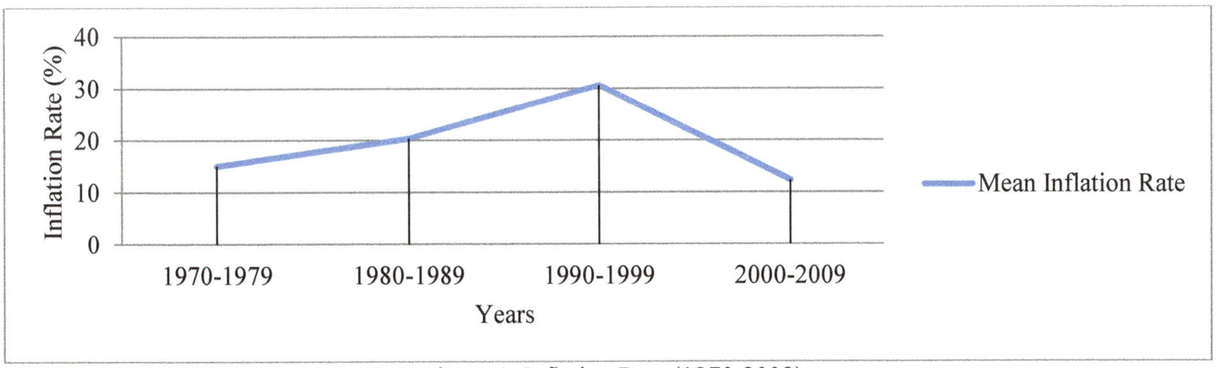

Figure 1: Inflation Rate (1970-2009)

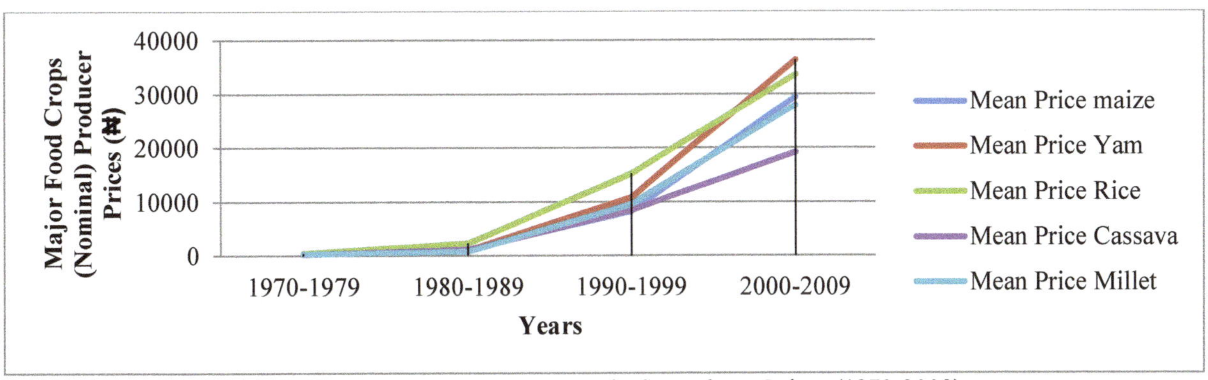

Figure 2: Major Food Crop (Nominal) Producer Prices (1970-2009)

Owing to the important role of agriculture in the Nigerian economy, government, over the years has pursued various policies to boost the agricultural sector. The policies include fiscal, monetary, pricing and exchange rate policies among others. The government went into direct agricultural production through associations such as the River Basin Development Authorities (RBDA), the Directorate of Food, Roads and Rural Infrastructure (DFRRI), and a number of food producing companies. A host of agricultural production programmes were also established at different times. These programmes included Operation Feed the Nation (OFN), Green Revolution (GR), National Accelerated Food Production Project (NAFPP) among others (Ukoha, 2007).

The controversy whether inflation affects RPV has generated vast empirical literature. Parks (1978) analysed the determinants of the variability in inflation across different components of the US consumer price index. Parks found a significant positive correlation with the aggregate inflation rate. Papers, like Parsley (1996), and Debelle and Lamont (1997) analysed the determinants of the variability of inflation across different US cities. Again they found a significant positive correlation with aggregate inflation. Furthermore, some empirical studies such as Domberger (1987) and Hercowitz (1982) found a positive correlation between the RPV and inflation rate consistent with the Menu cost models however, Silver and Ioannidis (2001) documented a negative one consistent with imperfect information model.

Empirical works such as Reinsdorf (1994) founds that RPV in the United States during the 1980s recession was monotonically decreasing in unanticipated inflation and that positive inflation shocks reduce RPV (that is, they found a negative relationship between inflation and measures of the dispersion of relative price change). Tosupport Reinsdorf (1994), Fielding et al., (2011) used data from inter-war Japan, their results indicated that in recessionary conditions a positive inflation shock does reduce RPV. However, this reduction is unlikely to correspond to higher consumer utility; this has implications for the conduct of monetary policy during a recession.

Garba (2000) analysed the impact and stability of Nigerian agricultural policies from 1970 to1993, the results show that volatility was most intense for most policy variables in the structural adjustment period, implying that agricultural policies were relatively more stable before than after adjustment. Ukoha, (2007) examined the relationships among the RPV of agricultural commodities, inflation and agricultural policy in Nigeria. The study showed that the effect of inflation on RPV among agricultural commodities is non-neutral, inflation had a significant positive impact on RPV in both long-run and the short-run, OFN, GR, Structural Adjustment Programme (SAP) and post-SAP policies had significant positive impact on RPV of food crops in the short run, but in the long run, none of these policies has significant impact on RPV of food crops. The study recommended that the OFN policy measures favoured food crops and therefore a selective application of the policy for food crops would ensure efficiency. The policy included liberal resource allocation to agriculture and the provision of infrastructural facilities. Furthermore, Akpan and Udoh (2009) estimated a functional relationship between RPV of grains and inflation rate in various agricultural policy periods in Nigeria. Empirical results revealed that inflation had a positive significant effect on RPV of grains. The result further showed that the SAP and civilian post SAP agricultural policy regimes brought about a positive significant shift in the coefficient of inflation which implies a increase in RPV of grains. Hence, the objective of this study is to establish quantitatively the relationships among agricultural policy, RPV of food crops and inflation in Nigeria.

MATERIAL AND METHODS

Scope of the study. The subject of RPV in Nigeria is wide and complex. However, this study focused on the role of inflation and agricultural policies on RPV of food crops from 1971 to 2009. The study covered the whole Nigeria.

Source of data. The data used for the study were secondary and obtained from publications of the Central Bank of Nigeria (CBN), Nigerian Bureau of Statistics (NBS), Food and Agricultural Organisation (FAO) and Nigerian Institute of Social and Economic Research (NISER). The following major food crops were considered for the analysis: Yam, Maize, Rice, Millet and Cassava (Ukoha, 2007). The data included annual producer prices (nominal), output of food crops and annual inflation rate, the data covered the period of 1970-2009.

Analytical technique

Measuring RPV. The nominal rate of price change consists of two aggregate components, the inflation and relative price component.

$$P_{i,t} = P_t + Z_{i,t} \qquad (1)$$

Where, $P_{i,t}$ = Nominal price of ith commodity in period t, P_t = the producer price index (PPI) in period t and $Z_{i,t}$ is the relative price of commodity i in time t. All variables are expressed in natural log.

The PPI was calculated using the Laspeyres price index as shown below:

$$P_t = \left[\frac{\sum P_n Q_o}{\sum P_o Q_o}\right] \times 100 \quad (2)$$

Where, P_t = PPI of 'i' commodity in period t, P_n = Price in the current year
P_o = Price in the base year, Q_o= Quantity in the base year

Inflation rate in period t can be defined as:

$$\Omega_t = \text{Ln}P_t - \text{Ln}P_{t-1} \quad (3)$$

RPV as defined by Parks (Parks, 1978) is:

$$V_t = \sqrt{\sum_{i=1}^{\mu} W_i(\Omega_{it} - \Omega_t)^2} \quad (4)$$

Where, V_t= Relative price variability, W_i= Weight of price 'i' in the price index (it reflects the relative importance of each commodity), Ω_{it}= nominal rate of price change in period t (inflation rate of price 'i') which equals:

$$\text{Ln}P_{i,t} - \text{Ln}P_{i,t-1} \quad (5)$$

The annual measure of RPV (RPV index) of food crops was computed from the Producer Price Index (PPI) computed from the data named above. Laspeyres price index was used to compute the PPI (Gupta, 1981). This study used 1/N as weight attached to each food crop (Akpan and Udoh, 2009), where N is the total number of observations in t_k (N=5, 't_k' represents 'each year'). The augmented dickey-fuller (ADF) test was carried out to determine the time series properties of the data. The variables RPV of food crop and inflation were integrated and therefore error correction models (ECM) and long-run static models were specified and estimated. The empirical models for the study were specified as shown below according to the objectives. All the equations were estimated by OLS method (See equations 6 and7).

OLS Estimation of objective 1: To derive the relationship between inflation and food crop RPV and determine the impact of inflation on food crop RPV.

$$V_{ft} = \alpha_0 + \alpha_1\lambda_t + U_t \quad (6)$$

Where, V_{ft}= RPV of food crops, α_0= constant, α_1= coefficient, λ_t= absolute value of inflation in period t (Annual inflation rate), U_t = error term

OLS Estimation of objective 2: To show the impact of different agricultural policies on inflation and RPV of food crops. The agricultural policies examined includes: the Price Control (PC)1971-1975, Operation Feed the Nation (OFN) 1976–1979, Green Revolution (GR) 1980–1985, Structural Adjustment Programme (SAP) 1986–1993, Military Post-SAP (MPSAP)1994 - 1999 and Civilian Post-SAP (CPSAP) 2000 to date (Akpan and Udoh, 2009).
The required equation is:

$$V_{ft} = \alpha_0 + \alpha_1\lambda_t + \alpha_2(PC * \lambda)_t + \alpha_3(OFN * \lambda)_t + \alpha_4(GR * \lambda)_t + \alpha_5(SAP * \lambda)_t + \alpha_6(MPSAP * \lambda)_t + \alpha_7(CPSAP * \lambda)_t + U_t \quad (7)$$

PC, OFN, GR, SAP, MPSAP and CPSAP are dummy variables for each agricultural policies respectively (1= policy period and zero otherwise). The product between each dummy variable and absolute inflation were found (for example PC*λ, OFN*λ, GR*λ etc.) to show the impact of inflation in different agricultural policy regimes on RPV of food crop. α_0 is the constant term, α's are coefficients, U_t is stochastic error term other variables are earlier defined (see equation 6).

RESULTS AND DISCUSSION

Table 1 reports the ADF test for RPV of food crop and absolute value of inflation. Comparing the ADF test statistic with the MacKinnon critical values, the ADF Statistic values are lesser than the MacKinnon critical values therefore that there is no unit root problem in the variables used. All the variables are stationary in their levels. This validates the use of OLS in estimating equations 6 and 7 without the risk of obtaining a spurious regression.

Table 2 reports Error Correction Model (Parsimonious / short run) estimate for equation 6. It revealed that equation 6 is significant at 1%; the model has a good explanatory power as shown by the value of the R-Square, The strong significance of the error correction term in the equation confirms that the residual of the static model is level stationary and that the series are co-integrated. The error correction term is negative and statistically significant at 1% probability level, therefore a quick convergence to equilibrium in each period, with intermediate adjustment captured by the differenced term. The Durbin–Watson statistics shows that autocorrelation is not a serious problem in the equation. The coefficient of inflation is positive at lag one.

Table 1: The Unit root test result (ADF- Statistic) for RPV of food crops and absolute value of inflation.

Variables	ADF Statistic	Critical Value of ADF 1%	5%	10%
V	-4.4320	-3.6171	-2.9422	-2.6092
λ	-4.3477	-3.6171	-2.9422	-2.6092

Table 2: Error Correction Model (Parsimonious / short run)†

Regressor	Coefficient	t-statistic
Constant	0.0551	0.4654
λ	-0.0290	-0.6835
D(λ(-1))	0.0310	0.8622
ECM(-1)	-0.5150***	-3.9930
R-squared	0.3729	
Adjusted R-squared	0.3159	
S.E. of regression	0.1706	
Sum squared residual	0.9605	
Log likelihood	15.0475	
Durbin-watson statistic	2.1569	
Mean dependent variable	-0.0284	
S.D. dependent variable	0.2063	
Akaike information criterion	-0.5972	
Schwarz criterion	-0.4230	
F-statistic	6.5410***	
Prob (F-statistic)	0.0014	

† estimated for equation 6

*** represent significance at 1% probability level.

Relationship between RPV of food crops and Inflation

The result on table 2 above shows that there is a positive relationship between RPV of food crops and inflation in Nigeria. This result is similar to those obtained by Caglayan and Filiztekin (2003), Smith and Lapp (1993), Jaramillo (1999) and Ukoha, (2007). Table 2 also shows that inflation explains 37% of the total variation in food crop prices. The result suggests that food crop producers in Nigeria are exposed to increased risks and uncertainties in their production activities during the period of high inflation. Also, there is high tendency of reallocation of resources from food crop sub-sector to other agricultural sub-sectors during the period of increasing rate of inflation. During the period of high inflation, due to high-anticipated probability of production risks, there will be reduced welfare for both producers and consumers of food crops in Nigeria.

Impact of different agricultural policies on inflation and RPV of food crops

In order to examine the impact of different agricultural policies on inflation and RPV of food crops, equation 6 was re-estimated by incorporating dummy variables that capture the government policies. The product of these dummy variables and inflation rates in different agricultural policy regimes were further found [that is, Agricultural policy (Dummy variable) * Annual inflation rate]. Table 3 reports the long run (static) impact of different agricultural policies on inflation and RPV of food crops. The table showed that inflation has a positive impact on RPV of food crops. PC Policy did not give any result; this may be because PC policy was more focused on cash crops than on food crops. OFN, GR, SAP and MPSAP, did not have any significant impact on inflation, this may be because many of the Nigeria policies were short-lived and suffered discontinuity from one political regime to another, and consequently, these policies did not have any significant impact on the RPV of food crops in Nigeria. This result is in line with Ukoha, 2007, who found that OFN, GR and SAP Policies did not have any significant impact on RPV of food crops in Nigeria in the long run. CPSAP caused a significant shift in the co-efficient of inflation causing a reduction in the co-efficient of inflation and consequently caused an decrease in RPV of food crops in Nigeria.

Table 3: Long run (Static) impact of different agricultural policies on inflation and RPV of food crops†

Regressor	Coefficient	T-value
λ	0.00020	0.03000
OFN*λ	-0.00367	-0.42000
GR*λ	-0.00603	-0.86000
SAP*λ	-0.00232	-0.38000
MPSAP*λ	-0.00264	-0.42000
CPSAP*λ	-0.02123***	-2.68000
Constant	0.35060	4.95000
R-square	0.23240	
F- statictic	1.62000	

† OLS estimation of equation 7.
Note: *** represent significance at 1% probability level.

CONCLUSION

The study shows that inflation has a positive impact on RPV of food crops in both the short run and the long run. The OFN, GR, SAP, MPSAP policies did not shift the coefficient of inflation and as a consequence did not have any significant impact on RPV of food crops. The CPSAP had a significant impact on inflation (at 1% level) thereby reducing the co-efficient of inflation and at the same time reducing RPV of food crops in Nigeria in the long run which further has implication for increased welfare. Inflation positively affects price variations among food crops. Thus, the need for policies that will buffer the food crop sub-sector from the effects of inflation. Policies that reduce the rate of inflation and minimise RPV among food crops are needed. Effective management of inefficiencies and misallocation of resources in the sub-sector should be explored.

REFERENCES

Ajibefun, I.A. 2004. Determinants of technical efficiency and policy implications in traditional agricultural production: Empirical study of Nigerian food crops farmers. A final research report, submitted to African Economic Research Consortium, Nairobi, Kenya.

Akpan, S.B. and Udoh, E.J. 2009. Estimating Grain Relative Price variability and Inflation Rate Movement in Different Agricultural Policy regimes in Nigeria. Humanity and Social Sciences Journal. 4(2): 107-113

Caglayan, M. and Filiztekin, A. 2003. Nonlinear impact of inflation on relative price variability. Economic Letters. 79(2): 213–18.

CBN 2010. Central Bank of Nigeria: Statistical Bulletin, 2010 Edition.

CBN/NISER 1992. The Impact of SAP on Nigerian Agricultural and Rural Life. Vol 1. The National report. A CBN/NISER National Study.

Cohen, B.H. 1999. Derivatives, volatility and price discovery. International Finance, 2: 167-202.

Debelle, G. and Lamont, O. 1997. Relative price variability and inflation: evidence from U.S. cities. Journal of Political Economy.105: 132-52.

Domberger, S. 1987. Relative price variability and inflation: A disaggregated analysis. Journal of Political Economy. 94: 547–67.

Fielding, D., Hajzler, C. and MacGee, J. 2011. Determinants of Relative Price Variability during a Recession: Evidence from Canada at the Time of the Great Depression. Economics Discussion Papers Series. No. 1107. University of Otago. Retrieved from http://hdl.handle.net/10523/1878

Garba, K.P. 2000. Analysis of the implementation and stability of Nigerian agricultural policies, 1970-1993. African Economic Research Consurtium. Nairobi paper, 101.

Gupta, C. B. 1981. An Introduction to Statistical Methods. New Delhi: Vikas Publishing House PVT Ltd.

Hercowitz, Z. 1982. Money and price dispersion in the United States. Journal of Monetary Economics. 9: 25-38.

Jaramillo, C.F. 1999. Inflation and relative price variability: Reinstating Parks results. Journal of Money, Credit and Banking. 31(3): 375–85.

Lapp, J.S. and Smith, V.H. 1992. Aggregate sources of relative price variability among agricultural commodities. American Journal of Agricultural Economics. 74: 1–9.

Loy, J.P. and Weaver, R.D. 1998. Inflation and relative price volatility in Russian food markets. European Review of Agricultural Economics. 25: 373–94.

Parks, R.W. 1978. Inflation and relative price variability. Journal of Political Economy.86: 79– 95.

Parsley, D. 1996. Inflation and relative price variability in the short and long run: New Evidence from the United States. Journal of Money, Credit and Banking.28: 323-41.

Reinsdorf, M. 1994. New Evidence on the Relation between Inflation and Price Dispersion. American Economic Review. 84(3): 720-731.

Silver, M. and Ioannidis, C. 2001. Intercountry differences in the relationship between relative price variability and average prices. Journal of Political Economy. 109(2):355-374.

Smith, V.H. and Lapp, J.S. 1993. Relative price variability among agricultural commodities and macroeconomic instability in the United Kingdom. Journal of Agricultural Economics. 44(2): 272–283.

Ukoha, O.O. 2007. Relative Price Variability and Inflation: Evidence from the Agricultural Sector in Nigeria. AERC Research Paper 171. African Economic Research Consortium, Nairobi.

18

Optimized Plant Tissue Culture Protocol for *In Vitro* Morphogenesis of an Endangered Medicinal Herb *Ceropegia Ensifolia* Bedd.

M. Chandrasekhara Reddy [1&2], P. V. Bramhachari [2] and K. Sri Rama Murthy [1]*

[1]*Department of Biotechnology, Montessori Mahila Kalasala, Vijayawada, Andhra Pradesh - 520 010, India*
[2]*Department of Biotechnology, Krishna University, Machilipatnam -521 001, Andhra Pradesh, India*
**Corresponding author*

SUMMARY

In the present study effect of different concentrations of sucrose and Thidiazuron (TDZ) on *in vitro* morphogenesis of *C. ensifolia* was investigated and rapid micropropagation protocol was developed from *in vitro* derived nodal explants. Among the sucrose concentrations tested, medium concentration (3%) of sucrose induced maximum number of healthy shoots (6.5±0.68). The shoots formed in this concentration are normal and elongated rapidly. Whereas at higher concentrations delayed shoot induction with stout shoots stunted in their growth was observed. Among different concentrations of TDZ tested healthy shoots with well developed leaves were formed in lower concentration i.e. 0.2 mg/l. the number of shoots formed in this concentration was limited to 3.6±0.18 shoots/ explant. The shoots raised *in vitro* were best rooted on MS medium supplemented with 0.1 mg/l α-Naphthaleneacetic acid (NAA) in combination with 0.05 mg/l 6-Benzyladenine (BAP). The maximum root induction (84%) with mean number of roots of 6.7±0.26 and with mean root length 3.5±0.28 cm was observed after six weeks of inoculation. Complete plantlets developed *in vitro* were acclimatized successfully with 85% survival in field conditions.

Keywords: *Ceropegia ensifolia;* Thidizuron (TDZ); sucrose; multiple shoots.

INTRODUCTION

It is a well-known fact that worldwide thousands of plant species are endangered and facing extinction with the current trend of their exploitation and destruction (Poi *et al.*, 2010; Wyse *et al.*, 2009). According to the International Union of Conservation of Nature (IUCN), it is anticipated that the current species extinction rate is between 1000 and 10,000 times higher than it would naturally be. Biodiversity conservation is of global concern which requires a holistic approach. Recognizing the need, the biodiversity convention was adopted by United Nations in 1992 (Sharma and Sharma, 2013). Convention on Biodiversity seeks to conserve the endangered plants at their genetic, species and ecosystem levels. Nonetheless it is now recognized that *ex situ* techniques can be efficiently used to complement *in situ* methods, and they represent conservation of highly endangered and rare species (Ramsay *et al.*, 2000). Recently, a combination of *ex situ* and *in situ* conservation methods of a critically endangered plant (*Ceropegia fantastica* Sedgw.) has been demonstrated (Chandore *et al.*, 2010). Nowadays *in vitro* multiplication has become an advanced tool for protecting medicinal plants, particularly rare and endangered species (Khan *et al.*, 2012).

The species of *Ceropegia* as a whole are under threat, owing to either destructive collection or habitat degradation. They are not only genetically depleted but also are scarcely available. These species are of economic importance (Jagtap and Singh, 1999) due to their starchy edible tubers with medicinal value (Murthy and Kondamudi, 2011). The tubers of this genus contain sugars, starch, fats, albuminoids, gums and crude fibers which are useful as nutritive tonic (Kirtikar and Basu, 1935). The active compound of tuberous roots is the alkaloid cerpegin which is active against diarrhea and dysentery inflammation of gums and delirious fevers of parturition (Nadkarni, 1976). Recently this genus has attracted the attention of several biologists for conservation, since many of the *Ceropegia* species are listed in RED data book of Indian medicinal plants (Srinivasarao *et al.*, 2010). Furthermore, the application of advanced tissue culture protocols is of economic use for the industrial production and consequently, conservation of *Ceropegia* species. However, for effective reintroduction of these plants, *in vitro* morphogenesis protocols must be established along with the habitat conservation and other appropriate *ex vitro* strategies. *Ceropegia ensifolia* Bedd. (Asclepiadaceae) is an endemic, perennial, tuberous twining herb distributed originally isolated from Peermadu Ghat of Kerala. This species is now extended its distribution in Tamil Nadu, i.e. at high wavy mountains in Theni district of Western Ghats, India. In recent days the population of this species is very rare due to habitat degradation. Conventional propagation of this plant species is apparently difficult, owing to low fruit and seed setting, less viability of seeds, poor seed germination and delayed rooting of seedlings. Therefore we propose an *in vitro* propagation method as an alternative technique for propagation of this endangered species and reestablishment of this plant population in natural habitat.

In vitro plant cells, tissues and organ cultures are not fully autotrophic, establishing a need for carbohydrates in culture media to maintain the osmotic potential and serve as energy and carbon sources for developmental processes (Chae, 2013). A variety of carbon sources viz. glucose; maltose, sucrose, dextrose etc are used in culture media depending upon genotypes, species and specific stages of growth (Bozena and Szczerba, 1991). However, sucrose is extensively used as a major transport-sugar in the phloem sap of many plants (Lemoine, 2000). In micropropagation systems, morphogenetic potential of plant tissues can greatly be manipulated by varying concentration of sucrose (Yaseen *et al.*, 2013). In addition to that TDZ (*N*-phenyl-*N'*-1,2,3-thidiazol-5-ylurea) is a substituted phenyl urea compound which was developed for mechanized harvesting of cotton bolls and emerged as a highly effective bioregulant of morphogenesis in plan tissue culture (Guo *et al.*, 2011). Unlike conventional phytohormones, it was proved that TDZ individually contented the requirements of various regenerative responses of many different plant species (Sanikhani *et al.*, 2006). However, in previous studies it has shown that TDZ exhibited both auxin and cytokinin like effects, although, it is chemically, different from auxins and cytokinins (Murthy *et al.*, 1998). It exhibits strong cytokinin like activity and promotes the proliferation of axillary shoots (Faisal *et al.*, 2005). In addition to that TDZ also releases the lateral bud dormancy and stimulates adventitious organ regeneration (Mroginski *et al.*, 2004). These interesting findings on TDZ raised many tantalizing questions for researching in plant morphogenesis. The purpose of the present work was to investigate the effects of sucrose and TDZ on morphogenesis of *C. ensifolia* using nodes with axillary buds as explants. So for the reports are not available on *in vitro* micropropagation for the regeneration of *C.ensifolia*.

MATERIALS AND METHODS

In the present study nodal segments derived from the *in vitro* grown plantlets were used as explants to study the effect of different concentrations of sucrose and TDZ on *in vitro* morphogenesis of *C. ensifolia.* In order to study the effect of sucrose on morphogenesis of *C. ensifolia,* single nodal segments were separated (1 cm long) and cultured on Murashige and Skoog (MS) media containing varying concentrations of sucrose (1 to 5%) along with BAP (3.0 mg/l) respectively. The pH of the media was adjusted to 5.6 to 5.8. Then the agar (0.8%) was added to the media and autoclaved for 20 min at 121 °C and15 lbs pressure. The cultivation was performed at 25±2°C with optimized room conditions for 6 weeks under fluorescent lightening (Philips) of 3000 lux (16/8, D/N cycle) at the top of the culture vessels. The plantlets grown *in vitro* were harvested from all replicate vessels of a treatment and evaluated their morphogenesis status. The parameters considered were a) percentage of shooting response, b) number of shoots per explants, c) shoot height (cm).

In order to study the influence of TDZ on shoot proliferation of *C. ensifolia*, nodal segments was inoculated on MS basal media augmented with TDZ at different concentrations (0.1 to 3.0 mg/l). The explants derived from sprouted shoots were transferred periodically to the fresh medium and in each subculture the transfer of explants were carried out for every six weeks. The frequency of explants sprouting, number of shoots formed per explants and shoot length after six weeks of culture were recorded as observations to know the optimum concentration of TDZ for effective shoot proliferation in *C. ensifolia*.

Attempts were made to root the cloned shoots under *in vitro* conditions by using various combinations of NAA, BAP and Kinetin (KN). The healthy shoots were harvested and cultured on 0.6% agar gelled MS medium supplemented with 0.1% NAA in combination with different concentrations of BAP and KN. Plantlets with well developed shoots and roots were taken out from the culture vials and were cleaned with distilled water to remove traces of agar. Roots of each plantlet were treated with 0.5% fungicide (carbendazim 50% WP) solution before planting into plastic cups contained sterile soil, sand and vermicompost in different ratios. The pots were covered with polythene covers; these hardened plantlets were irrigated with sterilized 1/4 MS salt solution devoid of sucrose for initial 2 to 3 days and were maintained in culture room at 25 ± 1°C with photoperiod of 16 h light. Subsequently after 4 to 6 weeks the covers were pored to allow free gaseous exchange with the outer environment and then removed to expose the plantlets to *ex vitro* conditions. Primary hardened plants after two weeks were transferred to earthen pots and irrigated with tap water and observed for their further growth.

Each experiment was repeated three times and twenty replicates were maintained. The data regarding number of shoots formed, shoot length, number of roots formed and root length was recorded periodically (six weeks) for shoot multiplication and rooting respectively. Raw data were subjected to analysis of variance (ANOVA) carried out by the SPSS 20 (SPSS Inc. Chicago, IL, USA). The values with $P \leq 0.05$ were considered to be statistically significant.

RESULTS AND DISCUSSION

The carbon sources (sucrose) serves as energy and osmotic agents to support the growth of plant tissues. In the present study different concentrations of sucrose were added in the media to know their effect on shoot initiation and proliferation. Different concentrations of sucrose showed varied responses during plant morphogenesis of *C. ensifolia*. The optimum response that is healthy shoot formation was observed when 3.0 mg/l sucrose was added to the medium (Table – 1). The explants inoculated in this medium showed fast sign of shoot sprouting when compared with the explants inoculated on other media. The shoots formed in this medium are good to observe with maximum number of shoots (6.5±0.68) and with maximum shoot length (8.4±0.25 cm) (Fig – 1B). These cultures survived for a longer duration in culture conditions. The shoots which are formed in the lower concentrations of sucrose are with linear stems with pale yellowish green leaves (Fig – 1A). Initially in the lower concentration (1.0 mg/l) the fast sign of shoot sprouting was observed, which developed into thin linear shoots very rapidly and latter these shoots under gone weathering and failed to survive for longer durations. The explants taken from these shoots were failed to sprout in further subcultures. Whereas at higher concentrations of sucrose the initial shoot sprouting was delayed for longer durations (nearly 2 weeks) and the shoots formed in these concentrations are stout and stunted in their growth (Fig – 1C). Browning of the medium was also observed due to leaching of phenolics from the base in to the medium. Swelling of the base of explants and tuber formation was also noticed when higher concentrations of sucrose present in the medium (Fig – 1F). The level of sucrose in tissue culture can affect the growth and development of the plants. To address this problem, several researchers showed that sucrose at 3% level was optimum for healthy shoot regeneration (Vinterhalter and Vinterhalter, 1999: Adjei, 2001, Gauchan, 2012).

In *C. ensifolia* lower concentrations of TDZ played a prominent role in shoot induction. The shoots that are formed at lower concentration of TDZ showed excellent results in their morphogenesis especially with healthy stem and broad dark green elongated leaves which resembled the plants of the natural environment. The maximum shoot sprouting (90) percentage with mean number of shoots (3.6±0.18) and mean shoot length (10.1±0.81 cm) was observed on MS medium supplemented with 0.2 mg/l TDZ (Table–2, Fig–1D). However, with the increase in concentration of TDZ, decrease in shoot sprouting percentage and shoot number and length was observed. The shoots that are formed at higher concentrations of TDZ showed reduction in leaf size (Fig–1E). Even though the shoots contain reduced leaf size, the nodal explants derived from these shoots are good in their shoot sprouting and produced healthy shoots in further subcultures. Some studies have previously reported the significant role of TDZ in shoot proliferation like cytokinins in a number of plant species (Lee, 2005; Shen *et al.*, 2010; Janarthanam *et al.*, 2012).

Rooting of *in vitro* derived shoots were carried out by transferring healthy shoots with 3 to 4 nodes to the root induction media which include various combinations of NAA, BAP and KN. The cumulative effect of BAP and NAA was clearly observed with the growth characteristics. Among all the combinations tested, NAA 0.1 mg/l in combination with BAP 0.05 mg/l was found optimum for *in vitro* rooting in *C. ensifolia* (Table – 3, Fig – 1G). The maximum root induction (84%) with mean number of roots (6.7±0.26) with mean root length (3.5±0.28 cm) was noticed in this medium after six weeks of transfer. Our data indicate that, combinations of NAA and KN did not assist in root induction of *C. ensifolia.* However when the medium was supplemented with NAA, a significant root induction was clearly noticed. Interestingly among all the combinations of NAA and KN treated, NAA 0.1 mg/l in combination with KN0.05 mg/l induced maximum rooting (66%) with mean number of roots (4.7±0.26) and with mean root length (2.8±0.46 cm). Previous publications have already addressed the similar feature that in most of the *Ceropegia* species viz. *C. spiralis* (Murthy *et al.*, 2010), *C. fantastica* (Chandore *et al.*, 2010), *C. attenuate* (Chavan *et al.*, 2011); individual auxins played an effective role in root induction. The prominent role of IAA and NAA in combination with BAP on root induction in *C. bulbosa* was reported by Goyal and Bhadauria (2006). The plantlets with well developed shoot and root system were successfully hardened off inside the growth room in sterile soil, sand and vermi compost in 1:1:2 ratios for about 4 to 6 weeks (Fig – 1H). The primary hardened plantlets were eventually established in earthen pots with nearly 85 percent survival rate.

Table 1. Effect of sucrose in combination of BAP (3.0 mg/l) on *in vitro* shoot proliferation of *Ceropegia ensifolia.*

Sucrose (%)	% of explant response	No of shoots/explant mean ± standard error	Shoot length (cm) mean ± standard error
1.0	93	4.0±0.06 [c]	6.7±0.26 [c]
2.0	98	4.5±0.27 [b]	7.1±0.23 [b]
3.0	100	6.5±0.68 [a]	8.4±0.25 [a]
4.0	92	4.6±0.20 [b]	4.6±0.20 [d]
5.0	87	3.8±0.16 [d]	4.1±0.16 [d]

Data indicate mean ± standard error of the 20 replicates per treatment in three repeated experiments. Mean followed by the same letter was not statistically significant at 0.05 % probability.

Table 2. Effect of TDZ on *in vitro* shoot proliferation of *Ceropegia ensifolia.*

TDZ	% of explant response	No of shoots/explant mean ± standard error	Shoot length (cm) mean ± standard error
0.1	70	2.1±0.20 [b]	6.1±0.25 [c]
0.2	90	3.6±0.18 [a]	10.1±0.81 [a]
0.3	82	3.2±0.18 [a]	7.2±0.94 [b]
0.4	80	2.6±0.32 [b]	7.1±0.75 [b]
0.5	83	2.5±0.28 [b]	6.4±0.93 [c]
1.0	75	2.1±0.20 [b]	6.1±1.01 [c]
1.5	75	2.0±0.06 [b]	4.7±0.48 [d]
2.0	78	1.6±0.35 [c]	4.4±0.11 [d]
2.5	74	1.4±0.30 [c]	4.8±0.22 [d]
3.0	71	1.1±0.21 [c]	3.8±0.33 [e]

Data indicate mean±standard error of the 20 replicates per treatment in three repeated experiments. Mean followed by the same letter was not statistically significant at 0.05 % probability.

Figure 1: Effect of sucrose and TDZ on *in vitro* morphogenesis of *Ceropegia ensifolia*

A. Induction of thin linear shoots on MS medium supplemented with 1.0 % sucrose + 3.0 mg/l BAP
B. Healthy shoots developed on MS medium supplemented with 3.0 % sucrose + 3.0 mg/l BAP
C. Short stouter shoots that developed on MS medium supplemented with 5.0 % sucrose + 3.0 mg/l BAP
D. Induction of multiple shoots on MS medium supplemented with 0.2 mg/l TDZ
E. Induction of multiple shoots on MS medium supplemented with 2.0 mg/l TDZ
F. Tuberization on MS medium supplemented with 5.0 % sucrose + 3.0 mg/l BAP
G. Induction of rooting on MS medium supplemented with 0.1 mg/l NAA + 0.05 mg/l BAP
H. Primary hardened plant ready to transfer in to field conditions

Table 3. Effect of auxins in combinations with cytokinins on *in vitro* rooting of *Ceropegia ensifolia.*

NAA	BAP	KN	% of explant response	No of shoots/explant mean ± standard error	Shoot length (cm) mean ± standard error
0.1	-	-	72	5.6±0.47 [b]	2.97±0.15[b]
0.1	0.01	-	76	5.9±0.31 [b]	3.2±0.14 [a]
0.1	0.05	-	84	6.7±0.26 [a]	3.5±0.28 [a]
0.1	0.10	-	60	4.0±0.06 [c]	2.1±0.60 [c]
0.1	-	0.01	61	4.3±0.20 [c]	2.4±0.30 [b]
0.1	-	0.05	66	4.7±0.26 [c]	2.8±0.46 [b]
0.1	-	0.10	58	2.1±0.60 [d]	1.5±0.2 [c]

Data indicate mean ± standard error of the 20 replicates per treatment in three repeated experiments. Mean followed by the same letter was not statistically significant at 0.05 % probability.

CONCLUSION

In the present study the effect of sucrose and TDZ on *in vitro* morphogenesis of *C. ensifolia* was investigated. It was observed that 3% sucrose was optimum for healthy shoot regeneration. Whereas amongst the concentrations of TDZ tested, 0.2 mg/l was good for inducing multiple shoots with broad elongated dark green leaves. This protocol can be helpful for large-scale multiplication of *C. ensifolia* plant species for effective conservation.

REFERENCES

Bozena, B. and Szczerba, J. 1991. Influence of different carbon sources on invertase activity and Growth of Sour Cherry (*Prunus cerasus* L.) Shoot Cultures. Journal of Experimental Botany, 42: 911-915.

Chae, S.C. 2013. Influence of carbon sources on shoot organogenesis in *Echinacea angustifolia* DC. Life Science Journal, 10: 1300-1303.

Chandore, A.N., Nimbalkar, M.S., Gurav, R.V., Bapat, V.A. and Yadav, S.R. 2010. A protocol for multiplication and restoration of *Ceropegia fantastica* Sedgw. A critically endangered plant species. Current Science, 99: 1593-1596.

Chavan, J.J., Mansingraj, S.N., Avinash, A.A., Sharad, S.K., Nikhil, B.G., Ghansham, B.D., Rajaram, V.G., Bapat, V.A. and Shrirang, R.Y. 2011. Micropropagation and *in vitro* flowering of endemic and endangered plant *Ceropegia attenuata* Hook. Journal of Plant Biochemistry and Biotechnology, 20: 276 - 282.

Faisal, M., Ahmed, N. and Anis, M. 2005. Shoot multiplication in *Rauvolfia tetraphylla* L. using thaidiazuron. Plant Cell Tissue and Organ Culture, 80: 187-190.

Gauchan, D.P. 2012. Effect of different sugars on shoot regeneration of maize (*Zea mays* L.). Kathmandu University Journal of Science, Engineering and Technology, 8: 119-124

Goyal, D. and Bhadauria, S. 2006. *In vitro* propagation of *Ceropegia bulbosa* using nodal segments, Indian Journal of Biotechnology, 5: 565-567.

Guo, B., Abbasi, B.H., Zeb, A., Xu, L.L. and Wei, Y.H. 2011. Thidiazuron: A multi-dimensional plant growth Regulator. African Journal of Biotechnology, 10: 8984-9000.

Janarthanam, B., Dhamotharan, R. and Sumathi, E. 2012. Thidiazuron (TDZ) - induced plant regeneration from internodal explants of *Santalum album* L. Journal of Biosciences Research, 3:145-153.

Khan, S., Al-Qurainy, F. and Nadeem, M. 2012. Biotechnological approaches for conservation and improvement of rare and endangered plants of Saudi Arabia. Saudi Journal of Biological Sciences, 19: 1-11

Kirtikar, K.R. and Basu, B.D. 1935. In Indian medicinal plants Vol.3 (Bishen Singh Mahendrapal, New Delhi, India). Pp. 1638.

Lee, S.W. 2005. Thidiazuron in the improvement of banana micropropagation. Acta Horticulturae, 692: 67-74.

Lemoine, R. 2000. Sucrose transporters in plants: update on function and structure. Biochimica et Biophysica Acta (BBA) – Biomembranes, 1465: 246–262.

Mroginski, E., Rey, H.Y., Gonzalez, A.N. and Mroginskin, L.A. 2004. Thidiazuron promotes *in vitro* plant regeneration of *Arachis correntina* (Leguminosae) via organogenesis. Plant Growth Regulation, 23:129-134.

Murthy, B.N.S., Murch, S.J. and Saxena, P.K. 1998. Thidiazuron: a potent regulator of *in vitro* plant morphogenesis. *In Vitro* Cellular and Developmental Biology-Plant, 34: 267-275

Murthy, K.S.R., Kondamudi, R. and Vijayalakshmi, V. 2010. Micropropagation of an endangered medicinal plant *Ceropegia spiralis* Wight. Journal of Agricultural Technology, 6: 179-191.

Murthy, K.S.R. and Kondamudi, R. 2011. Rapid shoot regeneration from thin cell layer explants of an endangered Asclepiad *Ceropegia spiralis*. Plant Tissue Culture and Biotechnology, 21: 63-73.

Nadkarni, K.M. 1976. Indian Materia Medica Popular Prakashan, Bombay, India. 1: 303-304.

Poi, C.D., Diss, G. and Freschi, L. 2010. Biodiversity matters in a changing world. Biology letters, rsbl 20100567.

Ramsay, M.M., Jackson, A.D. and Porley, R.D. 2000. A pilot study for the *ex situ* conservation of UK bryophytes. In: BGCI (ed) Eurogard-II European botanic garden congress. EBGC, Las Palmas de Gran Canaria: 52-57.

Sanikhani, M., Frello, S. and Serek, M. 2006. TDZ induces shoot regeneration in variousKalanchoe¨blossfeldianaPollen cultivars in the absence of auxin. Plant Cell Tissue and Organ Culture, 85:75-82.

Shen, X., Castle, W.S. and Gmitter, F.G. 2010. *In vitro* shoot proliferation and root induction of shoot tip explants from mature male plants of *Casuarina cunninghamiana* Miq. Hortscience, 45:797-800.

Sharma, D.K. and Sharma, T. 2013. Biotechnological approaches for biodiversity conservation. Indian Journal of Science Research, 4: 183-186.

Srinivasarao, C., Soni, A., Dulloo, M.E. and Naithani, S.C. 2010. Overcoming physiological dormancy in *Ceropegia odorata* seeds, an endangered rare species, with GA_3, H_2O_2 and KNO_3, Seed Science Technology, 38: 341-347.

Vinterhalter, D.V. and Vinterhalter, B.S. 1999. Hormone-like effects of sucrose in plant *in vitro* cultures. Phyton (Austria) Special issue: Plant Physiology, 39: 57-60.

Wyse, J.P., and Kennedy, K. 2009. The global strategy for plant conservation: a challenge and opportunity for the international community. Trends in plant science, 14: 578-580.

Yaseen, M., Ahmad, T., Sablok, G., Standardi, A. and Hafiz, I.A. 2013. Role of carbon sources for *in vitro* plant growth and development. Molicular Biology Reports, 40: 2837-2849.

19

An Overview on some Biotic and Abiotic Factors Affecting the Population Dynamics of Leucaena Psyllid, *Heteropsylla Cubana* Crawford (Homoptera: Psyllidae): Contributory Factors for Pest Management

Ahmed M. M. Ahmed[1,2] *; Luis Ramírez y Avilés[1];
Francisco J Solorio Sánchez[1], F.A. Al-Zyoud[3]
and Marcos Barros-Rodriguez[1]

[1]Department of Animal Nutrition, Faculty of Veterinary Medicine and animal science, University of Yucatan (UADY), Mérida, Yucatán, México, KM 15.5 carretera Mérida, Xmatkuil Apdo. P. O. BOX 4-116 Itzimna Merida. E-mail: saleh_moda@yahoo.com
[2]Department of Plant Protection, Faculty of Agriculture, Assiut University, Assiut P. O. BOX 71526, Egypt.
[3]Department of Plant Protection and Integrated Pest Management, Faculty of Agriculture, Mu'tah University, Karak P. O. BOX 61710, Jordan.
E-mails: raviles@uady.mx, ssolorio@uady.mx, firaszud@yahoo.com, ma_barrosr@yahoo.es
**Corresponding author*

SUMMARY

The present review was written with the objectives of providing basic information for pest control about leucaena psyllid, *Heteropsylla cubana* Crawford (Homoptera: Psyllidae) the most destructive insect pest of *Leucaena* sp. In the regard of pest control, the review focus on discussing the findings of the existing researches about the factors affecting the distribution and seasonal fluctuations of *H. cubana* in the world. These factors are biotic such: biological control agents, and abiotic such: climatic factors, leaves chemical composition and genetic control which significantly affect pest population. A basic understanding of the relationship of these factors with psyllid population is important in developing an integrated control strategy for psyllid in leucaena and determining the potential pest control needs under the given biotic and abiotic factors trend. Besides, develop some concepts of pest control which effect directly and indirectly on the survival and population dynamics of *H. cubana* around the world.

Keywords: Biological control; chemical composition; climatic factors; population fluctuation.

INTRODUCTION

Leucaena leucocephala (Lam.) de wit the fast growing tree has been known as a miracle leguminous tree for its long life, wood fuel source, shade crop for coffee and cacao. In the regard of animal production, leucaena is the important fodder crops around the world for its highly nutritional value compared to alfalfa (Shelton and Brewbaker, 1994; Tim, 2009).

In the past, the successful spread of the common leucaena in the tropical world may have been due to its vacancy of main diseases and pests behind, but recently leucaena psyllid, *H. cubana*: is a small yellow-green insect about 1-2 mm long and aphid-like sometimes called jumping plant lice adapted to feed on young growing shoots of Leucaena. The psyllid limits the continued use and expansion of leucaena trees. Moreover, it achieved international notoriety in the early 1980s when outbreaks and devastating defoliation of leucaena plantings were first detected in Florida and Hawaii with an extremely rapid rate of spread. Therefore, it was reported as a serious pest almost exclusively on *L. leucocephala* for the ability of completing its life cycle only on plants related to genus *Leucaena* and a few to a lesser extent on closely related mimosoid leguminous trees in different countries such as Florida and Hawaii (USA), Mexico, and Central America (Bray 1994). From there it spread rapidly and it is reasonable to assume that, sooner or later, all areas where leucaena is grown will be affected (Bray, 1994; Geiger *et al.*, 1995; Olckers, 2011).

Accordingly, results of pest control studies indicated that the variation in the extremely rapid or slow rate of psyllid population spread from area to another, is depending on a complex set of interactions between leucaena growth, climatic factors (particularly moisture), psyllid mortality factors, and other influence factors are largely responsible for its dispersal, together with movement by aircraft and other man-made transportation (Bray, 1994; Morris, 2000).

Therefore, many of psyllid's control programs concentrated on the contribution of factors affect the population of psyllid as a trend for pest control such as: predators, parasitoids, and entomopathogens; biology and behavior studies of psyllid; and producing new resistant accessions of leucaena (Geiger and Andrew, 2000; Singh, 2004; Finlay-Doney and Walter, 2005; McAuliffe, 2008; Lawrie, 2010; Shivankar, *et al.*, 2010).

This discussion of the existing researches to clarify the optimal use of the biotic and abiotic factors in controlling the population of leucaena psyllid. Besides, the explications of these effective factors, instead of using chemical control which can't be used with such animal fodder crop.

DISTRIBUTION OF LEUCAENA PSYLLID

In recent history, psyllid is a typical example of the risk of pest outbreaks in *L. leucocephala* plantations across the tropics and widely distributed by (Nair, 2007), known from its native habitat in Latin America, Florida in late 1983, Hawaii in April 1984, the Philippines and Taiwan, In 1986 it was noticed in Indonesia, Malaysia, Thailand, southern Myanmar, southern China and neighboring countries. Thus in less than 10 years, and subsequently this pest has spread from its native range in tropical America, across the Pacific to Asia and Africa – an unusual spread for an insect. In 1987 it appeared in the Andaman Islands in India and in Sri Lanka, and the next year in southern peninsular India. The westward movement continued, and in 1992 infestations were noticed in the African continent, in Tanzania, Kenya, Uganda, and Burundi and by 1994 in Sudan and Zambia (Geiger *et al.*, 1995; Ogol and Spence, 1997).

POPULATION DYNAMICS OF LEUCAENA PSYLLID

Psyllid populations are normally fluctuating quite widely over time. Different levels of pest abundance occurring in different parts in the same tree, according to the differences in the growth stages of *Leucaena* sp. Apparently, leucaena trees are vulnerable to high infestation of psyllids in the stage of producing new shoots and leaves because the young shoots of both native and giant varieties have been infested by high numbers of psyllid proportions in outbreaks (Fig. 1), with monthly mean numbers ranged from 0 to 44 nymphs and from 0 to 12.5 adults per shoot. Meanwhile, the ratios of nymphs per adult ranged from 3 to 13.4 and the population of psyllid has been found around the year (San Valentin, 1988), in India, the new shoots has been usually observed by heavy infestations; up to 3000 nymphs and adults per 15 cm of terminal shoot (Nair, 2007), and in Northern Thailand, the psyllid populations were highly boomed in juvenile leaves of leucaena, and then crashed when vegetative growth outpaced by Geiger and Andrew (2000).

BIOTIC FACTORS AND THE POPULATION DYNAMICS OF PSYLLID

Fecundity and other characteristics effecting on psyllids life cycle and its population dynamics

The population rates in all organisms are varies according to the interplay between ovipositing rates;

total number of eggs, adult's size, and periods of life cycle. These factors have been reviewed by Hodkinson (1974). The life cycle of *H. cubana* varies from location to another, with average (10 - 20 days) from egg to adult with several overlapping generations per year, and in tropical regions a population doubling time of 2.52 days to facilitate the massive population increase (Napompeth, 1994; Geiger *et al*., 1995; Olckers, 2011). In agreement with the observation that body size declined steeply during the population crash at the beginning of hot weather (NBCRC 1996; Napompeth 1998; Villacarlos *et al*., 1989). In Northern Thailand, there are a significant relationship between psyllid fecundity and various measures, such as the size of adult body and number of eggs laid over a 96-h period by Geiger and Andrew (2000). Average fecundity (eggs per day ± SD) was 51.0 ± 18.0 (n = 14) for non-desiccated females, with a maximum of 154. The regression during 96-h of total fecundity and measures of body size x (body length × head width^2) for 37 adult psyllid in laboratory studies was $y = 516.5x - 150.1$, $r^2 = 0.5042$, $P < 0.001$, $n = 37$, SE(y) = 102.9. The population dynamics of psyllid fall down under high temperatures, because it reduces both of body size (hence fecundity), and egg size by Geiger and Andrew (2000), and the total numbers of laid eggs on leucaena are the indicator of its upcoming arrival rates by Finlay-Doney and Walter (2005).

Biological control

Biological control agents such as predators, parasitoids, and entomopathogens are automatically regulating the population of any insect pests, and its absence cause severe damage to host plants especially for exotic pests in a new environment. The damage could be continuing until the populations of natural enemies build up. Exactly, that's what happened with psyllid at the time of its first spread, the psyllid caused severe and extensive damage to leucaena in Asia and Africa and in its native Central America, at a time when leucaena has been widely promoting in agroforestry (Rao *et al.,* 2000). Then, successive control by natural enemies (Bray, 1994; Shivankar *et al.,* 2010).

Predators and parasitoids

At any environmental system there are some natural enemies feeding on one or more stages in the life cycle of its pest. In this regard, there are considerable literatures on *H. cubana*'s natural enemies in native and exotic locations. Biological control efforts against psyllid were succeeded by using specific natural enemies such as the predators, *Curinus coeruleus* Mulsant and *Olla v-nigrum* Mulsant (Coleoptera: Coccinellidae), and the parasitoids, *Psyllaephagus yaseeni* Noyes (Hymenoptera: Encyrtidae) and *Tamarixia leucaenae* Boucek (Hymenoptera: Eupelmidae) by Shivankar *et al.* (2010). In contrast, some little countries rejected biological control due to its highly cost, and infective results on controlling psyllid populations (Shelton *et al*., 1998; Geiger and Andrew, 2000).

Figure. 1. Presents the infestation of high population numbers of Leucaena psyllids, *Heteropsylla cubana* Crawford infesting the new shoots of *Leucaena leucocephala* (Lam.) De Wit.

Successful trials of biological control agents. The larvae and adults of the common Ladybirds (Coleoptera: Coccenillidae) are good in this respect of psyllid control and greatly contributed in reducing psyllid populations in many localities (Funasaki, 1989). The predator, *C. coeruleus* is a handsome ladybird beetle about 4 mm long originally from Mexico. It is recognized by its iridescent blue-black color and orangish'Cheeks by Geiger *et al.* (1995). A significant efficiency in using *C. Coeruleus* was identified in Hawaii and many several countries as Southeast Asia in attacking psyllid larvae (Bray, 1994). In its native range it was found preying on psyllids, although *C. Coeruleus* was originally imported from Mexico to Hawaii in 1922 to control another pest, the coconut mealybug, *Nipacoccus nipae* Maskell (Homoptera: Pseudococcidae). But, it controlled another pest, leucaena psyllid and due to its entrance to a new environment it was rare and hardly found until the arrival of leucaena psyllid. Then, *C. Coeruleus*, has become the most abundant species of Coccinellidae in Hawaii and important predator for leucaena psyllid (Napompeth, 1994). Likewise, in Cuba a successful predatism model of *C. Coeruleus* had been done by Valenciaga *et al.* (1999) for controlling psyllid population with significant results presenting those Coccinellidae predators and other natural enemies are capable to keep the psyllid populations below the economic injury level (E.I.L.) in different agro ecosystems. In India, *C. coeruleus* was imported from Thailand to study the biological suppression of psyllids by Singh (2004), after well establishing of the predator in the fields, and about four months after release 20-30 adults per tree, the grubs consumed 10.630 eggs and 3.500 nymphs during their lifetime. The predator population starts building up in May and reached its peak in December to February, and the psyllid population declined from November and remained so up to March. Also, it has spread on its own to an area of more than 20 sq km in Bangalore and in about two years, after release, the population of psyllid was drastically reduced and the predator firmly established in the released sites. The reduction of pest population occurred on 20 marked trees per ha., after releasing 20 adults of predator per tree twice during July and October or about 1,000 to 5,000 beetles per hectare of leucaena trees, and weekly observation of *C. coeruleus* its superiority at released sites by Singh (2004).

The parasitoid, *Psyllaephagus* wasp which attacks psyllid with more specification to the genus *Heteropsylla*, but it hasn't released widely (Bray, 1994). Furthermore; the predator Ashy Gray Lady, *O. v-nigrum* was introduced from Mexico into Hawaii in 1908 for the controlling scale insects (Geiger *et al.*, 1995). The *O. v-nigrum* population was sparse, then significantly increased at the arrival of leucaena psyllid and actively attacked psyllids, but it wasn't abundant such the other Coccinellidae predator, *C. coeruleus*. The indigenous natural enemies *C. coeruleus* and *O. v-nigrum* suppressed psyllid populations during the early years of its presence in Hawaii, although they considered general pest feeders (Napompeth, 1994).

Furthermore, the parasitoid, *P. yaseeni*, was reared in Thailand at laboratories for the future field release of leucaena psyllid mummies. It has firmly established and widespread in all areas with significant results during the season of psyllid's peak from October to March, and the parasite densities were much higher than psyllid (Winotai, 1989; Napompeth, 1994).

Both of *C. coeruleus*, *P. yaseeni*, in Thailand, covered most of adjoin countries to Thailand that never introduced these biological control agents for the highly abilities of fast spread. Therefore, it can be concluded that *C. coeruleus* and *P. yaseeni* could widespread throughout continental southeastern Asia throughout the introduction (Napompeth, 1994).

The biological control was accepted to control psyllid in different countries for a number of reasons. Historically, most pests in order Homoptera easly can be defeated by biological control (Greathead, 1989; Waage, 1989). Benefits / cost ratios for biological control programs are high, rating from 1.5:1-150:1 (Norgaad, 1988; Tisdell, 1990). In case of its success, it could be a permanent control having the most ecologically sound option with an equal benefits at all economic starts.

Non successful trials of biological control agents. Biological control was rejected in little countries such as Vietnam for its highly cost in obtaining the specific natural enemies, and the efforts for moving these agents from country to country, in spite of identifying about 30 species of natural enemies including three species of spiders, and six species of fungi (Shelton *et al.*, 1998). Although, *C. coeruleus* had a partial success in Indonesia (Mangoendihardjo and Wagiman, 1989; Wagiman *et al.*, 1989), but it was failed to be established in many seasonal-dry areas (Funasaki *et al.*, 1989; Oka, 1989; Wagiman *et al.*, 1989).

The Coccinellid predator, *Olla v-nigrum* has also been introduced widely but it appears ineffective to control psyllid (Chazeau *et al.*, 1992). Also, the parasitoid *P. yaseeni* never succeeded in regulating psyllid population, because the high numbers of psyllid were fluctuated normally when *P. yaseeni* are existence and there was no evidence of its ability to control psyllid populations by Geiger and Andrew (2000).

In conclusion, the exploration of the psyllid's native area for further predators and parasites should continue. It is not an easy task, and any new organism will need to be carefully tested before release (Bray, 1994).

Entomopathogens agents in controlling psyllid

Fungi, as entomopathogen scored high percent of reduction (82%), in Taiwan, among the populations of psyllid during heavily infested leucaena plantation by using the fungicide *Beauveria bassiana* (Liu *et al.*, 1990). Another, epizootics of fungi *Entomophthora* sp., *Entomophaga* sp., and *Fusarium* sp. were observed in Northern Thailand causing significant reductions in psyllid populations (91% mortality), during two weeks in the period of maximum rainfall precipitation by Geiger and Andrew (2000).

ABIOTIC FACTORS AND THE POPULATION DYNAMICS OF PSYLLID

Climatic factors and the activity of psyllid

There were a number of attempts to assess the effects of environmental factors on population dynamics of psyllid (McAuliffe, 2008). The psyllid pressure In a humid-tropical site (1800 mm annual rainfall) in Indonesia, was negatively correlated with rainfall, and positively correlated with solar radiation and wind velocity (r^2= 0.51, 0.52 and 0.68, respectively) by Mangoendihardjo *et al.* (1990). The same result was emphasized in Australia, Queensland, when psyllid populations reduced during the periods of intense rain (McAuliffe, 2008).

In wet seasons, rainfall deterred the populations whereas strong winds during the dry season helped in disseminate and increase the numbers of psyllid (Mangoendihardjo *et al.*, 1990), because it was noticed in Kenya that the physical effects of rain washing the mature and immature stages of psyllids. Therefore, lowered psyllid damage occurred during wet seasons by Wandera and Njarui (1998). In the sub-humid, seasonally dry tropics of Thailand, psyllid peak numbers occurred at the end of the dry season and the beginning of the wet season (Napompeth, 1989). In cooler climates, psyllid numbers were high throughout the year at Southeast Queensland and upland regions in Hawaii (Austin *et al.*, 1996, Castillo *et al.*, 1997).

In the drier sub-humid environments of Central Queensland (650 mm annual rainfall), psyllids aren't seen during dry or windy weather, but populations build up quickly during rainy periods with high relative humidity. A 3-years evaluation of a large collection of *Lecaena* accessions at Los Banos, Philippines (humid tropical site with 2100 mm annual rainfall) and at Brisbane, Australia (sub-tropical with summer dominant rainfall of 1500 mm annually) provided some insights into the effects of climatic parameters on psyllid damage. Psyllid pressure was assessed by psyllid damage to susceptible *Leucaena* accessions. Plant responses to climatic changes may also affect the severity of psyllid damage, thereby confounding studies based solely on psyllid damage. However, psyllid damage scores (Wheeler, 1988) are highly correlated with psyllid populations and the minimum scores only occurred in the absence of psyllids by Bray and Woodroffe (1988).

At subtropical Brisbane, psyllid pressure was high throughout summer, autumn and winter but was consistently low during the spring, a season associated with low rainfall and relative humidity, cool nights and warm days. At this site, significant positive correlations were obtained between psyllid damage and mean minimum temperature (r^2= 0.30), mean maximum temperature (r^2= 0. 17) and mean daily temperature (r^2= 0.22), indicating that psyllids may be favored warm temperatures. The study also showed that there was little or no psyllid damage when mean minimum temperatures were less than 10 °C. These findings confirmed those of (Austin *et al.*, 1996) in Florida where low mean daily temperatures below 12 °C were associated with low psyllid populations. At tropical Los Banos, a negative significant relationships was obtained between psyllid damage and mean daily maximum temperature (r^2=0.62), mean daily radiation (r^2=0.46) and mean daily temperature (r^2=0.42). The results showed that psyllid damage was low at maximum temperatures above 33°C and confirmed the findings of (Patil *et al.*, 1992) who identified an upper developmental temperature range for psyllids of 30-35 °C under laboratory conditions. However, relative humidity had no discernible effect on mortality (Baker *et al.*, 1993), none psyllid matured in 2 replicates when mean temperatures inside the field cages were 29.5 and 29.1 °C and maximum temperatures over 36 °C. A regression of development rates on mean temperatures yields a lower temperature threshold of 9.68 °C. Furthermore, psyllid populations at valley sites fell dramatically at the onset of the hot season, when mortality and desiccation of adult psyllids were widespread.

In tropical countries such as Mexico, Thailand, Papua New Guinea and northern Australia, the psyllid is most active during the cooler months (Napompeth, 1994; Geiger *et al.*, 1995; Geiger and Gutierrez, 2000). The insect also has a distinct upper thermal limit and numbers often decline substantially during warmer periods (Napompeth, 1994).

Generally, psyllid population is affected by temperature, moisture, humidity and exposure to wind

(Geiger and Andrew, 2000; McAuliffe, 2008) and the ups and downs of the psyllid populations are related to an optimum cooler temperature range and the availability of tender shoots in Hawaii. Also, psyllid damage is a result of interaction between several different climatic factors (Napompeth, 1994). But in fact, there was a little agreement between different studies in the effective climatic factor among weather factors on the dynamics of psyllid during correlation studies between climatic factors and psyllid pressure (Mullen and Shelton, 1998). Furthermore, the cool or dry seasons aren't the reason behind the highly psyllid damage, as a result of the observations of two experimental years in Kenya that the high level of psyllid pressure occurred due to the decline in natural eneimes numbers or the site climatic factors (Wandera and Njarui, 1998).

Chemical compositions of leaves and shoots

The effects of mimosine, tannins, phenols, and fibers as chemical compositions on animal food production were taken in consideration by researchers, especially mimosine and tannins (Wheeler *et al.*, 1995) because, it was noticed in Australia that *Leucaena* was free from insect pests infestation in the past due to the insecticidal properties of the mimosine content in the growing young leaves. The mimosine content varies from tissue to tissue inside the same tree from 8-12% in actively growing shoots, from 4-6% in young leaves, and 4-5%, in young pods and seeds. These mounts of mimosine enable *Leucaena* cultivars to show different levels of resistance (Norton *et al.*, 1995; Shelton and Jones, 1995), as well as tannins protect plants from psyllid attack, but the immediate mechanism for psyllid resistance in some leucaena species still need further studies to be understood (Griffith, 1991; Wheeler *et al.*, 1995).

Therefore, psyllids resistance in Leucaena accessions is generally related to high tannin and fiber contents in Australia (Elder *et al.*, 1998; Shelton and Jones, 1995), because psyllid-tolerant species such *L. pallid* and *L. diversifolia* contain high levels of tannins and fibres than the susceptible *L. leucocephala* (Shelton and Jones, 1995). There isn't a wide variation between accessions within a species in tannin content (e. g. *L. diversifolia*), but tannin content vary throughout the year within an accession, and the varieties which have high content of tannins such as *L. lanceolata* shows different levels of resistance and wasn't vulnerable to psyllid infestation, but in case of *L. collinsii*, which has a little tannins content shows moderate or high susceptibility to psyllid attack in Vietnam (Wheeler *et al.*, 1995).

Leucaena species and variety selection

Few studies included details in using techniques for pests and diseases control (Schroth *et al.*, 2000). Most of researches results indicate that psyllid infestation problems were particularly severe for the narrow genetic base of different leucaena accessions (Rao *et al.*, 2000). Therefore, responses of *Leucaena* sp. toward psyllid infestation are varied from highly susceptible to highly resistant (Mullen *et al.*, 1998). The apparent resistance between psyllid and Leucaena was investigated in some Leucaena varieties (Finlay-Doney and Walter, 2005), and several Leucaena accessions (Ibrahim *et al.*, 1998, Jones; 1998, Mullen and Shelton, 1998).

In general, *leucaena* species are equally vulnerable to psyllid infestation because of the narrow genetic base of the 'Hawaiian-type' stocks that comprised most stands in Hawaii (Morris, 2000; Olckers, 2011), but it was found during a complete range study of psyllid responses to different Leucaena species that the highly resistant species were (*L. collinsii subsp. collinsii, L. conferticapitula, L. esculenta subsp. esculenta and L. matudae*) and the highly susceptible were (*L. leucocephala and L. multicapitula*), with considerable variation, both between and within species (Shelton *et al.*, 1998). The best traits of all leucaena commercial varieties in The United States of America was Wondergraze combines for its excellent growth under psyllid insect attack (Lawrie, 2010). However, in Hawaii results of crossing *L. leucocephala* with *L. pallida* produced a resistant hybrid (KX2- Hawaii) tolerates cool weather, has psyllid resistance, and low mimosine content by (Shelton, 1998, Tim, 2009). Therefore, it was confirmed that, KX2-Hawaii was resistance for psyllid by the Australia Council of International Agricultural Research (Mullen and Shelton, 1998; Shelton and Brewbaker, 1994; Brewbaker, 2008). The experimental trials for introducing a resistant *Leucaena* sp. was conducted in Queensland, when psyllid resistance genes have been moved from *L. pallida* into an elite *L. leucocephala* intraspecific hybrid through a process of cross pollination and repetitive backcrossing. Then, the variety exhibits high levels of resistance to psyllid damage. Also, there are a relationship between psyllid resistance and the volatile substance (Caryophyllene) in three species of Leucaena (*L. leucocephala*, *L. pallida* and their hybrids) in Queensland, (Finlay-Doney and Walter, 2005). The difference between three varieties when caryophyllene proportions dropped dramatically was significant. Another study on resistant accessions, elsewhere in Australia, reported that Tarramba (K636) was the least resistant cultivar of any of *L. leucocephala* cultivars, as there is no evidence that the new cultivar Tarramba (K636) was more psyllid resistance than other commercial *L.*

leucocephala cultivars (e.g. Cunningham and Peru). This contrast with the results of (Bray, 1994; Gieger *et al.*, 1995) who used damage rating in their assessment method. The lake of Tarramba (K636) resistance returns to its greatest ability to growth by lateral branching, this observation indicates that Tarramba can be severely attacked by psyllid in coastal central Queensland. Cunningham variety is a well adapted to the dry seasonally tropical environment under heavy attacks of psyllids, because it has high leaf: stem ratio and high leaf density and should be considered as a useful parent for crossing with other species to produce adapted, psyllid-tolerant and nutritious hybrids in Australia (Jones, 1998; Jones *et al.*, 1998) whom stated that, Cunningham variety gave higher steer gains and psyllid-resistant and showed more psyllid tolerant than cv. Tarramba. Nevertheless, the magnitude of the superior steer performance on cv. Cunningham was greater in the absence of any psyllid damage. Good resistance was also found in some, but not all, accessions of *L. collinsii*, *L. pallida* and *L. trichanadra*, unfortunately, there was no evidence of resistance in *L. leucocephala* although there was variation in degree of susceptibility, and the mechanisms of psyllid resistance in Leucaena remain unresolved (Shelton *et al.*, 1998). The serious damage of psyllid occurred in the susceptible lines with large infestation differences among *Leucaena* accessions the most tolerant lines were *L. trichandra* OFI 53/88, CPI 46568; *L. diversifolia* CPI 33820; *L. esculeuta* OFI 47/87; *L pallida* CSIRO composite, OFI 79/92, and the *L. pallida* x *L. leucocephala* hybrid UQ118. At these times, the two *L. leucocephala* cultivars were severely affected, with cv. Cunningham the most susceptible accession, having scores 1 to 2 units higher than cv. Tarramba, in Australia by (Jones, 1998).

CONCLUSION

This review discusses the findings of the existing researches about Leucaena psyllid, *Heteropsylla cubana* Crawford (Homoptera: Psyllidae) partly in the tropical regions in order to protect the most important forage tree, *L. leucocephala* which play a valuable role in the world of agriculture over a long period of time due to their multifaceted value, increasing and diversifying use of this wonderful species. However, the future use of *L. leucocephala* is in dangerous for the invasion of psyllids. On an optimistic view, several observations worthwhile remembering according to the discussed researches. Firstly, the use of biological control agents predators, parasitoids and entomopathogens in order to increase populations of natural enemies is required. The adults of the predator *C. coeruleus* greatly contributed to population reduction of psyllid in many localities, 20-30 adults per tree consumed 10.630 eggs and 3.500 nymphs of psyllids during their lifetime. Furthermore, using parasitoids such as *P. yaseeni* has firmly established and widespread in all areas with significant results during the season of psyllid's peak. Additionally, the epizootics of entomopathogenic fungi caused significant reductions in psyllid populations (91% mortality), and 82% of reduction in psyllid populations occurred during heavy infestations after applying the fungicide of *Beauveria bassiana* for biological control. Secondly, using the resistant hybrids with taking in consideration the amounts of leaf chemical composition content for each hybrid, especially tannin content which decreases psyllid numbers. Thirdly, planting the suitable cultivar or hybride under the optimum climatic factors as climatic conditions have the authority affect on psyllid population dynamics.

REFERENCES

Austin, M.P., Williams, M.J., Hammond, A. E., Frank, J.H., Chambliss, C.G. 1996. Psyllid population dynamics and plant resistance of leucaena selections in Florida. Tropical Grasslands 30: 223-228

Baker, P.S., Patil, N.G., Ooi, P.A.C., Murphy, S.T., Cross, A.E. 1993. Demographic parameters of the leucaena psyllid *Heteropsylla cubana* (Crawford) under various constant temperature and relative humidity regimes. CAB, Walllingford, UK.

Bray, R.A. 1994. The leucaena psyllid. In: Gutteridge R.C. and Shelton, H.M. (eds.). Forage Tree legumes in Tropical Agriculture. CAB International, Oxford. pp. 283-291.

Bray, R.A., Woodroffe, T.D. 1988. Resistance of some Leucaena species to the leucaena psyllid. Tropical Grasslands 22, 11–16.

Brewbaker, James, L. 2008. Registration of KX2-Hawaii, Interspecifi c-Hybrid Leucaena. The Journal of Plant Registrations 2, 190–193.

Castillo, A.E., Cuyagan, O.E., Fogarty, S., Shelton, H.M. 1997. Growth, psyllid resistance and forage quality of *Leucaella leucocephala , L. pallida, L. diversifolia* and the Fl hybrid of *L. leucocephala* x *L. pallid* .Tropical Grasslands 31, 188-200.

Chazeau, J., Capart, I., de Larbogne, L.B. 1992. Biological control of the leucaena psyllid in NewCaledonia: The introduction of *Curinus coeruleus* (Coccinellidae). Leucaena Research Reports.13: 59-61.

Elder, R.J., Middelton, C.H., Bell, K.L. 1998. *Heteropsylla cubana* Crawford (Psyllidae) and *Coccus longulus* (Douglas) (Coccidae) infestation on lucaena species and hyprids in costal centeral Queensland. Australian Journal of Entomology 37: 52-56.

Finlay-Doney, M., Walter, G.H. 2005. Discrimination among host plants (Leucaena species and accessions) by the psyllid pest *Heteropsylla cubana* and implications for understanding resistance. Agricultural and Forest Entomology 7: 153–160.

Funasaid, G. 1988. Status of *Psyllaephagus* sp. nr. *rotundiforinis* (Howard) (Hymenoptera: Encyrtidae) in Hawaii. Leucaena Research Reports. 9:14

Funasaki, G.Y., Lai, P.Y., Nakahara, L.M. 1989. Status of natural enemies of *Heteropsylla cubana* Crawford (Homoptera: Psyllidae) in Hawaii. In: Napompeth B. and MacDicken, K.G. (eds.). Leucaena Psyllid: Problems and Management. Proceedings Bogor, Indonesia. Winrock International, Bangkok, Thailand.

Geiger, C.A. and Andrew, P. Guterrez. 2000. Ecology of *Heteropsylla cubana* (Homoptera: Psyllidae): Psyllid Damage, Tree Phenology, Thermal Relations, and Parasitism in the Field. Environmental Entomology. 29: 76-86.

Geiger, C.A., Napompeth, B., Van Den Beldt R.J. 1995. An update on the status of the leucaena psyllid in Southeast Asia. In: Shelton H.M., Piggin, C.M. and Brewbaker, J.L. (eds.). Leucaena: Opportunities and Limitations. Canberra: Australian Centre for Agricultural Research. pp. 8-125.

Greathed, D.J. 1989. Biological Control as introduction phenomenon: a preliminary examination of programmes against the Homoptera: The Entomologist 108, 28-37.

Griffith, D.W. 1991. Considered tannins. In: D`Mello, Duffs, J.P.F., C. M. and Duffs, J.N. (eds.). Toxic substances in crop Plants. Royal Society of chemists, Cambridge, UK. pp. 181-201.

Hodkinson ID. 1974. The biology of psylloidea (Homoptera): A review. Bull. Entomol. Res., 64, 325-339.

Ibrahim, T.M., Shelton, H.M., Mullen, B.F., Horne, P.M. 1998. Leucaena –Adaptation, Quality and Farming systems. In: Shelton H.M., Gutteridge, R.c., Mullen, B.F. and Bray, R.A. (eds.). Leucaena Germplasm Evaluation in North Sumatra. Proceedings in workshop held in Vietnam, 9-14 February. pp. 131

Jones, R.J. 1998. Leucaena – Adaptation, Quality and Farming systems. In: Shelton H.M., Gutteridge, R.c., Mullen, B.F. and Bray, R.A. (eds.). Evaluation of Leucaena Species in the Dry Tropics of N.E. Australia for Yield and Psyllid Tolerance. Proceedings in workshop held in Vietnam 9-14 February. pp. 158.

Jones, R.J., Galgal, K.K., Castillo, A.C., Palmert, B., Deocareza, A., Bolam, M. 1998. Leucaena – Adaptation, Quality and Farming systems In: Shelton H.M., Gutteridge, R.c., Mullen, B.F. and Bray, R.A. (eds.). Animal Production from Five Species of Leucaena. Proceedings in workshop held in Vietnam 9-14 February. pp. 251.

Lawrie Andrew. 2010. The Leucaena Network News. [Accessed 14 March 2014.] Available from URL: http://www.leucaena.net/leucaenanews.pdf.

Liu, S.D., Chang, Y.C., Huang, Y.S. 1990. Application of entomopathogenic fungi as biological control of Leucaena psyllid, *Heteropsylla cubana* Crawford (Homoptera: Psyllidae) in Taiwan. Plant Protection Bulletin Taipei 32: 49–58.

Mangoendihardjo Soeprapto, Wagiman, F.X., Andi Ihsyono, Y., Sujono Mohamad. 1990. Seasonal abundance of Leucaena psyllid population in Yogyakarta, Indonesia. In: Napompeth B. and MacDicken, K.G. (eds.). *Leucaena Psyllid: Problems and Management*) Proceedings of an international workshopheld in Bogor, Indonesia. Winrock International, Bangkok, Thailand. pp. 111-138.

Mangoendihardjo, S., Wagiman, F.X. 1989. Some experiences in mass breeding, release and evaluating the establishment of *Curinus coeruleus* Mulsant in Yogyakarta and Central Java. In: Napompeth B. and MacDicken, K.G. (eds.). *Leucaena Psyllid*: Problems and Management. Proceedings of an international workshopheld in Bogor, Indonesia. Winrock International, Bangkok, Thailand.

McAuliffe, M. 2008. Investment Opportunity: New PBR *Leucaena* variety. A full information package is available from Melissa McAuliffe. [Accessed 13 March 2014.] Available from URL: http://www.leucaena.net/pr_info_pack.pdf.

Morris, C. 2000. The leucaena psyllid finally appears (and disappears) in South Africa. Leucnet News 7: 28.

Mullen, B.F., Shelton, H.M. 1998. Leucaena – Adaptation, Quality and Farming systems In: Shelton H.M., Gutteridge, R.c., Mullen, B.F. and Bray, R.A. (eds.). Accession evaluation, Agronomic Evaluation of the Leucaena Foundation Collection: 1. Subtropical Australia. Proceedings in workshop held in Vietnam 9-14 February. pp. 107.

Mullen, B., Frank, V.E., Date, R.A. 1998. Specificity of rhizobial strains for effective N-fixation in the genus leucaena. Tropical Grasslands, 32: 110-117.

Nair KSS.2007. Tropical Forest Insect Pests. The United States of America by Cambridge University Press, New York, pp. 263.

National Biological Control Research Center (NBCRC).1996. Annual Report. Bangkok: National Biological Control Research Center of Thailand, Kasetsart University, Thailand.

Napompeth, B. 1994. Leucaena psyllid in the Asia-Pacific region: implications for its management in Africa. In: LeucaenaPsyllid: a Threat to Agroforestry in Africa. FAO Corporate Document Repository Food and Agriculture Organization of the United Nations, Rome, Italy. pp. 1–15.

Napompeth, B. 1998. Leucaena psyllid in Thailand – a country report. In: Napompeth B. and MacDicken K.G. (eds.). *Leucaena Psyllid: Problems and Management*. Proceedings of an International Workshop held in Bogor, Indonesia, 16-21 January, 1989. Winrock International, Bangkok, Thailand. pp. 45-53.

Norgaard, R.B. 1988. Economics of the Cassava mealy bug (*Phaenacoccus manihoti*, Homoptera: Pseudococcidae) biological control in Africa. Entomophaga 33: 3-6.

Norton, B.M., Lowry, B., McSweeny, C. 1995. Leucaena - opportunities and limitations In: Shelton H.M., Piggin, C.M., Brewbaker, J.L. (eds.). The Nutritive Value of Leucaena Species. pp. 103-108.

Ogol CKPO, Spence, J.R.1997. Abundance, population dynamics and impact of the leucaena psyllid *Heteropsylla cubana* Crawford in a maize-leucaena agroforestry system in Kenya. Insect Science and its Application 17: 183–92.

Oka, I.N. 1989. Progress and future activities of the leucaena psyllid research program in Indonesia. In: Napompeth B. and MacDicken, K.G. (eds.). *Leucaena psyllid: problems and management*. Proceedings of an international workshop held in Bogor, Indonesia, January 16-21, 1989. Winrock International, Bangkok, Thailand.

Olckers, T. 2011. Biological Control of *Leucaena leucocephala* (Lam.) de Wit (Fabaceae) in South Africa: A Tale of Opportunism, Seed Feeders and Unanswered Questions. African Entomology 19(2): 356-365.

Patil, N.G., Baker, P.S., Pollard, G.V. 1992. Demographic parameters of the Leucaena psyllid *Heteropsylla cubana* Crawford under various temperature and relative humidity regimes. Trinidad, CAB International Institute of Biological Control.

Rao, M.R., Singh, M.P., Day, R. 2000. Insect pest problems in tropical agroforestry systems: Contributory factors and opportunities for management. Kluwer Academic Publisher, printed in Netherlands. Agroforestry Systems 50: 243-277.

San Valentin, H. 1988. Seasonal abundance, distribution and host range of psyllid infesting ipil-ipil. [Accessed 15 March 2014.] Available from URL: http://erdb.denr.gov.ph/enr/i4d/i4d_fo_psy.pdf

Schroth, G., Krauss, U., Gasparotto, L., Duarte Aguilar, J.A., Vohland, K. 2000. Pests and diseases in agroforestry systems of the humid tropics. Agroforestry Systems 50: 204.

Shelton, A.M., Vandenberg, J.D., Ramos, M., Wilsey, W.T. 1998. Efficacy and persistence of Beauvaria bassiana and other fungi for control of diamondback moth on cabbage seedlings. Journal Entomol. Science 33: 142-151.

Shelton, H.M. , Jones, R.J. 1995. Opportunities and limitations in leucaena. In: Shelton H.M., Piggin C.M., Brewbaker J.L. (eds.). Leucaena - Opportunities and Limitations pp. Proceedings of workshop held in Bogor, Indonesia. ACIAR Proceedings No. 57. .pp16-23.

Shelton, H.M. 1998. Leucaena – Adaptation, Quality and Farming systems. In: Shelton H.M., Gutteridge, R.c., Mullen, B.F. and Bray, R.A. (eds.). The Leucaena Genus: New Opportunities for Agriculture (A Review of Workshop Outcomes). Proceedings in workshop held in Vietnam 9-14 February. pp.16.

Shelton, H.M., Brewbaker, J.L. 1994. *Leucaena leucocephala* – the Most Widely Used Forage Tree Legume In: Ross C., Gutteridge and Shelton, H.M. (eds.). Forage Tree Legumes in Tropical Agriculture. CAB, Walllingford, UK. pp 15-29.

Shelton, H.M., Gutteridge, R.C., Mullen, B.F., Bary. 1998. Leucaena – Adaptation, Quality and Farming systems. (eds.). Shelton H.M., Gutteridge R.c., Mullen B.F. and Bray, R.A. ACIAR Proceeding No. 86, ISBN 1-86320-232-3. pp.17-18.

Shivankar, V.J., RAO, C.N. 2010. Psyllids and Their Management. Pest Management in Horticultural Ecosystems 16: 1- 4.

Singh, S.P. 2004. Some Successes Stories in Classical Biological Control of Agriculture Pests in India. pp. 27 Asian-Pacific Association of Agricultural Research Institutions (APAARI) FAO-RAP, Maliwan Mansion 39 Phra Atit Road Bangkok 10200, Thailand.

Tim Motis .2009. Leucaena Tree and Sweet Corn Breeding Programs. EDN, ECHO development Notes. [Accessed 13 February 2014.] Available from URL:http://c.ymcdn.com/sites/www.echocommunity.org/resource/collection/CAFC0D87-129B-4DDA-B363-B9733AAB8F1/edn99.pdf.

Tisdell CA. 1990. Economic impact of biological control of weeds and insects. In: Mackauer M., Ehler, L.E. and Roland, J. (eds.). Critical Issue of in Biological control. . Intercept Press, Andover, Hants, UK. pp.301-316.

Valenciaga, N., Barrientos, A., Mora, C.1999. Performance of the beneficial entomofauna in

Leucaena leucocephala (Lam) de Wit areas. Cuban Journal of Agricultural Science 33: 319–24.

Villacarlos, L.T., Paglinawan, R.V., Robin, R.P. 1989. Factors affecting leucaena psyllid populations in Leyte, Philippines. In: Napompeth, B and MacDicken K.G. (eds.). Leucaena psyllid: problems and management. Proceedings in Bogor, Indonesia. Winrock International - F/FRED, Bangkok, Thailand. pp. 122-129.

Wagaa, J. 1989. Exploration for Biological Control agents of leucaena Psyllid in tropical America. In: Napompeth, B. and MacDicken K.G. (eds.). Leucaena Psyllid: Problems and Management. Proceedings in Bogor, Indonesia. Winrock International- F/FRED, Bangkok, Thailand. pp. 144-152.

Wagiman, F.X., Mangoendihardjo, S., Maahrub, E. 1989. Performance of *Curinus coeruleus* Mulsant as a predator against leucaena psyllid. In: Napompeth, B. and MacDicken K.G. (eds.). Problems and Management. Proceedings in Bogor, Indonesia. Winrock International- F/FRED, Bangkok, Thailand. pp. 163-165.

Wandera, F.P., Njarui, D.M.G. 1998. Leucaena-Adaptation, Quality and Farming systems. In: Shelton H.M., Gutteridge, R.c., Mullen, B.F. and Bray, R.A. (eds.). Psyllid Tolerance and Productivity of Leucaena Species and Hybrids in Semi- Arid Kenya. Proceedings in workshop held in Vietnam 9-14 February. pp. 138.

Wheeler, R.A. 1988. Leucaena psyllid tri al at Waimilnalo, Hawaii. Leucaena Research Reports 8: 25:29.

Wheeler, R.A, Norton, B.M., Shelton, H.M. 1995. Leucaena - opportunities and limitations. In: Shelton H.M., Piggin, C.M., Brewbaker, J.L. (eds.). Considered Tannins in Leucaena Species and Hybrids and Implications for nutritive Value. Proceedings of workshop held in Bogor, Indonesia. ACIAR Proceedings No. 57. pp. 112-116.

Winotai, A. 1989. Biological control of leucaena psyllid, *Heteropsylla cubana* Crawford (Homoptera: Psyllidae), in Thailand. Ph.D. Thesis, Kasetsart University, Bangkok. 128 pp.

In Vitro Culture of Rumen Ciliate Protozoa based on *Parmentiera Aculeate* Kunth Medium

Alejandro Ley de Coss[1], Juan F. Aguirre Medina[1]*, Consepción Arce Espino[1], Willians de León de León[2], Saúl Posada Cruz[1], René Pinto Ruíz[3], Francisco Guevara Hernandez[3], Susana G. Gayón Amaro[4]

[1]Faculty of Agronomic Sciences, Campus IV, Autonomous University of Chiapas (UNACH), Coast Highway and Station Huehuetán Railway, Huehuetán, Chiapas, Mexico. CP. 36670. Email: juanf56@prodigy.net.mx.
[2]Master in Tropical Agropecuary Production Student at UNACH.
[3]Faculty of Agronomic Sciences, Campus V, UNACH. Villaflores, Chiapas, Mexico.
[4]Faculty of Veterinary Medicine and Zootechny, Autonomous University of the State of Mexico, Toluca, Estado de Mexico.
**Corresponding author*

SUMMARY

The present research aims to measure the ciliated protozoa concentration, pH, the concentration of volatile fatty acids (VFA) and the *in-vitro* dry matter degradation in a culture environment based *Parmentiera aculeata* Kunth. The treatments were: T1, 300 µL of *Avena sativa*; T2, T3 and T4 had 150, 300 y 450 µL of *P. aculeata* Kunth, respectively; and T5, 300 µL of *Chenopodium ambrosioides* L. Every treatment was inoculated with ruminal fluid and incubated for 3, 6, 12, 24, 48 and 72 hours to 38±0.5 °C. Based on a randomly experimental design, the data was analyzed by GLM and the means were compared with Tukey ($p \leq 0.05$) while the protozoa concentration was analyzed with a rank test. There were differences in the protozoa concentration and pH between treatments; T3 and T4 had the largest amounts ($p \leq 0.05$) compared with the other treatments, especially with T5. There was no difference ($p \leq 0.05$) in the acetic acid concentration between T1, T2 and T3 compared with T4 at 72 hours and with T5 from 6 hours of incubation, besides for the concentration of butyric acid T5 had the lowest amount ($p \leq 0.05$) at 72 hours. The smallest concentration of total VFA was for T4 and T5 from 48 hours. *P. aculeata* Kunth used like subtract helps to preserve for 3 days a ciliated protozoa population of 10^4 without alterate the fermentation process.

Keywords: Agroforestry; Livestock; Animal nutrition; Rumen protozoa.

INTRODUCTION

In tropical and subtropical regions of Mexico and Central America are many species of forage plants (FP) which can increase the amount and availability of some of the nutrients of a ruminant diet (Sosa *et al.,* 2004). In previous studies with FP the balance energy/protein, the presence and consequences of anti-nutritionals factors (saponins, tannins, alkaloids, phenols, etc.) in ciliate protozoarian populations, the effect in the voluntary intake, palatability, digestibility of dry matter and the animal response has been shown (Sosa *et al.,* 2004).

The protozoarian species and its ruminal concentration is related with the nature of the diet, physiological conditions of the animal and the nutritional stress (Göçmen *et al.,* 2001); in addition of the time of the sample and the circumstances associated with the *in-vitro* process. Because of all this factors the amounts of ciliates that had been reported showed a variation between 10^5 to 10^6 cells mL^{-1} of the culture environment or ruminal fluid (Ley de Coss *et al.,* 2011; Kumar *et al.,* 2009). Concentrations greater than 10^4 ciliates per mL has been found in environments where the energy source was the *Avena sativa* L flour (Ley de Coss *et al.,* 2011a; 2011b), *Manihot esculenta* Crantz, *Musa* spp L., *Oryza sativa* L. (Coleman, 1981), *Triticum aestivum* L. hay (Micha£owski *et al.,* 2001) and *Dactylis glomerata* L. while the bacterial population was regulated ($<10^8$ cells mL^{-1}) by the addition of antibiotics against gram negatives (2,000 U mL^{-1} penicillin and 130 U mL^{-1} streptomycin) (Dehority, 2008). By other side, the positive effect of the addition of minerals based in K_2HPO_4, KH_2PO_4, $(NH_4)_2SO_4$, NaCl, $MgSO_4$, $CaCl_2{\cdot}H_2O$, Na_2CO_3 8%, L–cystein (dissolved in 2N NaOH) + Na_2S-$9H_2O$ with oxide-reduction indicators like resazurin and CH_3COONa 1.5%; to the environment was evaluated (Dehority, 2008; Ley de Coss *et al.,* 2011a; 2011b). Therefore, the objective of this study was to evaluate if a culture environment, whose only source of energy was given by the extract of the fruit of *P. aculeata* Kunth and *Chenopodium ambrosioides* L., would allow *–in vitro-* the viability of ciliates to a concentration (10^4) similar to the ruminal natural conditions, and make future evaluations about the capability of some plants to eliminate protozoarians.

MATERIALS AND METHODS

This investigation was conducted by the Animal Nutrition and Biotechnology Laboratory of the Faculty of Agriculture Science, Campus IV, UNACH, Chiapas, Mexico.

Culture medium and treatments

For the anaerobic culture environment (AMC, table 1) were added 5 mL of CH_3COONa 1.5 %. For the witness treatment (T1), 54 mL of AMC were added in vials of 100 mL with 150 µL of *A. sativa* L every 9 mL of AMC, 20000 UI of penicillin and 25 mg of streptomycin. For the treatments T2, T3 and T4, AMC was added with the same doses of antibiotics but, instead of *A. sativa* L; 150, 300 and 450 µL of the soluble extract (weight: volume) of 2.0 g of *P. aculeata* Kunth in 10 mL of distilled water were incorporated to the system of culture. In the T5 treatment, 300 µL of the soluble extract (weight: volume) of 2.0 g of *Ch. ambrosioides* L in 10 mL of distilled water were added to the medium.

Table 1. Anaerobic cultive medium (AMC) for the maintenance of ruminal ciliate protozoa.

Substance	Amount per 100 mL
Distillated water, mL	47.6
Clarified ruminal fluid (CRF), mL (1)	30.00
Mineral solution I, mL (2)	5.00
Mineral solution II, mL (3)	5.00
Sodium carbonate 8 %, mL (4)	5.00
Sulfide-cystein solution, mL (5)	2.00
Rezarsurin 0.1% solution, mL (6)	0.1
Sodium acetate 1.5 % solution, mL (7)	5.00
Trypticase-pepton, g	0.20
Levadure extract, g	0.10

1) Clarified ruminal fluid was filtrated and centrifuged to 17664 g for 15 minutes, and sterilized 20 minutes to 21 °C at 15 psi. (2) In every 1000 mL; 6 g K_2HPO_4. (3) In every 1000 mL of H_2O; 6 g KH_2PO_4, 6 g $(NH_4)_2SO_4$, 12 g NaCl, 2.45 g $MgSO_4$ and 1.6 g $CaCl_2{\cdot}H_2O$. (4) 8 g Na_2CO_3 per 100 mL of distillated H_2O. (5) 2.5 g L–cystein (in 15 mL 2N NaOH) + 2.5 g Na_2S-$9H_2O$ (in 100 mL of H_2O). (6) 0.1 mL resazurin in 100 mL suggested by Ley de Coss *et al* (2011a, 2011b). (7) 5 mL of CH_3COONa 1.5 % suggested by Dehority (2008).

Soluble extract obtention

P. aculeata Kunth fruits and *Ch. ambrosioides* L. forage were collected in the Soconusco area, Chiapas, Mexico. The samples were dehydrated by natural conditions and dried in a stove (VWR International, Sheldon, U.S.A.) at 40°C until 13% of moisture; grinded in a Willey mill (ED-5, Tomas-Willey Mill, U.S.A.) with 1 mm sieve and stored in amber glass containers. To get the soluble extract, 2.0 g dry matter of each plant suspended in 10 mL of distillated water were deposited in tubes of 18 x 150 mm; then this suspension was mixed 3 minutes in a vortex (Science

MED, MX-S, U.S.A.) and let it repose 5 minutes. The supernatant fluid represents the soluble extract.

Protozoa concentrate obtention

The ruminal fluid (FR) was collected through an esophagi cannula of the nearest reticule-rumen of a F1 bovine (zebu x Sweden; 450 kg LW) nourished with *Cynodon plectotaschyus* K. (Schum) Pilger and 3.0 kg d^{-1} of a supplement with 13% protein made with maize, sorgum, soybean paste, urea and a mix of: 1) macrominerals (%), Ca (17), (12), Mg (5), Na (7), Cl (10.59, K (0.04) and S (0.05); 2) vitamins (UI), A (350,000), D (150,000), E (150); 3) microminerals (ppm), Mn (4,000), F (2,939), Zn (6,000), Cu (1,000), I (500), Se (40), Co (60). This diet was offered for 7 days before collecting the ruminal fluid which was filtered using a 75 mm-diameter-filtration-conical-funnel (Mod. TS5190N11, Kimax, Mexico) with cotton crapes; and incubated at 38 ± 0.5°C under CO_2 conditions. After that, the protozoa concentrate was taken through a 500 mL-separation-glass-funnel (Mod. K29034f, Kimax, Mexico) for 60 minutes to had the protozoa sedimentation which were dissolved in the cultivate medium without carbohydrate source.

Inoculation and protozoa calculation

The treatments were inoculated under aseptic conditions in presence of CO_2 (95 % purity, INFRA®, Mexico) with 6.0 mL of protozoa concentrate which was obtained with the previous described method. The inoculated mediums were incubated (Felisa FE-133A, Feligneo, Mexico) to 38 ± 0.5 °C. When the incubation period ended (3, 6, 12, 24, 48 y 72 h), the inoculated mediums were homogenized and 0.5 mL were taken out with a pipette; the direct protozoa alive count was made with a phase-contrast microscope (40X, Biológico BX51, Olympus, U.S.A.) using the technique and calculation formulas described by Ley de Coss *et al.* (2011a; 2011b). Moreover, a sample of 1.0 mL of medium was extracted under anaerobic conditions to determinate pH with a pH-meter (Orion A250, Orion Research, Inc., U.S.A.); the samples were taken conserving temperature and anaerobic conditions with the purpose of reducing protozoa death. For the VFA analysis, 2.5 mL of cultivate medium were retired at 3, 6, 12, 24, 48 and 72 h of incubation and were centrifuged for 10 min at 17,664 g; 2.0 mL of supernatant fluid were mixed with 1 mL of 25%-metaphosphoric acid; the VFA were determinated by gasses-chromatography (Claurus 500, Perkin-Elmer™, U.S.A.) with the conditions described by Ley de Coss *et al.* (2011a; 2011b). Dry matter *in-vitro* degradability (DMiD) of fruits of *P. aculeata* Kunth was calculated with cultivate tubes (18 x 150 mm) where 0.3 ± 0.01 g DM were added and sterilized with an autoclave (Felisa, FE-397, Mexico), then 9.0 mL of cultivate medium were added to each tube under anaerobic conditions (presence of CO_2). The tubes were inoculated with 1 mL of CRF and incubated for 3, 6, 12, 24 and 48 h at 38 ± 0.5 °C; by the end of the incubation period, the amount of DM was estimated.

Experimental design and statistical analysis

The experimental design was completely randomized and the data of DMiD, VFA concentration and pH of the cultivate mediums was analyzed with GLM procedure of SAS, while protozoa concentration in the cultivate medium was studied through Kruskal-Wallis proof with independents ranges of Wilcoxon. All the means were compared with Tukey proof ($p \leq 0.05$) and the statistical package (13-11) was used in the complete study.

RESULTS

Protozoa concentration and dry matter degradability

Differences had been found ($p \leq 0.05$) in protozoa concentration (Table 2) between treatments; at 3h of incubation, the treatments T3 (300 µL) and T4 (450 µL) observed the largest amounts of alive cells, compared with the control (T1), T2 and T5. At 6 and 12 h post-incubation, the protozoa concentrations were bigger ($p \leq 0.05$) in T1 (control) and T2, while at the end of the incubation period; only T3 and T4 had the largest protozoa concentrations ($p \leq 0.05$). So, doses of 300 and 400 µL of extract of *P. aculeata* Kunth kept, under *in-vitro* conditions and 72 h, the biggest amounts of protozoa appertaining to any taxonomical order of ruminal protozoa.

The pH in the cultivate mediums were different ($p \leq 0.05$) since 12 h of incubation between treatments, the lowest pH kept in T3 until 48 h; however in the treatments whose been the greater protozoa amount, pH was higher phenomenon which involve that pH affected the increase of those microorganisms. By the way, the effect of *Ch. ambrosioides* L (T5), pH in the medium was lower ($p \leq 0.05$) affecting the growth of protozoa.

Dry matter *in-vitro* degradability (DMiD) of *P. aculeata* Kunth was of 24.21, 45.87 y 89.76 % at 6, 12 y 24 h of incubation respectively; after 24 h there was no residual material, so apparently was a total degradability.

Volatile Fatty Acids (VFA) concentration

The amount of acetate through the first 6 h of incubation did not change ($p > 0.05$) between treatments (Table 3); although, it was observed the indirect proportion between the amount of *P. aculeata* Kunth extract and the production of acetate, to greater amounts of *P. aculeata* Kunth extract, less acetate production. While the propionic concentration was similar ($p > 0.05$) between treatments, there was difference at 12 h between the control and treatments with *P. aculeata* Kunth about the butyric acid concentration; although, with the defauning plant was a smaller amount of that VFA for 72 h ($p \leq 0.05$), just at the ending of the incubation period. This information shows that *P. aculeata* Kunth extract reduces acetic and butyric concentration but does not modify the amount of propionic. The total concentration of VFA at 72 h was smaller ($p \leq 0.05$) in treatments 4 and 5; the last one had the *Ch. ambrosioides* L.

DISCUSSION

The concentrations and protozoa species were similar through the entire study; that is why the fruit that had been evaluated has potential as nutrients source, similar at the data reported by Coleman (1981) with the development *in vitro* of Order Entonidios protozoa in flour-based mediums of *Musa* spp. L. and *Oryza sativa* L.; when those mediums were substituted for *O. sativa* L. starch (4.6 x10^3 mL^{-1}) and *T. aestivum* L., observed the development of *Entodonium caudatum* Stein, *Epidinium ecaudatum* F. y *Ophryoscolex caudatus* Eberlein while in mediums made with de *Canavalia ensiformis* (L.) DC flour, growth (2.9 x10^3 ciliates mL^{-1}) *E. caudatum* y *E. ecaudatum caudatum* for 9 months (Williams y Coleman, 1992). According to Teferedegne *et al.* (1999), the ruminal microorganisms are adaptable to chemical substances with defauning capabilities, which involve the action of ruminal bacteria which at the same time, have repercussions on the FVA production because of the adaptation and metabolism of the chemical secondary substances (Ivan *et al.*, 2003).

The defauning effect of *Ch. ambrosioides* L. (T5) was reported by Ley de Coss *et al* (2011B); also mentioned 2.0 x10^4 protozoa mL^{-1} of medium when *A. sativa* L. and some tuber starch were used (Dehority, 2008; Ley de Coss *et al.,* 2011a). In this investigation, protozoa of all the taxonomic orders reported in the rumen were observed while Dehority, 2008 mentioned a medium where only growth *Epidinium cadatum* y *Entodonium caudatum.* All this information is important because one of the elementary factors for protozoa *in-vitro* cultivate is the change between saturated medium and a fresh one; besides from addition antibiotics to regulate growth and bacterial population; this allows realize studies about nutritional requirements and metabolic activity of this microorganisms (Onodera y Henderson, 1980; Martin *et al.,* 1999; Dehority, 2008).

Table 2. Protozoa concentration (10^4) in cultivate mediums with *P. aculeata* Kunth y *Chenopodium ambrosioides* L.

Treatment	Incubation period (h)					
	3	6	12	24	48	72
T1, Control	3.00b	4.40a	4.23ab	2.50c	2.00a	1.17b
T2, 150 μL‡	3.00b	3.53d	4.27a	2.17c	1.93a	1.20b
T3, 300 μL‡	3.40a	4.23c	3.87bc	3.90a	1.27bc	1.96a
T4, 450 μL ‡	3.83a	4.30b	3.30cd	2.77b	1.96ab	2.30a
T5, 300 μL§	2.87c	1.00e	0.67d	0.03d	0.02c	0.00d
SEM€	1.33	1.00	1.30	1.08	2.16	1.48

[a, b, c, d] Means with different letter in the same column are differents ($p \leq 0.05$); [€] Standard Error; [‡]Solution made with 2 g *P. aculeata* Kunth in 10 mL of water; § Solution made with 2 g *Ch. ambrosioides* L. in 10 mL of water.

Table 3. Volatile Fatty Acids (VFA) concentration in cultivate mediums with *P. aculeata* Kunth y *Ch. ambrosioides* L.

Time, hours	0	3	6	12	24	48	72
	Acetic acid, mM L^{-1}						
T1, control	22.90^{a}	23.75^{a}	23.93^{ab}	25.34^{a}	34.36^{ab}	31.63^{b}	42.25^{ab}
T2, 150 μL‡	22.51^{a}	24.95^{a}	23.91^{ab}	23.78^{a}	40.18^{a}	38.85^{ab}	46.42^{ab}
T3, 300 μL‡	23.96^{a}	26.08^{a}	25.04^{a}	21.34^{a}	42.31^{a}	42.61^{a}	55.77^{a}
T4, 450 μL ‡	18.25^{a}	23.80^{a}	22.70^{b}	23.12^{a}	25.33^{b}	22.77^{c}	25.30^{b}
T5, 300 μL§	20.23^{a}	19.89^{a}	18.67^{c}	16.78^{c}	16.22^{c}	15.55^{c}	19.23^{b}
SEM$^{€}$	4.08	1.06	0.73	2.98	4.45	3.97	8.23
	Propionic acid, mM L^{-1}						
T1, control	4.50^{a}	4.71^{a}	4.79^{a}	4.93^{a}	4.53^{a}	4.21^{a}	4.71^{a}
T2, 150 μL‡	4.41^{a}	4.86^{a}	4.67^{a}	4.14^{a}	4.04^{a}	4.09^{a}	3.03^{a}
T3, 300 μL‡	4.59^{a}	5.04^{a}	4.74^{a}	3.81^{a}	3.97^{a}	3.40^{a}	3.33^{a}
T4, 450 μL ‡	3.77^{a}	4.91^{a}	4.69^{a}	4.78^{a}	4.41^{a}	4.25^{a}	3.28^{a}
T5, 300 μL§	4.00^{a}	4.56^{a}	4.56^{a}	4.54^{a}	4.23^{a}	4.65^{a}	3.89^{a}
SEM$^{€}$	0.85	0.20	0.14	0.52	0.26	0.41	0.90
	Butyric acid, mM L^{-1}						
T1, control	1.87^{a}	1.97^{a}	1.97^{a}	2.02^{a}	2.03^{a}	3.63^{a}	4.96^{a}
T2, 150 μL‡	1.75^{a}	2.02^{a}	1.90^{a}	1.65^{a}	1.79^{ab}	2.68^{ab}	3.81^{ab}
T3, 300 μL‡	1.78^{a}	2.06^{a}	2.01^{a}	1.55^{a}	1.67^{b}	1.71^{bc}	3.49^{ab}
T4, 450 μL ‡	1.59^{a}	2.05^{a}	1.97^{a}	2.04^{a}	1.84^{ab}	1.27^{c}	1.68^{b}
T5, 300 μL§	1.76^{a}	2.34^{a}	1.00^{b}	1.02^{b}	1.34^{b}	1.12^{c}	0.78^{c}
SEM$^{€}$	0.37	0.09	0.12	0.21	0.33	0.41	1.23
	Total VFA, mM L^{-1}						
T1, control	29.27^{a}	30.43^{a}	30.71^{a}	32.3^{a}	40.93^{ab}	40.14^{ab}	56.26^{a}
T2, 150 μL‡	28.67^{a}	31.48^{a}	30.48^{a}	29.41^{a}	46.01^{a}	49.29^{a}	52.59^{a}
T3, 300 μL‡	33.33^{a}	33.19^{a}	31.80^{a}	26.71^{a}	47.95^{a}	47.72^{a}	67.26^{a}
T4, 450 μL ‡	23.62^{a}	30.76^{a}	29.36^{b}	29.94^{a}	31.59^{b}	31.63^{b}	30.26^{b}
T5, 300 μL§	25.99^{a}	26.79^{a}	24.23^{c}	22.34^{b}	21.79^{c}	21.32^{c}	23.9^{b}
SEM$^{€}$	5.29	1.35	0.90	3.76	4.44	4.62	8.68

$^{a, b, c,}$ Means with different letter in the same column are differents ($p \leq 0.05$); $^{€}$ Standard Error ‡Solution made with 2 g *P. aculeata* Kunth in 10 mL of water; § Solution made with 2 g *Ch. ambrosioides* L. in 10 mL of water.

Different nutrients sources (energy, protein and mineral) had been used in *in-vitro* cultivation technics of ruminal ciliate protozoa (Ivan y Entz, 2004; Dehority, 2008). In the present research the cultivate was obtained when the fruit of *P. aculeata* Kunth, *A. sativa* L. and a mineral mix were added (Table 1); in other reports, nutritional broths based in *Triticum aestivum* L. 1.5%, *Dactylis glomerata* L. 1%, solutions of *T. aestivum* L. 1.5% plus *Medicago sativa* L. 1 %, *Sorghum vulgare, Sorghum bicolor* (L.) Moench 5% plus *Panicum maximun* Jacq. 5% (Dehority, 1998; 2008) had been used. The antibiotics doses that were added to the cultivate medium allowed the development of Holotrica class protozoa (10^4 cells mL^{-1}), while Lee *et al.* (2000) mention that a dose of 0.1 mg mL^{-1} of streptomycin sulfate, penicillin and cloramphenicol keeps stable pH to 6.53, but the metabolic activity of ciliate protozoa was inhibited for the presence of fungi and bacteria in cultivate mediums.

About DMiD of *Parmentiera aculeata* Kunth fruit, rates from 87.8 to 100% of degradability has been reported (Moctezuma *et al.,* 1993); in the other hand García-Castillo *et al.* (2008) mention that *in-situ* degradability of *Parmentiera aculeata* Kunth fruit was about 33.3% at the first 24 h of introduced the samples in the rumen; although there is a significant difference between results because of the ripeness conditions of the fruit utilized: in this study, a mature fruit was used; while García-Castillo *et al.* (2008) used a green *P. aculeata* Kunth fruit, that is why the ripeness affected the dry matter degradability including the ruminal protozoa growth. The amounts obtained in this study can be related to the solubility of the dry-grinded fruit and the high content of organic matter (OM), especially soluble fiber. In relation to, García-Castillo *et al.* (2008) mention that *P. aculeata* Kunth fruit contents 14.7% DM, 91.2% OM, 2.8% CP and 82.5% fiber; about this results, Sosa *et al.* (2004) indicate that there is not a direct relation between the variation in the NDF content and DM*iv*D in forage plants; just the lignification rate could reduce the degradability of the dry matter.

When the amount of *P. aculeata* Kunth fruit extract increased in the medium, the production of acetic acid reduced, about this Ley de Coss *et al.,* (2011a) shown that defauning plants like *Ch. ambrosioides* L. reduces the acetic concentration in the medium, similar results were reported by Nagaraja *et al.* (1992) y Mendoza *et al.* (1993); but this result does not coincide with Ivan *et al.* (2004) whose study indicates that *Enterolobium cyclocarpum* (Jacq.) Griseb, defauning plant does not affect VFA concentration but reduces the amount of ciliate protozoa like a transitory effect. Abreu *et al.* (2004) indicate a smaller amount of acetic and larger of propionic acid in ovines nourished with *Sapindus saponaria* L. fruit, besides increase 67% protozoa concentration with a rate of 121 mM L^{-1} total VFA which are superior at the rate found in this study. On that score, Ivan *et al.* (2000) show the FVA concentration depends of the ciliate protozoa specie and if *Dasytricha* predominate, the amount of total VFA would be 73.5 mM L^{-1} and 88.1 mM L^{-1} if the predominant specie would be *Isotricha* in diets based in corn forage.

In tropical regions, in important to analyze nutritional alternatives for the balance of glucogenics and acetogenics fatty acids in animals because 90% of glucose comes from neogenesis, specially from propionic acid, and 70% of the energy comes from the VFA that are produced in the rumen (Dehority, 2003), using fruits provided of forage plants has allowed increase the amount of propionic, improving the balance of gluconeogenics fatty acids according to Navas-Camacho *et al.* (1994).

CONCLUSION

In conclusion, the hydro-soluble fraction of *P. aculeata* Kunth fruit allows to keep an approximate amount of 2.3 x10^4 protozoa per mL of medium without alterations on the pH range when the extract dose is superior to 300 µL. Only the largest extract applied dose reduced the total VFA concentration, particularly acetic and butyric acids without changes in propionic. The DM*iv*D was high from 24 h of incubation, which allowed disposing nutrients for the microorganisms. The medium would allow the evaluation of defaunating capability of forage plants which secondary compounds

Acknowledgments

To National Council of Science and Technology (CONACYT) for the investment of this investigation as a part of the project named "Estimation and ambient impact of carbon capture in *Elaeis guineensis* Jacq plantations in the State of Chiapas, Mexico" which is part of the Scientific Development to attend National Problems Project (PDCPN2013-01) and the scholarship of Willians de León de León (Scholarship holder 284222) to make his studies in the program of Master in Tropical Farming Production Sciences at the Autonomous University of Chiapas.

REFERENCES

Abreu, A., Carulla, J. E, Lascano C. E., Díaz, T. E., Kreuzer, M. and Hess, H. D. 2004. Effects of *Sapindus saponaria* fruits on ruminal fermentation and duodenal nitrogen flow of sheep. Journal of Animal Science; 82:1392-1400.

Coleman, G. S. 1981. Cultivation of rumen Entodiniomorphid protozoa on tropical animal feeds. Tropical Animal Production; 6:11-14.

Dehority, B. A. 2008. Improved in vitro procedure for maintaining stock cultures of three genera of rumen protozoa. Journal of Animal Science; 86:1395-1401.

Dehority, B. A. 2003. Rumen microbiology. Nottingham, UK. Nottingham: Univ. Press.

García C. G., Martínez T. J., Montañez V. O., Sánchez O. l., Posada C. S., Izaguirre F. F., Martínez P. G., 2008. Degradación ruminal de la materia seca del fruto cuajilote (*Parmentiera edulis*); Zootecnia Tropical. 26:1-8.

Göçmen, B., Dehority, B. A., Talu, G. H., Raslegdi, S. 2001. The rumen ciliate fauna of domestic sheep (*Ovis ammon aires*) from the Turkish Republic of Northern Cyprus. Journal of Eukaryotic Microbiology; 48(4):455-459.

Ivan, M., Entz, T., Mir, P. S, Mir, Z. Mcallister, T. A. 2003. Effects of sunflower seed supplementation and different dietary protein concentrations on the ciliate protozoa population dynamics in the rumen of sheep. Canadian Journal of Animal Science; 83:809–817.

Ivan, M., Koenig, K. M., Teferedegne, B., Newbold, C. J., Entz, T., Rode, l. M. Ibrahim M., 2004. Effects of the dietary *Enterolobium cyclocarpum* foliage on the population dynamics of rumen ciliate protozoa in sheep. Small Ruminant Research; 52:81-91.

Ivan, M., Neill, l., Forster, R., Alimon, R., Rode, l. M. Entz, T. 2000. Effects of *Isotricha*, *Dasytricha*, *Entodinium*, and total fauna on ruminal fermentation and duodenal flow in wethers fed different diets. Journal of Dairy Science; 83:776–787;

Kumar, S., Puniya, A. K., Puniya, M., Dagar, S. S., Sirohi, S. K., Singh, K. Griffth, G. W. 2009. Factor affecting rumen methanogens and methane mitigation strategies. World Journal of Microbiology and Biotechnology; 25:1557-1566.

Lee, S. S., Ha, J. K. and Cheng, K. J. 2000. Relative contributions of bacteria, protozoa, and fungi to in vitro degradation of orchard grass cell walls and their interactions. Applied and Environmental Microbiology; 66(9):3807–3813.

Ley de Coss A., Martínez T, J., Marroquín A, F. J., García C. C. G., Montañez V, O. D., Guerra M, E. 2011a. La capacidad desfaunante del extracto de plantas. Revista Científica FCV-LUZ; XXI (5):414-420.

Ley de Coss, A., Cobos P, M., Hernández S, D., Guerra M, E. 2011b. Formulación de un medio de cultivo anaerobio para protozoarios ruminales y valuación *in vitro* en la capacidad desfaunante del extracto de plantas. Revista Científica FCV-LUZ; XXI (1):43-49.

Martin, C., Devillard, E. Michalet D, B. 1999. Influence of sampling site on concentrations and carbohydrate-degrading enzyme activities of protozoa and bacteria in the rumen. Journal of Animal Science; 77:979–987.

Mendoza, G. D., Britton, R. A., Stock, R. A. 1993. Influence of ruminal protozoa on site and extent of starch digestion and ruminal fermentation. Journal of Animal Science; 71:1572-1578.

Micha£owski, T., Rybicka, K., Wereszka, k., Kasperowicz, A. 2001. Ability of the rumen ciliate *Epidinium ecaudatum* to digest and use crystalline cellulose and xylan for *in vitro* growth. Acta Protozoological; 40:203-210.

Moctezuma, S. A., Lerma D. C. E., Martinez, D. R., Salinas C, J., Yado, P. R. 1993. *In vivo* digestibility of complete sheep diets with various amounts of chote fruit (*Parmentiera edulis*). Reunión Nacional de Investigación Pecuaria. Jalisco, México; 27-30.

Navas C. A., Laredo, M. A., Cuestas, A., Ortega, O., Romero, M. 1994. Evaluation of tropical trees with hing or medium saponin content as dietary alternative to eliminate ciliate protozoa from the rumen. Proceedings of the Society of Nutrition Physiology; 3:204-212.

Nagaraja, T. G., Towne, G., Beharka, A. A. 1992. Moderation of ruminal fermentation by ciliated protozoa in cattle fed a high-grain diet. Applied and Environmental Microbiology. 58: 2410-2414.Norton, B. W. 1994. Anti-nutritive and toxic factors in forage tree legumes. En: Gutteridge RC, Shelton HM. Forage Tree Legumes in Tropical Agriculrure. Wallingford, UK: CAB international.

Onodera, R. and Henderson, C. 1980. Growth factors of bacterial origin for the culture of the rumen oligotrich protozoon *Entodinium caudatum*. Journal of Applied Bacteriology; 48:125-134.

Sandoval, C. A., Herrera, F. 2001. Cambios en la población de protozoarios ruminales por efecto de la inclusión de *Canavalia ensiformis* a la dieta de bovinos. Revista Biomédica 12:166-171. [En línea]; http://www.revbiomed.uady.mx/pdf/rb011233.pdf. [Fecha de acceso el 27 de agosto del 2012].

SAS. 2009. Statics. Versión 8.0 Edition. Cary (NC): SAS institute Inc.

Sosa, R. E., Pérez, D., Ortega, L., Zapata, G. 2004. Evaluación del potencial forrajero de árboles y arbustos tropicales para la alimentación de ovinos. Técnica Pecuaria en México; 42:129-144.

Teferedegne, B., Mcintosh, F., Osuji, P. O., Odenyo, O., Wallace, R. J., Newbold, C. J. 1999. Influence of foliage from different accessions of the sub-tropical leguminous tree, *Sesbania sesban,* on ruminal protozoa in Ethiopian and Scottish sheep. Animal Feed Science Technology; 78:11–22.

Williams, A. G., Coleman, G. S. 1992. The Rumen Protozoa. Brock Springer Serie in Contemporary Bioscience. Springer-Verlag, New York Inc. pp: 133-164.

Genetic Diversity and Sex Identification in *Genipa Americana* L.

Ana Veruska Cruz da Silva[1]*, Evandro Neves Muniz[1], Camila Santos Almeida[2], Marina Ferreira da Vitória[2], Ana da Silva Ledo[1], Marília Freitas de Vasconcelos Melo[3], Allivia Rouse Carregosa Rabbani[4]

1. Researcher at Embrapa Coastal Tablelands, 49025-040, Aracaju-SE, Brazil. Email: ana.veruska@embrapa.br
2. Federal University of Sergipe, 49100-000, São Cristóvão-SE, Brazil
3. Faculty of Agricultural Sciences, State University of Sao Paulo, 18610-307, Botucatu-SP, Brazil
4. Federal Institute of Education, Science and Technology of Bahia, 45810000, Porto Seguro-BA, Brazil
**Corresponding Author*

SUMMARY

Ten genotypes of *Genipa americana* L. located in Arauá (Sergipe, Brazil) were analyzed using RAPD markers. The genotypes were selected according to sex - five females and five males - in an attempt to find a molecular marker associated with the sex prediction. 74 DNA fragments were generated by 20 primers, 85% of which were polymorphic. Results show a quantitative genetic diversity value of 0.28 and a Shannon index of 0.42 in the population. The similarity among the specimens according to Jaccard's coefficient ranged from 0.22 to 0.75. The genetic diversity was relatively large among these individuals.

Key words: Tropical fruits; Genetic resources; Ecosystems.

INTRODUCTION

Genipa americana L. – Rubiaceae, commonly known as genipap, is a fruit species naturally occurring between Mexico and Brazil, with ecological, social and cultural importance. Its distribution is common in riparian areas, withstands long periods of immersion during the flood season the rivers. The species is highlighted by its multiplicity of uses. It is recommended for recovery of degraded areas, and its fruits are edible, with important quality attributes for foods industry, dye (Penalber *et al.*, 1996) and traditional medicine to treat anemia, icterus, asthma, and liver and spleen problems (Conceição *et al.*, 2011).

The species has high ecological plasticity occurring in several forest formations throughout tropical America, suggesting genetic diversity among individuals and populations, since differentiation between species is based on both genetic and environmental factors.

The sex determination in some fruits is still not well understood. The development of mechanisms of sex determination are extremely useful in plant breeding (Alstrom-Rapaport *et al.*, 1998). In papaya, Souder *et al.* (1996) constructed a genetic linkage map using RAPD and defined two markers flanking for determining locus about 7 cM. Has a many farmers dissatisfied by the high rate of male plants in the areas of genipap, but research on the species is still limited.

Besides the analysis of genetic diversity, the techniques based on PCR have used to detect markers associated with sex in dioecious and bisexual plants. RAPD has been used in several studies, as in Guggal (*Commiphora wightii* R.) (Sanghamitra *et al.*, 2010), *Hancornia speciosa* Gomes (Costa *et al.*, 2011), *Myrciaria tenella* O. Berg (Pinheiro *et al.*, 2011), *Cattleya labiata* (Pinheiro *et al.*, 2012) among many other species.

The goal of this study was to evaluate the genetic diversity and try to identify the sex of brands associated genipap in a natural population located in Arauá, Sergipe, Brazil.

MATERIAL AND METHODS

Previously characterized as to sex by local farmers, located in a natural population in the city - five male subjects (279, 281, 283, 285, 287) and five female individuals (280, 282, 284, 286, 288) were randomly selected of Arauá (latitude 11 ° 17'23" and longitude 37 ° 38'31"), Sergipe, Brazil. All subjects were georeferenced, and the healthy and ripe fruits were transported to the Seed Laboratory of Embrapa Coastal Tablelands (CPATC) in Aracaju-SE for the formation of seedlings in nursery.

DNA was isolated from 1g young leaves per plant as described by Doyle and Doyle (1990). For PCR-ISSR, 20 primers from Operon (Operon Technologies, EUA) and IDT (Integrated DNA Technologies, Germany) were used to screen for polymorphisms (Table 1). Polymerase chain reaction (PCR) amplifications were performed using a PTC-100 thermocycler (Programmable Thermal Controller, MJ Research Inc., Waltham, MA, United States), with reactions consisting of 5 min initial denaturation at 95°C, followed by 45 cycles of denaturation at 94°C for 1 min, primer annealing at 52°C for 1 min and extension at 72°C for 45 s, and one final extension step at 72°C for 10 min.

Fragments were visualized on 2% agarose gel [1 x TBE – 89 mM Tris, 89 mM boric acid, 2.5 mM ethylenediamine tetraacetic acid (EDTA), pH 8.3] in a horizontal electrophoresis system (Sunrise, Gibco BRL) run at a constant voltage of 100 V for 60 min. The gel was stained with ethidium bromide solution (5 mg/mL) for 30 min. ISSR amplification products were visualized under ultraviolet light using a Gel Doc L-Pix imaging system (Loccus Biotecnologia, Brazil).

Only data from intensely stained, unambiguous and clear fragments were used for statistical analysis. ISSR markers were scored as present (1) or absent (0). The polymorphic information content (PIC) was calculated according to Ghislain (1999) and the marker index (MI) was determined as described by Zhao *et al.* (2007). A data matrix of ISSR scores was generated, and similarity coefficients were calculated using Jaccard's arithmetic complement index. A dendrogram was constructed using the Unweighted Pair Group with Arithmetic Mean (UPGMA) cluster algorithm. To determine dendrogram robustness, the data were bootstrapped with 10.000 replicates using FreeTree software (Pavlicek *et al.*, 1999). Principal coordinates analysis method (PCoA) was performed using the software package XLSTAT (http://www.xlstat.com/) based on the calculated Jaccard's similarity coefficients. The Shannon diversity index (I) and genetic diversity (G) (Brown and Weir, 1983) were calculated using Genalex v.6.3 (Peakall and Smouse, 2005).

RESULTS AND DISCUSSION

The primers used allowed the molecular characterization and study of the genetic diversity among individuals of genipap. However, no molecular markers associated with sex was observed. The 20 primers used in generating a total of 74 bands, of which 63 were polymorphic (85%) (Table 1). The primer showed the highest number of polymorphic bands was the X01 (6), while the A15 and W2 showed no polymorphic bands. The PIC value ranged from 0.00 to 0.48 and IM from 0.41 to 2.00. For the values of genetic structure, the Shannon index (I) averaged 0.42, and the genetic diversity index (G) presented a mean value of 0.28.

The percentage of polymorphism was found to consistently with Rabbani *et al.* (2012), using 12 primers in a natural population in the riparian area of the Lower São Francisco (Sergipe State) gave 74 polymorphic fragments, a total of 88 (84%).

Table 1. Number of polymorphic fragments (NPF), % polymorphism, polymorphic information content (PIC), marker index (MI) and genetic diversity (G) from 10 Genipa Americana L. individuals from Sergipe, Brazil.

Primer	Sequence	NPF	%P	PIC	MI	I	G
A2	TGC CGA GCT G	5	60	0,27	0,54	0,39	0,27
A3	AGT CAG CCA C	4	50	0,23	0,46	0,33	0,23
A4	AAT CGG GCT G	5	100	0,35	1,76	0,52	0,35
A8	GTG ACG TAG G	4	100	0,48	1,92	0,67	0,48
A9	GGG TAA CGC C	5	100	0,30	1,50	0,47	0,30
A10	GTG ATC GCA G	5	100	0,40	2,00	0,57	0,40
A11	CAA TCG CCG T	5	60	0,19	0,57	0,30	0,19
A13	CAG CAC CCA C	3	33	0,14	0,14	0,20	0,14
A15	TTC CGA ACC C	0	100	0,00	0,00	0,00	0,00
A16	AGC CAG CGA A	3	100	0,28	0,84	0,44	0,28
A18	AGG TGA CCG T	3	100	0,29	0,87	0,45	0,29
A20	GTT GCG ATC C	3	100	0,44	1,31	0,63	0,44
IDT18	GGA GGA GAG G	3	33	0,12	0,12	0,22	0,12
B18	CCA CAG CAG T	4	100	0,41	1,65	0,60	0,41
X01	CTG GGC ACG A	6	100	0,27	1,60	0,43	0,27
X03	TGG CGC AGT G	4	100	0,34	1,37	0,52	0,34
W2	ACC CCG CCA A	0	100	0,00	0,00	0,00	0,00
W4	CAG AAG CGG A	4	100	0,32	1,28	0,48	0,32
W13	CAC AGC GAC A	4	100	0,45	1,79	0,64	0,45
K20	GTG TCG CGA G	4	100	0,33	1,34	0,50	0,33
Total		4	87	0,28	1,05	0,42	0,28

The Shannon index (I ≈ 0.42) can be considered low, as it was below 0.5. The 'I' varies from 0 to 1, and it is considered that the closer the value is to zero, the lower the diversity. The values found in this study are similar to those obtained by Allnut et al. (1999) for other tree species *Fitzroya cupressoides* with 0.42 to 0.56 and *Swietenia macrophylla*, with 0.27 to 0.41 (Gillies *et al.*, 1999). The mean genetic diversity (G ≈ 0.28) obtained are in agreement with those found for other species. For *Trichilia pallida* Swartz ranged from 0.27 to 0.33 (Zimback *et al.*, 2004) and *Aspidosperma polyneuron*, the average range is 0.28 (Torezan *et al.*, 2005). The nonzero valuesfor L indicate that there is an excess of homozygotes or heterozygotes in the sample group (Barreira *et al.*, 2012).

The high degree of polymorphism identified among the individuals studied is the heterozygosity for this marker, which can contribute to infer the genetic relationships (Table 2). According to Jaccard coefficient, the lowest genetic similarity was found between 285 and 279 individuals (0.22) and highest among 288 individuals and 287 (0.78).

There was the formation of three groups. The first consisted of four individual (280-283), the second with six (284-288) and third with only one individual (279), which was the most divergent within the population (Figure 1).

The similarity data are in accordance with the ACoP. In this study, the efficiency of RAPD markers was confirmed, because it was found that the percentage of variability accumulated in the first two axes was 52.26%, which confirms the genetic diversity among the individuals in this study. A bidimensional graphic was generated, and the parameters defined in the analysis reinforced the positioning of the different genotypes (279) (Figure 2).

It was possible to verify a natural clustering structure, depending on the genetic similarity between them evaluated. The use of more than one clustering method, due to differences in the ranking, optimization and ranking the groups, allows classification is complemented by the criteria that each technique uses, and prevents erroneous inferences are adopted in the allocation of materials, within a given genotype subgroup (Silva *et al.*, 2012). To genipap, the most divergent individual was 279.

Table 2. Genetic similarity (%) (below diagonal) and estimated error (above diagonal) among 10 *Genipa americana* L. individuals.

	279 (M)	280 (F)	281 (M)	282 (F)	283 (M)	284 (F)	285 (M)	286 (F)	287 (M)	288 (F)
279 (M)	-	0,33	0,25	0,27	0,32	0,25	0,22	0,28	0,25	0,23
280 (F)	0,33	-	0,74	0,63	0,50	0,50	0,60	0,46	0,52	0,56
281 (M)	0,25	0,74	-	0,64	0,51	0,51	0,58	0,47	0,53	0,60
282 (F)	0,27	0,63	0,64	-	0,75	0,60	0,65	0,52	0,57	0,66
283 (M)	0,32	0,50	0,51	0,75	-	0,50	0,54	0,46	0,57	0,58
284 (F)	0,25	0,50	0,51	0,60	0,50	-	0,66	0,63	0,61	0,67
285 (M)	*0,22*	0,60	0,58	0,65	0,54	0,66	-	0,53	0,58	0,71
286 (F)	0,28	0,46	0,47	0,52	0,46	0,63	0,53	-	0,68	0,60
287 (M)	0,25	0,52	0,53	0,57	0,57	0,61	0,58	0,68	-	0,78
288 (F)	0,23	0,56	0,60	0,66	0,58	0,67	0,71	0,60	*0,78*	-

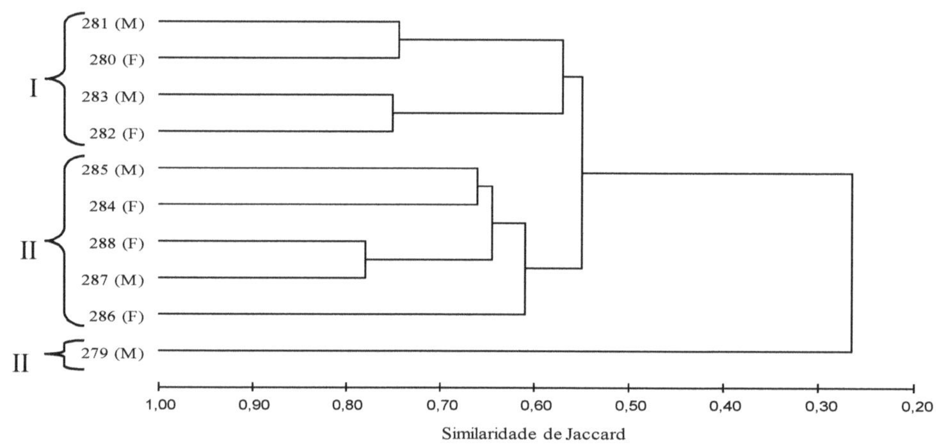

Figure 1. Genetic similarity dendrogram generated by Jaccard's similarity coeficiente and bootstrap analysis (10.000 x) and groups (I, II, III) for the 10 *Genipa americana* L. individuals using RAPD markers.

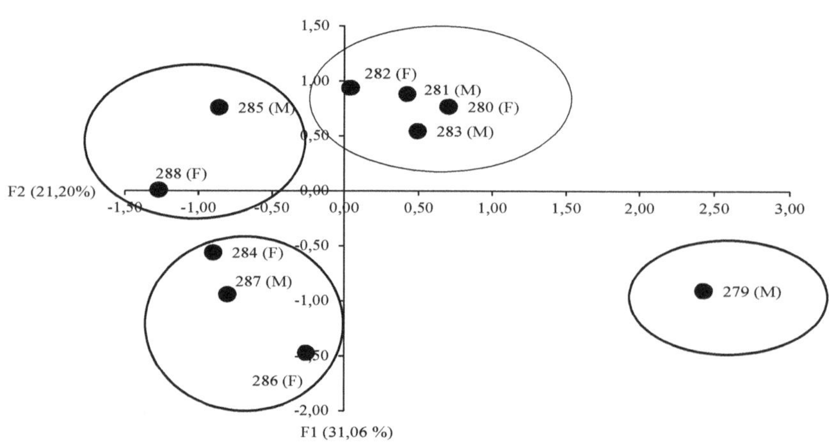

Figure 2. Principal Coordinates Analysis (PCoA) of 10 Genipa Americana L. individuals collected from natural population in Sergipe, Brazil based on RAPD markers data./HV

CONCLUSION

This is the first report to predict sex genipap reported in northeastern Brazil. Unable to identify a specific molecular tag to sex using the primers tested, but our study is a contribution to the characterization of native genotypes. The genetic variation and the genetic relationships among wild genipap were efficiently determined using RAPD markers. The discrimination of genipap from Sergipe (Brazil) and identification of genotypes more genetically divergent may contribute to development of strategies for the implementation of conservation, breeding programs efforts and commercial exploitation.

REFERENCES

Allnut, T.R.; Newton, A.C.; Lara, A.; Premoli, A.; Armesto, J.J.; Vergara, R. and Gardner, M. , 1999. Genetic variation in *Fitzroya cupressoides* (alerce), a threatened South American conifer. Molecular Ecology, Malden, 8 (6): 975-87.

Alstrom-Rapaport, C; Lascoux, A.; Wang, Y.Y.C.; Roberts, G. and Tuskan, G.A. 1998. Identification Of A Rapd Marker Linked To Sex Determination In The Basket Willow (*Salix Viminalis* L.). Journal Of Heredity, 89 (1): 44 49.

Barreira, S.; Sebbenn, A.M.; Scolforo, J.R.S. and Kageyama, P.Y. 2006. Diversidade genética e sistema de reprodução em população nativa de *Eremanthus erythropappus* (DC) MacLeish sob exploração. Scientia Forestalis, 71: 119-130.

Brown, A.H.D. and Weir, B.S. 1983. Measuring genetic variability in plant populations, in Isozymes in Plant Genetics and Breeding, Part A (Tanksley SD, Orton TJ, Editors). Elsevier Science Publishing, Amsterdam, p. 219-239.

Conceição, A.O.; Rossi, M.H.; Oliveira, F.F.; Takser, L. and Lafond, J. 2011. *Genipa americana* (Rubiaceae) fruit extract affects mitogen-activated protein kinase cell pathways in human trophoblast-derived BeWo cells: implications for placental development. Journal of Medicinal Food, 14: 483-494.

Costa, T. S.; Silva, A.V.C.; Ledo, A.S.; Santos, A.R.F. and Silva Júnior, J.F. 2011 Diversidade genética de acessos do banco de germoplasma de mangaba em Sergipe. Pesquisa Agropecuária Brasileira, 46 (5): 499-507.

Doyle, J.J. and Doyle, J.L. 1990 Isolation of plant DNA from fresh tissue. Focus, 12: 13-15.

Ghislain, M.; Zhang, D.; Fajardo, D.; Huamán, Z. and Hijmans, R.J. 1999. Marker assisted sampling of the cultivated Andean patato Solanumphureja collections using RAPD markers. Genetic Resources and Crop Evolution, 46: 547-555.

Gillies, A.C.M.; Navarro, C., Lowe, A.J., Newton, A.C.; Hernández, M.; Wilson, J. and Cornelius, J.P. 1999. Genetic diversity in Mesoamerican populations of mahogany (*Swietenia macrophylla*), assessed using RAPDs. Heredity, 83 (6): 722-32.

[Pavlicek, A.; Hrda, S. and Flegr, J. 1999. FreeTree – freeware program for construction of phylogenetic trees on the basis of distance data and bootstrap/jackknife analysis of the tree robustness. Application in the RAPD analysis of genus *Frenkelia*. Folia Biologica, 45: 97-99.

Peakall , R. and Smouse, P. E. 2005.GenAlEx 6: Genetic analysis in Excel., population genetic software for teaching and research. Australian National University, Canberra, Australia.

Penalber, T.J.A.; Sadala, M.A.C.; Castro, M.S. and Faria, L.J.G. 1996. Ensaios de extração e aplicação de corantes do fruto do jenipapeiro (*Genipa americana*). Revista Brasileira de Corantes Naturais, 2:129-135.

Pinheiro, L.R.; Almeida, C.S. and Silva, A.V.C. 2011. Diversidade genética de uma população natural de cambuizeiro e avaliação pós-colheita de seus frutos. Scientia Plena, 7 (6): 1-5.

Pinheiro, L.R.; Rabbani, A.R.C.; Silva, A.V.C.; Lédo, A.S.; Pereira, K.L.G.; Diniz, L.E.C. 2012. Genetic diversity and population structure in the Brazilian *Cattleya labiata* (Orchidaceae) using RAPD and ISSR markers. Plant Systematics and Evolution, 12.

Rabbani, A.R.C.; Silva-Mann, R. and Ferreira, R.A. 2012.Variabilidade genética de *Genipa americana* L. pertencente ao baixo curso do rio São Francisco. Revista Árvore, 36 (3): 401-409.

Sanghamitra S.; Geetha, K.A.; Hidayath, K.P. and Maiti, S. 2010. Identification of RAPD markers linked to sex determination in guggal [*Commiphora wightii* (Arnott.)] Bhandari. Plant Biotechnology Reports, 4 (1): 95-99.

Silva, A. V. C.; Santos, A.R.F.; Lédo, A.S.; Feitosa, R.B.; Almeida, C.S.; Silva, G.M. and Rangel, M.S.A. 2012. .Moringa genetic diversity from Germplasm Bank using RAPD markers. Tropical and Subtropical Agroecosystems, 15 (1): 31-39.

Souder, S.N.; Manshardt, R.M. and Stiles, J.I. 1996. A genetic linkage map of papaya based on randomly amplifies polymorphic DNA markers. Theoretical Applied Genetics, 93: 547-553.

Torezan, J.M.D; Souza, R.F.; Ruas, P.M.; Ruas, C.F.; Camargo, E.H. and Vanzela, A.L.L.

2005.Genetic variability of pre and post-fragmentation cohorts of *Aspidosperma polyneuron* Muell. Arg. (Apocynaceae). Brazilian Archives of Biology and Technology, 48 (2): 171-180.

Zhao, K.; Zhou, M.Q. and Chen, L.Q. 2007. Genetic diversity and discrimination of Chimonanthuspraecox (L.) link germplasm using ISSR and RAPD Markers. Hortscience, 42: 1144–1148.

Zimback, L.; Kageyama, P.Y.; Veiga, R.F.A. and Mello Júnior, J.R.S. 2004. Estrutura genética de populações de *Trichilia pallida* Swartz (Meliaceae) por marcadores RAPD. Scientia Forestalis, 65: 114-119.

Effect of Feeding Dinning Room and Kitchen Waste on Growth Performance of Growing Pigs

G. Ramírez-Zúñiga[1], R. F. García-Castillo[1*], J. Salinas-Chavira[2], A. Vega[3], M.H. Ruiloba[4], Juan D. Hernández-Bustamante[1], A. Valdéz-Oyervides[1] and J. M. Fuentes-Rodríguez[1]

*[1]Universidad Autónoma Agraria Antonio Narro, Buenavista, Saltillo, Coahuila, México. CP. 25315. Email: gacr430421@gmail.com**; guadalupe12_88@hotmail.com; david_busta@hotmail.com; antoniovaldezo@hotmail.com. jesusfuentes@hotmail.com*
[2]Universidad Autónoma de Tamaulipas. Facultad de Medicina Veterinaria y Zootecnia, Cd. Victoria, Tamaulipas, México. Email: jsalinasc@hotmail.com.
[3]Universidad Autónoma de Chiriquí. Facultad de Recursos Naturales. Email: Aravega@cwpanama.net.
[4]Universidad de Panamá. Facultad de Ciencias Agropecuarias. David, Chiriquí, República de Panamá. Email: mruiloba@hotmail.com.
**Corresponding author*

SUMMARY

This research used 41 growing backyard piglets (11.47 ± 1.2 kg BW) fed for 22 d and assigned at random to three treatment (T) groups, respectively (T1 to T3) with three repetitions. The proportion of commercial concentrate (CC) to kitchen waste (DW) was: T1, 100:0; T2, 50:50 and T3, 0:100. Diets contained: T1, 17.3, 13.6 and 16.3% CP and 3,321, 3,526 and 4,011 McCall/kg of ME, respectively. Weight gain, carcass characteristics, minerals and metabolites in blood serum were evaluated. The weight gain, slaughter weight, hot and cold carcass weight, hot and cold carcass yield, rib eye area and back fat thickness were not affected by DW ($P > 0.05$). The cuts of leg shoulder and rib were not affected by treatment ($P > 0.05$) for dry matter, ash and crude protein content. The increase in DW reduced ether extract content of leg and rib ($P < 0.05$) but not in shoulder ($P > 0.05$). The concentration of Ca decreased with DW inclusion ($P < 0.05$). No treatment effect was observed in blood metabolites ($P > 0.05$). It is concluded that feeding with DW the backyard growing pig do not affect growth performance or quality of meat.

Key words: Dining Waste; growth; carcass; blood serum; piglets.

INTRODUCTION

Backyard pig production is a very familiar practice, where the activity operates as a system of saving in the rural field. In addition, this activity has had great acceptance and generalization for its beneficial results. In this activity the domestic and productive units are physically integrated. This means a social activity and important source of income in the rural family (Santamaría et al. 2011). They produce most of the backyard pig inventory under rustic conditions and are producers with low income (SAGARPA, 2009).

The shortage in feed as well as the reduced supply and high cost of ingredients used in animal feeding; forced to look for promising alternatives for the care of livestock production species. On the other hand, the accumulation of garbage or waste increases and becomes ecological problem in the localities, there which, in many Asian countries, the accumulation of waste and nutrients causes a great imbalance. Therefore, the technological development in the use of wastes of food for humans and feed for animals will contribute to improve the self-sufficiency of food, which would correct the imbalance by the accumulation of nutrients and make a more sustainable agriculture (Kawashima, 2004; Chittavong *et al.* 2012; Yi, 1999).
Accumulation and environmental pollution caused by these waste requires eliminating the waste of food through the feeding of animals. Similarly, results from the use of waste, plant by-products and animals in animal feed have been reported. As an example in the Caribbean (Domínguez, 1997) recycling this material to be an attractive option (Westendorf *et al.* 1996); in the American continent (Aguilar, 2011; Kornegay *et al.* 1970; Myer *et al.* 1999; Pérez, 1997). The backyard pig production is distributed for the supply of micro-regional markets or to self-sufficiency (SAGARPA, 2009). Therefore, the objective of this research was to evaluate the behavior of backyard growing piglets fed with waste from dining room and kitchen (DW) through the daily weight gain, measures zoometric, carcass traits, mineral and metabolites in blood serum.

MATERIALS AND METHODS

This study was conducted in the metabolic unit for research of the Universidad Autónoma Agraria Antonio Narro, Buenavista, Saltillo, Coahuila, Mexico. Which is geographically located at 25 ° 22 ' North latitude and 101 ° 22 ' length West, with an altitude of 1742 meters above sea level, average temperature of 17.7 ° C and average annual precipitation of 225 mm (García, 1987).

We used nine pens of concrete equipped with automatic waterers and feeders that were disinfected with chlorine and lime (calcium oxide) 15 days prior to initiation of the investigation. It was used 41 piglets in the stage of growth with 11.47±1.2 kg of LW from the Backyard Pigs Program of the University. At the beginning of the evaluation, the piglets were weighed and this practice was repeated every 8 days until the end of the evaluation. The previously identified animals were distributed at random in nine groups and formed three treatments with three repetitions each.

It was evaluated as a feed a commercial concentrate (CC) and dining room and kitchen waste (DW), forming three treatments with three replications. Each repetition was considered an experimental unit. Treatments contained T1 100:0; T2 50:50 and T3, 0:100 CC:DW, respectively. These diets were established according to the system used by small backyard pig producers. Feed was offered such as collected (CBT) by animal: T1, 1.2; T2, 0.600 + 2.0; T3, 4.0 kg, all in wet basis. The DW had 30% of DM in average. According to the concentration of dry matter in the diets was calculated daily feed consumption in dry basis of 1.2 kg per day per animal for treatments T1, T2 and T3. The feed was offered once per day and contained (% CP and Mcal/kg of ME); T1, 17.3% and 3,261; T2, 13.6% and 3,526; T3, 16.3% and 3,620. The average daily gain (ADG) was calculated taking into account the difference between final and initial weight divided by the days of the experiment.

Feed samples were dried at 60° C in an oven until constant weight; then samples were ground in a Willey mill with 2 mm diameter mesh. These samples were analyzed to determine dry matter (105° C), crude protein (N x 6.25), ether extract, crude fiber and extract free nitrogen, according to AOAC (1997). The metabolizable energy (ME) was obtained by the equation reported by National Research Council (NRC, 1998).

At the end of the experiment the piglets were measured (cm) for: height at the cross, (from the ground to the top of the cross); length of tubers, (from the joint escapula-humeral to the top of the rump); chest circumference (from the base of the cross going through the sternum ventral base and then to the base of the cross), forming a circle straight around the flat sacks.

Subsequently, two piglets in each replicate chosen at random (6 per treatment) were sacrificed humanely to record the weight at sacrifice and hot carcass weight (without considering the weight of the head, viscera, legs and blood, according to the system practiced in the region) and determine the hot carcass yield (hot carcass weight as percent of the live weight); cold carcass weight and cold carcass yield (cold carcass weight as percent of the live weight at sacrifice); the rib eye area using grid of IOWA and back fat thickness between the seventh and eighth rib at a seven centimeters of the spine according to the instructions of operation of the Draminski back fat scanner. From the carcass were obtained samples to perform chemical analysis of dry matter, crude protein, ash and ether extract (AOAC, 1997).

At the end of the experiment, blood samples were obtained. For this sampling were obtained at random two piglets from each repetition (6 per treatment). These samples were centrifuged at 2000 g for 15 minutes. The obtained serum was analyzed in a Spectronic ® 20 GenesysTM model 4001 - 4 spectrophotometer, USA; to determine metabolites preparations; Glucose, method Trinder. GOD-POD; creatinine, Jaffe method without deproteinization; cholesterol, method CHOD-PAD and hemoglobin, ICSH standardized method and an absorption spectrophotometer Atomic Varían series AA-1275, Australia, Were identified minerals previously prepared; P, method phosphomolybdate; CA, Orto method Cresoftaleina; Mg, Xilidil blue method; CU and Zn, serum diluted in water).

The data were analyzed in a completely randomized design for three treatments and an equal number of repetitions (3) each as an experimental unit (Stell and Torrie, 1986). Applying the system of statistical analysis (SAS ® version 9.0), with the procedure General Linear Models (GLM).

RESULTS AND DISCUSSION

The DW was variable in its nutrient content (PC, EE, FC, ashes, ELN) and energy, showing wide ranges between the evaluated feed samples. The average daily gains (ADG) of the growing piglets are showed in table 1. The DW did not affect ($P \geq 0.05$) daily weight gain. The average daily gain (ADG) during the experiment was 0.261, 0.310 and 0.264 kg/day for T1, T2 and T3 respectively. Although there was no significant difference, numerically the highest GDP was for 50:50 CC:DW T2, had 47 g more on increase of weight than piglets fed diets containing 100 CC and 100 DW.

The consumption CP was 0.190, 0.163 and 0.196 kg and energy was 3.586, 4.231 and 4.344 Mcal ME/day per animal for treatments T1, T2 and T3 respectively. Consumption of CP obtained in this study is lower than those reported by (NRC, 1998); which animals of this weight and age should consume 0.209 kg of CP. However, ME consumption per day was exceeded in all diets. Therefore, the feeding system commonly used by the backyard pig farmer offers lower CP and higher ME concentration (NRC, 1998).

The variable content of nutrients in DW was its limitation. Diets containing DW (T2 and T3) used in this work were high in fat 7.7 and 14.5%. Situation that raises the ME content in the diet. In addition, this situation can alter the energy:protein relationship and it may be necessary to increase the intake of dietary protein. Because it is necessary to maintain an appropriate relationship energy:protein for suitable responses in growth performance (Westendorf et al. 1998). Especially with young pigs at growing stage, and may be risk of protein restriction if diets are not well complemented (Allee and Hines, 1972).

However, the dried residue of restaurants as feed has the potential to obtain a nutritious product for pigs, and also offers a viable option as solid waste (Myer et al. 1999). On the other hand, a diet adequate in protein and amino acids is critical when using grain distillery (Lynn et al., 2009) or dehydrated waste of restaurants (Myer et al., 1999) for an optimal response in pigs.

The statistical analysis for carcass characteristics, the results do not show significant difference ($P \geq 0.05$) among treatments. However, it was noted that animals fed with 50:50 ratios were heavier, followed by piglets fed with T1, 100:00 and finally 00:100 CC:DW (table 1).

The weight of pigs at slaughter was 17.8, 22.7 y 16.3 kg, hot carcass weights were 10.3, 12.8, and 9.3 kg; as a result, the hot carcass yield was: 57.9, 56.4, and 57.1%. The weight of the cold carcass was 9.7, 12.0, and 8.0 kg, while the cold carcass yields were 54.5, 53.0 and 49.1%, respectively. It is noted that the behavior of piglets is largely dependent on the quality of the diet (Colina-Rivero *et al.* 2010).

Something important needed have mention is that the diet CC: DW 50:50 offered to piglets from T2. These piglets in general had the best performance. It is probable that the wet condition of DW mixed with CC improved the feed utilization or by associative effect among them. Since the animals eat to meet caloric requirements, but if the diet is diluted in water, pigs will consume a greater volume (Church *et al.* 2010).

In animals consuming waste of cafeteria, the dietary water intake was significantly higher in the group fed with waste of cafeteria and they drank much less water, but total intake did not differ. The pig shows ability to regulate their balance of water, regardless of the water content of the diet (Westendorf *et al.*, 1998).

The area of rib eye (cm^2) and back fat thickness (mm) were not statistically different (P≥0. 05). However, the largest area in rib eye was animals fed with 50:50 CC:DW as opposed to piglets fed diets with CC or DW. The back fat thickness presents the following values 16.4, 15.3 and 13.9 mm for T1, T2 and T3 respectively (table 1). In a study involving 39 native without hair male pigs (CCP) with an initial BW of 6.9 ± 2.38 kg and approximately 40 days of age; the weight of carcass and back fat measured before and after slaughter increased as the slaughter weight increased (Santos-Ricalde *et al.* 2011).

To statistically assess the chemical composition of the carcass was no significant difference (P≥0. 05) on the determination of DM, ash and CP of leg, rib, and arm. However, the content of EE showed significant difference (P≤0. 05) between treatments in the leg and rib cuts, but without significant difference (P≥0. 05) on the arm cut. The significance of the EE content in leg and rib in the T1 100:00 CC:DW is not biologically explained. Since the higher content of EE in the diet was in the diets of T2 and T3 (table 2).

This regard Braña, (2011), believes that the main components that make up the empty weight of a pig vary with age; so, to get from the birth to weight of sale, a pig will need to increase your size 80 times. Therefore the growth of a part of the body of the pig always holds proportion with the rest of the body. For example, bone mass always keeps proportion with muscle mass. However, fat is the exception to the rule. The growth of fat is extremely variable and does not keep any proportion with muscle mass. Corroborating this, pigs with 1.4, 7, 28 and 112 kg BW will have 2, 15, 12, and 25% of fat; but CP remains constant at 17.0% on average. Very different from what happens with lean tissues, the proportion of fat in the body, shows greater variability between pigs (Lance *et al.* 2001).

Since Westendorf *et al.* (1998) and Kornegay *et al.* (1970), indicate that the carcass quality of pigs fed with waste of cafeteria is similar to the pigs who receive traditional diets.

External measurements in cm (height, length of tubers and chest circumference) of piglets at the end of the research were not affected (P ≥ 0.05) by the addition of the DW in the ration (table 3). For the treatments (100:0), (50:50) and (0:100) CC:DW the following measures, (cm) were obtained, respectively: height to the cross, 38.5, 37.3 and 36.4; length of tubers, 39.0, 39.8 and 36.4; thoracic circumference, 57.2, 54.4 and 55.1. In this regard the formation of bone mass of the pig is related and keeps proportion with the formation of muscle mass. Unless pigs of different breeds being compared. But the performance of the young pig is more effect of maternal genetics than the own genetic (Braña, 2011).

Table 1. Growth performance of piglets from backyard fed with waste of dining room and kitchen

	Waste of dining room and kitchen (%)				
	0	**50**	**100**		
Determination	**T1**	**T2**	**T3**	**SEM**	**P≥F**
Number of repetitions	3	3	3		
Average daily weight, kg	0.261	0.301	0.264	0.035	0.61
Weight at slaughter, kg	17.8	22.7	16.3	3.99	0.541
Hot carcass weight, kg	10.3	12.8	9.3	2.179	0.543
Cold carcass weight, kg	9.7	12.0	8.0	2.4	0.537
Hot carcass yield (%)	57.9	56.4	57.1	3.568	0.976
Cold carcass yield (%)	54.5	53.0	49.1	5.02	0.757
Rib eye area (cm^2)	24.3	27.0	23.3	4.815	0.829
Back fat thickness (mm)	16.4	15.3	14.9	1.341	0.539

SEM = standard error of the mean. P≥F = Probability

Table 2. Chemical evaluation of the carcass from backyard piglets fed with waste of dining room and kitchen.

	Waste of dining room and kitchen (%)				
	0	**50**	**100**		
Determination	**T1**	**T2**	**T3**	**SEM**	**P≥F**
Leg					
Total dry matter	23.6	23.7	22.3	1.23	0.691
Ash	1.1	0.9	1.1	0.11	0.417
Crude protein	18.2	19.6	18.4	2.40	0.910
Ether extract	7.3	4.8	1.7	0.99	0.020
Rib					
Total dry matter	24.1	25.8	22.8	2.51	0.718
Ash	1.0	1.0	0.9	0.08	0.787
Crude protein	19.8	19.5	19.0	1.53	0.929
Ether extract	17.3	8.8	2.3	2.46	0.015
Arm					
Total dry matter	24.2	22.2	23.1	1.44	0.642
Ash	1.0	1.2	1.0	0.06	0.062
Crude protein	21.2	19.7	18.9	0.66	0.128
Ether extract	5.2	6.7	3.7	1.33	0.354

SEM = standard error of the mean. P≥F = Probability

The concentration of minerals: P, Mg, Cu, and Zn in blood serum had no significant difference (P≥0. 05) among treatments receiving diets based on C, C: DW and DW; with the exception of the Ca which had significant difference (P≤0. 05; Table 4). It is important to highlight that the concentrations of all measured electrolytes were within the reference range for pigs and that the change mentioned in the electrolyte status did not affect physiological homeostasis of animals.

Bone tissue is the major user of these three elements Ca, P and Mg, because it contains 23% of mineral substances. Nevertheless, these requirements are particularly important during growth, last month of gestation and lactation. However, the lack of these elements in piglets presents a detention of growth and rickets by ossification disorders (Zert, 1969).

Most natural feeds are deficient in one or more mineral elements or they are not in balance; diets of pork almost always require specific elements that are rich in calcium or phosphorus. Virtually existing in all supplements calcium is readily absorbed from the gastrointestinal tract of the pig. Conversely, the phosphorus present in the majority of plant products is in the form of phytate (phytic acid) that are not decomposed effectively in the gastrointestinal tract (Rincker *et al.* 2005).

Table 3. Body development in backyard piglets fed with waste of dining room and kitchen.

	Waste of dining room and kitchen (%)				
	0	**50**	**100**		
Variables (cm)	**T1**	**T 2**	**T3**	**SEM**	**P≥F**
Height to the cross	38.5	37.3	36.4	1.982	0.759
Length of tubers	39.0	39.8	36.2	3.2	0.721
Chest circumference	57.2	54.4	55.1	3.086	0.804

SEM = standard error of the mean. P≥F = Probability

Table 4. Mineral concentration in serum of backyard piglets fed with waste of dining room and kitchen.

	Waste of dining room and kitchen (%)					
	0	**50**	**100**			
Determination (mg/dl)	**T1**	**T2**	**T3**	**SEM**	**P≥F**	**Normal levels (mg/dl)***
Phosphorus	3.30	2.86	3.96	0.09	0.073	1.8 – 3
Calcium	9.83	7.88	9.35	0.37	0.023	9.3 –11.5
Magnesium	4.96	4.77	4.94	0.3	0.894	2.3 – 3.5
Copper	0.05	0.06	0.06	0.01	0.751	0.7 – 1.4
Zinc	0.12	0.11	0.15	0.01	0.118	0.5 – 1.2

SEM = standard error of the mean. P≥F = Probability
*Merck, (1991)

The concentration of metabolites (glucose, creatinine, cholesterol and hemoglobin) in blood serum of growing piglets fed diets based on DC and DW were not statistically different (P≥0. 05; Table 5). Therefore, the concentration of metabolites was not affected by the addition of DW.

In this study all metabolites in blood serum concentrations were within the normal range except hemoglobin (Merk, 1991). Similar result was obtained by Aguilar, (2011), by feeding piglets in initiation with waste of dining room and kitchen (DW); this suggest that the DW does not affect the concentration of metabolites and hence does not affect the metabolic pathways of these animals.

Other researchers like Colina *et al.* (2009) used meal of pijiguao (*Batris gasipaes* H:B.K.) (HP) plus synthetic lysine; they evaluated hematological parameters in pigs. The addition of 25% HP and lysine to diets did not affect the hematological parameters.

CONCLUSIONS

According to results obtained in this research it is concluded: feeding growing pigs with concentrated (CC) or waste from dining room and kitchen (DW), does not affect the behavior of the piglets. Similarly, the concentration of metabolites and minerals in blood serum were not different with exception of Ca with lower concentration in blood serum in the group 50:50 (T2). Investigations indicate that DW has nutritional value as indicated by the profile of nutrients; the quality of meat from pigs fed with DW has adequate chemical quality. The limiting of DW is its high moisture content, which could reduce the dry matter intake. These results indicate that waste from dining room and kitchen has good nutritional value and may be useful in the diet. The positive effects of the DW in the phase of growth of pigs and its influence on biochemical parameters were demonstrated in this work.

Table 5. Concentration of metabolites in serum of backyard piglets fed with waste of dining room and kitchen.

	Waste of dining room and kitchen (%)					
	0	**50**	**100**			
Variables (mg/dL)	**T1**	**T2**	**T3**	**EE**	**P≥F**	**Normal levels* (mg/dL)**
Glucose	75.6	82.2	75.0	19.4	0.959	66.4-116.1
Creatinine	2.1	2.2	2.2	0.1	0.719	0.8 – 2.3
Cholesterol	121	92.5	109.7	22.0	0.676	81.4-134.1
Hemoglobin	14.8	12.7	16.7	2.11	0.533	9 – 13

EE = Error estándar de la media. P≥F = Probabilidad
*Merck, (1991)

REFERENCES

Aguilar, V.A.B. 2011. Desempeño, desarrollo corporal y evaluación sanguínea de lechones de traspatio alimentados con desperdicio de comedor y cocina. Tesis de Licenciatura de Ingeniero Agrónomo Zootecnista. Universidad Autónoma Agraria Antonio Narro. Buenavista, Saltillo, Coahuila, México.

Allee, G.L., Hines, R.H. 1972. Influence of fat level and calorie:protein ratio on performance of young pigs. Journal Animal Science 35:210. (Abs).

AOAC. 1997. Official methods of analysis. Association of Official Analytical Chemist. (15th Ed.). Washington, D.C. 1018 pp.

Braña, V.D. 2011. Aspectos prácticos sobre el crecimiento en cerdos. En: Menorías. Taller sobre crecimiento y control de calidad en cerdos y sus productos. AMENA, INIFAP, AMEXITEC Y CONACYT 40 AÑOS. Guadalajara, Jalisco, México. Pp. 1-8. www.amena.mx

Colina-Rivero, J., Araque-Molina, H., Jerez-Timaure, N., Rico-Barreto, D. 2010. Crecimiento y características de la canal de cerdos en engorde alimentados con harina de pijiguao (Bactris gasipaes H.B.K.) y lisina. Revista Científica Facultad de Agronomía (LUZ). 27: 251-269

Colina-Rivero, J., Risini-Valenti M., Tovar, C., Araque-Molina, H. 2009. Peso de órganos y parámetros hematológicos de cerdos en engorde alimentados con harina de pijiguao (Bactris gasipaes H.B.K.) y lisina. XXI Reunión Bienal de la Asociación Latinoamericana de Producción Animal. San Juan, Puerto Rico. Pp27.

Chittavong, M., Linberg, J.E. Janson J. 2012. Feeding regime and management of local Lao pigs Central Lao PDR. Tropical Animal Health and Production. En línea: Unbound medline.

Church, D.C., Pond, W.G., Pond K.R. 2010. Nutrición animal aplicada. 2da. Ed. Editorial Limusa Wiley. Fundamentos de nutrición y alimentación de animales. Cap.14. México. Cap. 14:289-305.

Domínguez P.L. 1997. Desperdicios procesados y subproductos agroindustriales y de pesca en la alimentación porcina en Cuba. En: Tratamiento y utilización de residuos de origen animal, pesquero y alimenticio en la alimentación animal. Estudio FAO Producción y Sanidad Animal. La Habana, Cuba, Pp. 161-178.

García E. 1987. Datos meteorológicos de las estaciones empleadas en el presente trabajo actualizados en 1980. Segunda parte 4ta Ed. Instituto de Geografía. Modificaciones al sistema de clasificación climatológica de Köppen. UNAM. México. Pp. 87-88.

Kawashima, T. 2004. The use of food waste as a protein source for animal feed - current status and technological development in Japan. In: Protein Sources for the Animal Feed Industry. FAO Corporate Document Repository. Agriculture and Consumer Protection. Bangkok. 303-309.

Kornegay, E.T., Vander Noot, G.W., Barth, K.M., Graber, G., Macgrath, W.S., GIlbreath, R.L., Bielk, F.J. 1970. Nutritive evaluation of garbage as a feed for swine. Bull. No. 829. N.J. Agric. Exp. Sta., Rutgers, State University of New Jersey, New Brun.Swick.

Lance De, C.F., Birkett, S.H., Morel C.H. 2001. Protein, fat and bone tissues growth in swine. In: Swine nutrition. Edited by Austin & Southern. 2nd Edition. CRC Press. Boca Raton. USA. Pp65-81.

Lynn, B.A., Wohlt, J.E., Schoknecht, P.A., Westendorf, M.L. 2009. Variation in nutrient content of microbrewers grain and effect on performance and carcass composition when fed wet to growing-finishing swine. Professional Animal Science 25:426-433.

Merck. 1991. Valores y procedimientos clínicos. Parte III. El manual Merck de veterinaria. Cuarta Ed. Merck & Co., Inc. Océano/Centrum. España. Pp1075-1122.

Myer, R.O., Brendemuhl, J.H., Johnson, D.D. 1999. Evaluation of dehydrated restaurant food waste products as feedstuffs for finishing. Journal Animal Science 77:685-692.

NRC, 1998. Nutrient requirements of swine. National Research Council. 10th Edition. Nutrient requirements of domestic animals. National Academy Press. Washington DC. USA. Pp111-112.

Pérez, V.M. 1997. Política cubana de recuperación de todo de desperdicio y subproductos para la producción porcina y saneamiento ambiental. En: Tratamiento y utilización de residuos de origen animal, pesquero y alimenticio en la alimentación animal. Estudio FAO Producción y Sanidad Animal. La Habana, Cuba. Pp137-160.

Rincker, M.J., Hill, G.M., Link, J.E., Meyer, A.M., Rowntree, J.E. 2005. Effects of dietary zinc and iron supplementation on mineral excretion, body composition, and mineral status of nursery pigs. Journal Animal Science 83:2762-2774. En línea: http://jas.fass.org/content/83/12/2762

SAGARPA. 2009. Situación actual y perspectivas de la producción de carne de porcino en

México. Coordinación General de Ganadería. En Línea: http//:www.sagarpa.gob. Mx/ganadería.

Santamaría, G. J., Mariano, I., Domínguez, M., Palacios, E., Thomas, G., Jiménez, B., Montezuma, V., Santos, U, Palacios, C. 2011. Aceptación y apropiación de tecnologías para la innovación tecnológica de sistemas de producción de la agricultura familiar Ngäbe Bugle. Memorias. LVI Reunión del PCCMCA, Sonsonate, 04/25-29. El Salvador, CA. Pp169.

Santos-Ricalde, R.H., Trejo-Lizama, W, Osorio-Hernández W. 2011. Rendimiento de la canal y desarrollo de los órganos torácicos y abdominales de los 25 a los 45 kg en cerdos criollos pelones. Revista Científica, FCV-LUZ / Vol. XXI, N° 5, 396–402.

Steel, R. G. D., Torrie, J. H. 1986. Análisis de la varianza I: Clasificación de una vía (Cap. 7); Comparaciones múltiples (Cap. 8). Bioestadística: Principios y Procedimientos. 2da. Ed. (primera en español). McGraw-Hill, New York, EUA. Pp 132-165, 166-185.

Westendorf, M.L.; Dong, Z.C.; Schoknecht, P.A. 1998. Recycled cafeteria food waste as a feed for swine: Nutrient content, digestibility, growth, and meat quality. Journal Animal Science 76:2976-2983.

Westendorf, M.L., Zirkle, E.W., Gordon, R. 1996. Feeding food or table waste to livestock. Professional Animal Science 12:129-137.

Yi, R. 1999. Experiencia en el uso de hierbas medicinales para mejorar la salud y el desempeño animal en China. En: Biotecnología en la Industria de la Alimentación Animal. Alltech México. Vol VII. Pp217-234.

Zert. P. 1969. Vademécum del productor de cerdos. Ed. Acriba. Zaragoza, Espino.

23

Dry Matter and Nutrient Intake and Diet Composition in *Leucaena Leucocephala* – Based Intensive Silvopastoral Systems

Cesar A Cuartas C.[1], Juan F Naranjo R.[2], Ariel M Tarazona M.[3], Guillermo A Correa L.[4] and Rolando Barahona Rosales [3]

[1]*Doctorado en Ciencias Animales, Universidad de Antioquia.*
[2]*Grupo INCA-CAES, Facultad de Medicina Veterinaria y Zootecnia, Universidad CES, Medellín. Colombia.*
[3]*Departamento de Producción Animal, Facultad de Ciencias Agrarias, Universidad Nacional de Colombia, Facultad de Ciencias Agrarias, Departamento de Producción Animal, AA 1779, Sede Medellín, Colombia.*
Email: rbarahonar@unal.edu.co
[4]*Departamento de Ingeniería Agronómica, Facultad de Ciencias Agrarias, Universidad Nacional de Colombia, Sede Medellín, Colombia.*
**Corresponding author*

SUMMARY

Knowledge of forage intake and grazing selectivity is vital to improve the use of feed resources by grazing animals. To estimate forage selectivity and intake by steers grazing in intensive silvopastoral systems (ISS) under tropical dry forest (bs-T) conditions, intake was evaluated using the n-alkane technique in: 1) a *Leucaena leucocephala* (over 10000 shrubs/ha) and *Megathyrsus maximus* ISS system and, 2) a *M. maximus* monoculture system. Eight steers with an average weight of 248±23.1 kg live weight were evaluated. Animals grazing in the ISS system had greater dry matter intake (26.1 vs. 20.4 g kg^{-1} d^{-1} of BW, P = 0.044) and had greater intake (g d^{-1}; $P \leq 0.02$) of crude protein (954 vs. 499), calcium (62.1 vs. 36.2) and fat (94.2 vs. 69.6) than those grazing in the *M. maximus* monoculture system, respectively. Compared to conventional pastures, DMI of animals grazing in ISS increases, contributing to greater animal productivity.

Key words: beef cattle; legume; n-alkane; silvopastoral system; voluntary intake

INTRODUCTION

The mechanisms that regulate dry matter intake (DMI) and forage selectivity are complex and highly variable among species and individuals (Gordon and Prins, 2008) and are affected by a diversity of factors including sward height and density (Barahona and Sanchez, 2005). Understanding how animals select their diet is essential to improve their utilization of natural resources, especially in consideration to the high diversity of forages that can be used in animal feeding (Poppi *et al.*, 2000). One approach for assessing forage selectivity and DMI is the use of the alkane technique (Mayes *et al.*, 1986), which uses alkanes in the cuticle of forages as internal markers. This technique has been tested in different grazing systems in the tropics (Hendricksen *et al.*, 2002; Gaviria *et al.*, 2015) and in different animal species such as cattle (Unal and Garnsworthy, 1999), sheep (Dove *et al.*, 2000), goats (Ferreira *et al.*, 2005) and buffaloes (Hendricksen *et al.*, 2002). The main difficulties reported for this technique is that forages contain varying concentrations of alkanes, which may be very low in some tropical grasses and as C_{33} alkane is not present in sufficient quantities, estimates of dry matter intake using the pair C_{32}/C_{33} in these forages may not be suitable (Laredo *et al.*, 1991; Rodríguez *et al.*, 2007).

Intensive silvopastoral systems (ISS) are a promising alternative to intensify livestock production in the tropics, improving the competitiveness of livestock enterprises and significantly improving forage and animal productivity (Tarazona *et al.*, 2013; Cuartas *et al.*, 2014). An ISS is a form of agroforestry with high densities (from 5000 to 30000 shrubs/ha) of fodder shrubs such as *Leucaena leucocephala* cv. Cunningham and *Tithonia diversifolia* associated with high-biomass producing grasses and timber trees, fruit trees or palms (from 50 to 500 trees/ha). The ISS incorporates intensive rotational grazing by dividing the paddocks with electric fences, and offering a constant water supply and mineralized salt ad libitum, thus allowing high stocking rates and high milk and meat production per hectare (Cuartas *et al.*, 2014).

As little information is available thus far, the aim of this study was to evaluate the selectivity and intake of forages of beef steers grazing in intensive silvopastoral systems with different forages and relate them to their nutrient intake and nutritional requirements.

MATERIALS AND METHODS

Location

The study was conducted on a beef farm located in Sucre, Colombia (09°51'24.86"N; 75°26'02.44"W). The life zone corresponds to tropical dry forest (T-df). The ISS is at 134 m of altitude, receives an average rainfall of 1000 mm/year, with a relative humidity of 83.5% and an average temperature of 27.1 °C.

Grazing systems evaluated

Two grazing systems were included in the study: (a) an intensive silvopastoral system (ISS) with *Leucaena leucocephala* (over 10000 shrubs ha^{-1}) and *Megathyrsus maximus*, associated with 500 timber trees of the species *Cordia gerascanthus* and *Tabebuia rosea* and (b) a *M. Maximus* pasture also associated with 500 timber trees of the species *C. gerascanthus* and *T. rosea*. The *M. maximus* pasture included some *Guazuma ulmifolia* trees, but dry matter (DM) yield estimates showed that the contribution of this legume to the diet did not exceed 10%. Thus, this system provided a good contrast to the ISS. The timber trees associated to both systems are not normally part of the diet, but their presence affects the forage production in both systems.

Animals and grazing management

Two animal groups, one for each pasture system, were utilized, and each was composed of four steers Zebu steers of different crosses of 248± 23.1 kg live weight (LW). The daily grazing area was 600 m^2, further divided into two strips with 12 hours of occupation in the ISS or used as a single strip with 24 hours of occupation in the *M. maximus* pasture. Hence, grazing strips had 44,5 and 44 days of rest, respectively. The comparative yield technique (Haydock and Shaw, 1975) was used to provide weekly estimates of edible biomass in both systems.

Estimation of DMI with the n-alkane technique

Alkanes were offered in gelatin capsules with 250 mg of a mixture with known amounts (75 mg) of n-dotriacontane (C_{32}) and n-hexatriacontane (C_{36}). Animals were first used to the presence of the persons supplying the capsules. Two of such capsules were administered daily for 15 days. The first ten days, corresponded to an adaptation period and starting on day 11th, fecal samples were collected during five consecutive days in the morning and evening. Fodder samples were collected on days 1, 10 and 15 of the experiment, which were kept frozen until the end of the experimental period, when they were dried at 55°C in a forced air oven. Dry samples were ground to pass a 1 mm sieve and stored in glass vials for subsequent analysis.

Alkanes were extracted from duplicate fecal (0.1 g) and fodder (0.2 g) samples following the protocol of Dove and Mayes (2006). Both n-docosane (C_{22}) and n-tetratriacontane (C_{34}) served as internal standards. N-alkanes were determined in a gas chromatograph fitted with a 20 m x 0.25 mm x 0.25 µm capillary column. Injector and detector temperatures were 280 and 320°C, respectively. Column temperature started at 200°C and was increased at a rate of 10°C min^{-1} until reaching a final temperature of 300°C, which was maintained for 20 min. Hydrogen (20 mL min^{-1}) was used as the carrier gas.

To calculate forage intake (kg DM d^{-1}), the equation of Dove and Mayes (2006) was used:

$$DMI\left({}^{kg}/_{d}\right) = \frac{\frac{F_i}{F_j} * D_j}{H_i - \left[\left(\frac{F_i}{F_j} * H_j\right)\right]}$$

Where:
F_i = concentration (mg kg DM^{-1}) of odd-chain n-alkane in faeces;
F_j = concentration (mg kg DM^{-1}) of even-chain n-alkane in faeces;
D_j= amount of synthetic n-alkane offered (mg d^{-1})
H_i = odd-chain *n*-alkane concentration (mg kg DM^{-1}) in forage;
Hj = even-chain *n*-alkane concentration (mg kg DM^{-1}) in forage.

In turn, diet composition was estimated using the approach of Dove and Mayes (2006):

$$Calculated\ fecal\ alkane_i = \alpha Grass + \beta Legume$$

In this equation, constants *α* (*grass*) and (*β*) *legume* are the recovery-corrected fecal concentrations of alkane$_i$ evaluated as unknowns. For determining diet composition, the following equation was used:

$$x = \frac{\alpha}{\alpha + \beta}; y = \frac{\beta}{\alpha + \beta}$$

Estimation of enteric methane (CH_4) emissions

Enteric methane emissions were estimated using the Ym (the percentage of gross energy intake that is emitted as enteric methane) values obtained in the *in vivo* (Molina *et al.*, 2015a, 2015b) and *in vitro* (Rivera *et al.*, 2015) work of the Fundación Centro para la Investigación en Sistemas Sostenibles de Producción Agropecuaria – CIPAV and Universidad Nacional de Colombia Sede Medellín – UNAL, for animals consuming diets representative of an ISS (74 g/100 g *Cynodon plectostachyus* and 26 g/100 g *L. leucocephala*) and a grass-only pasture (100 g/100 g *C. plectostachyus*). Methane emissions were also estimated using the Ym values for grazing cattle suggested by the IPCC (2006).

Data Analysis

To compare variables, a completely randomized design ($y_{ij}=\mu + \alpha_i + \epsilon_i$) was used, where y_{ij} is the trait measured; μ is the overall average; α_i is the treatment effect at the ith level and ϵ_i is the experimental error. Treatments consisted of the grazing systems evaluated. Means were compared across treatments using Tukey's test at a significance level of 0.05.

RESULTS

N-alkane concentration

Total concentration of odd-chain alkanes reached values of 850.8, 860.9 and 442.8 mg kg^{-1} for *M. maximus*, *L. leucocephala* and *G. ulmifolia* respectively (Table 1). The most abundant odd-chain alkane was either C_{33} or C_{31}.

Table 1. Odd-chain alkane concentration of the forages included in this study (mg kg^{-1} dry matter)

Alkane	*Megathyrsus maximus*	*Leucaena leucocephala*	*Guazuma ulmifolia*
C_{27}	62.91	32.25	77.23
C_{31}	285.06	288.06	154.41
C_{32}	14.69	11.25	14.76
C_{33}	315.61	299.73	126.75
C_{34}	------	------	------
C_{35}	187.21	240.86	84.44
C_{36}	6.70	7.92	3.81

Dry matter intake

Estimates of DMI are shown in Table 2, which in average were adequate for the animals. However, great variability was observed in individual DMI, with intakes (g kg LW^{-1}) being as low as 16 and as high as 33. On average, DMI of animals grazing on SSP was 5.7 g kg LW^{-1} greater than that of animals grazing on the *M. maximus* pastures ($p<0.05$). Similarly, fecal output was 1.30 times greater in animals grazing in the ISS ($p<0.01$). This was due to the increased DMI of animals in ISS as the apparent digestibility coefficient of dry matter of both diets was similar (0.534 vs. 0.538 for the ISS and *M. maximus* pastures, respectively).

Forage selectivity

The diet of animals grazing in the ISS contained a greater proportion of legume than that of those grazing in the *M. maximus* pasture (30.2 vs 9.0%, respectively, $p<0.01$; Table 3). This obeyed to the greater availability of *L. leucocephala* in the ISS compared to that of *G. ulmifolia* in the *M. maximus* pasture.

One animal grazing on the ISS only consumed 1.0 kg DM of leucaena (18.6% of the diet), while the other three animals consumed on average 2.23 kg (34.8% of the diet). In the *M. maximus* pasture, given the limited supply of *G. ulmifolia*, there was less variability in selectivity. However, the diet of two animals grazing in the *M. maximus* pasture *G. ulmifolia* composed of 11.2% of the diet, while that of the other two animals was comprised of only 5.8% of the diet.

Table 2. Estimated dry matter intake of animals grazing in an intensive silvopastoral system (ISS) with *Leucaena leucocephala* and *Megathyrsus maximus* and in a monoculture system based on *M. maximus*.

Parameter	ISS	Monoculture	RMSE	*P value*
Intake, kg d^{-1}	5.99	5.39	2.68	0.35
Intake, g kg LW^{-1}	26.1	20.4	4.9	0.04
Fecal production, kg d^{-1}	2.71	2.40	0.28	0.14
Fecal production, g kg LW^{-1}	11.9	9.1	0.60	<0.01
Digestibility, coefficient	0.534	0.538	0.057	0.90
Live weight (LW), kg	231	266	40.9	0.33

Table 3. Selectivity on dry matter Intake of animals grazing in an intensive silvopastoral system (ISS) with *Leucaena leucocephala* and *Megathyrsus maximus* and in a monoculture based on *M. maximus*.

Parameter	ISS	Monoculture	RMSE	*P* value
Grass intake, % total diet	68.8	91.3	1.00	<0.01
Grass Intake, kg day^{-1}	4.12	4.92	1.81	0.13
Legume intake, kg day^{-1}	1.87	0.47	0.70	<0.01
Grass intake, g kg LW^{-1} d^{-1}	17.8	18.6	2.90	0.72
Legume intake, g kg LW^{-1} d^{-1}	8.2	1.8	0.15	<0.01

Dietary nutrient composition and nutrient intake by grazing animals

The nutrient content of the diet consumed in both systems is shown in Table 4. Animals grazing in the ISS consumed a diet with greater content of crude protein, ether extract and lignin and lower content of NDF, ADF and cellulose. The protein content of the diet consumed in the ISS (159.4 g kg^{-1}) was sufficient to meet the requirements of these animals.

As shown in Table 5, animals in ISS consumed 1.9 times (950 vs 500g) more protein than those grazing in the *M. maximus* pasture ($p<0.01$). Similarly, intake of ether extract was 1.35 times (94 vs 70 g) greater in these animals than in the *M. maximus* pasture ($p = 0.02$). Concerning fiber (NDF, ADF and cellulose) intake, there was no difference between both animal groups.

Table 4. Nutrient content (g kg^{-1} DM, unless indicated otherwise) of the diet of animals grazing in an intensive silvopastoral system (ISS) with *Leucanea leucocephala* (Ll) and *Megathyrsus maximus* (Mm) and in a system based on *M. maximus* with some inclusion of *Guazuma ulmifolia* (Gu).

Fraction	ISS		Monoculture		Diet composition, %	
	Mm	Ll	Mm	Gu	ISS	Monoculture
Nitrogen	17.0	44.3	14.4	19.1	25.5	14.8
Crude protein	106.1	276.8	89.9	119.1	159.4	92.4
NDF	663.8	324.2	675.7	290.2	557.7	642.7
ADF	362.6	123.0	365.2	171.4	287.8	348.6
Lignin	63.0	77.0	63.0	63.0	67.4	63.0
Cellulose	299.6	46.0	302.2	108.4	220.4	285.6
Hemicellulose	301.2	201.2	310.5	118.8	270.0	294.1
Calcium	8.6	14.3	5.9	15.2	10.4	6.7
Phosphorus	1.5	2.1	1.7	2.7	1.7	1.8
Ether extract	12.4	2.3.1	12.4	18.1	15.7	12.9
Ash	95.0	69.2	109.0	117.0	86.9	109.7
Organic matter	905.0	930.8	891.0	883.0	913.1	890.3

Table 5. Average nutrient intake (kg d^{-1}) by animals grazing in an intensive silvopastoral system (ISS) with *Leucaena leucocephala* and *Megathyrsus maximus* and in a monoculture system based on *M. maximus*.

Parameter	ISS	Monoculture	RMSE	*P* value
Crude protein	0.95	0.50	0.07	<0.01
NDF	3.34	3.46	0.92	0.74
ADF	1.72	1.88	0.26	0.43
Lignin	0.40	0.34	0.01	0.14
Cellulose	1.32	1.54	0.17	0.18
Hemicelullose	1.62	1.58	0.21	0.85
Calcium, g	62.10	36.20	<0.001	<0.01
Phosphorus, g	10.10	9.60	<0.001	0.69
Ether extract, g	94.20	69.6	<0.001	0.02
Ash	0.52	0.59	0.03	0.25
Organic matter	5.47	5.80	2.19	0.25

Estimated enteric methane (CH_4) emissions

Using the Ym values obtained in *in vivo* (Molina *et al.*, 2015a, 2015b) and *in vitro* (Rivera *et al.*, 2015) experiments by CIPAV-UNAL, there was no difference in estimated methane emissions of animals grazing in both systems, which were in average 54 kg $head^{-1}$ $year^{-1}$. Given the greater energy intake of steers in ISS, the use of the Ym values suggested by the IPCC (2006) led to estimations of methane emissions that were 1.22 greater in ISS than in the *M. maximus* pasture.

With average daily weight ga|ins (g) of 818 and 300 (Cuartas *et al.*, 2013), to gain 300 kg of live weight, steers would have to graze for 12.0 and 32.8 months in a typical ISS and *M. maximus* pasture, respectively. With an average daily DMI of 10.6 and 8.32 kg/animal (Cuartas *et al.*, 2013) throughout that period, based on the Ym values of CIPAV-UNAL (Molina *et al.*, 2015a, 2015b, Rivera et al., 2015) methane emissions associated with gaining those 300 kg of LW were estimated to be 92.5 and 232 for animals in the ISS and *M. maximus* pastures, respectively. Hence, it can be estimated that in producing a kilogram of live weight, enteric methane emissions would be 7.09 and 17.79 kg of CO_2 equivalent in ISS and *M. maximus* pastures, respectively.

Table 6. Estimates of methane production by animals grazing in an intensive silvopastoral system (ISS) with *L. leucocephala* and *M. maximus* and in a pasture based on *M. maximus* using the Ym values of CIPAV-UNAL (Molina *et al.*, 2015a, 2015b, Rivera et al., 2015) and those of IPCC (2006).

Parameter	ISS		Monoculture	
	CIPAV-UNAL	IPCC 2006	CIPAV-UNAL	IPCC 2006
Intake, kg	5.99		5.39	
GE intake, MJ	110.54		91.13	
Y_m[a]	7.21% ± 0.65%	6.5% ± 1.0%	9.20% ± 0,20%	6.5% ± 1.0%
MJ CH_4[b]	7.97	7.20	8.38	5.94
CH^4 g d^{-1} [c]	143	129	151	106
Kg CH_4 an^{-1} $year^{-1}$	52.27	47.10	55.98	38.90

[a] Ym: methane conversion factor; [b]Mega Joules (MJ) of gross energy (GE) going to CH_4; [c] Grams of CH_4 calculated given that one kg of methane contains 55.65 MJ of GE.

DISCUSSION

Great variability has been reported with regard to the content of alkanes in forages, which could be related to soil and climatic factors and/or the laboratory technique used. For example, the concentration of odd–chain n-alkanes in *M. maximus* ranged between 16.8 and 22.3 mg kg DM^{-1} and 13.4 and 13.7 mg kg DM^{-1} for *L. leucocephala* (Sanchez *et al.*, 2009). In turn, in 25 species of plants used for grazing in Sudan, the concentrations of odd–chain n-alkanes in *Chloris gayana*, *Cenchrus ciliaris* (blooming) and *Stylosanthes guianensis* were 478, 571 and 406 mg kg DM^{-1}, respectively (Ali *et al.*, 2005), values within the ranges reported in this study.

Delgado *et al.* (2000) reported that the largest differences in content (mg kg DM^{-1}) of long chain n-alkanes between *M. maximus* and *L. leucocephala* occurred in C_{31} (69-83 and 31-33, respectively) and in C_{33} (48 and 16, respectively). In our study, this occurred for C_{27} and C_{35} (*L. leucocephala* vs. *M. maximus*), but differences of n-alkane content of these two forages with *G. ulmifolia* were much more contrasting. As it was also observed in this study, Delgado *et al.* (2000) reported that the most abundant odd–chain n-alkanes in *M. maximus* were C_{31} and C_{33}, which disagrees with earlier reports that in tropical forages, C_{33} is not present in sufficient quantities (Laredo *et al.*, 1991). Clearly, as with most plant metabolites, the concentration of n-alkanes in forages changes with the time of year (Delgado *et al.*, 2000). Therefore, when using n-alkanes to determine DMI, it is necessary to determine the content of n-alkanes in representative samples of the forages consumed by animals.

Daily DMI intakes (g kg BW^{-1}) found in this study are similar to those reported by Gaviria et al., 2015. In this study, we observed greater DMI (1.28 times) in animals grazing in ISS than those grazing in a *M. maximus* pasture. Forage intake is greatly influenced by its chemical and nutritional quality and the nutritional quality of forages in ISS is superior to that of most tropical pastures (Cuartas *et al.*, 2013). With greater nutritional quality and biomass availability (Cuartas *et al.*, 2014), greater DMI is to be expected in ISS.

The maximum DMI occurs when the diet digestibility reaches values between 66 and 68% (Faria and Mattos, 1995). Thus, the greater DMI observed in ISS could obey to changes in rumen residence times, since there were no differences in total tract DM digestibility (Table 2). At the same degree of digestibility, legume intake by ruminants is greater than that of grasses (Niderkorn and Baumont, 2009), because legume particles are cuboidal whereas those of grasses are long and thin, leading to greater rates of passage for legumes (Barahona and Sanchez, 2005). Additionally, management practices in ISS ensure that animals are offered forages closer to their optimal nutritional value (less mature), with low fiber content and thus, increased digestibility (Boval and Dixon, 2012). In turn, protein degradability can be low in *L. leucocephala* due to condensed tannin content (Barahona *et al.*, 2003), with a fraction of the protein protected from hydrolysis in the rumen, favoring protein bypass to the duodenum (Rodriguez *et al.*, 2010).

Another factor that can contribute to greater DMI in ISS is that grasses subjected to moderate shading regimes, as occurs in ISS, may have greater protein and lower fiber content those growing at full exposition (Murtagh *et al.*, 1987). Additionally, DMI relates to the ability of grazing animals to regulate body temperature under tropical conditions, with high environmental temperatures leading to decreased DMI (Tarazona *et al.*, 2013). The shade provided by trees in ISS could have contributed to the results reported in this study. With greater DMI, there is greater productivity and steers grazing in ISS had daily gains between 832 and 856 g/d and reached harvest weight at around 540 days of age (Mahecha and Angulo, 2012).

As observed in this study, forage intake is a parameter of high individual variability. The DMI of steers in the *M. maximus* and ISS systems (20.4 and 26.1 g kg LW^{-1} d^{-1}, respectively), seem low when compared to the DMI of lactating cows. For example, Smith *et al.* (2005) estimated a DMI of 31.5 g kg LW^{-1} d^{-1} in Holstein cows with a live weight of 549 kg and an average milk production of 35 L. This is probably due to the greater nutritional requirements of lactating animals compared to meat producing animals. Estimates of DMI in grazing beef cattle of the Angus and Simmental breeds ranged between 14 and 15 g kg LW^{-1} d^{-1} (Estermann *et al.*, 2003).

Fecal DM excretion was greater in ISS, and this related to the greater DMI. Fecal excretion plays an important role in the maintenance of soil fertility and nutrient recycling. Theoretically, the deposition of fecal matter in pastures allows nutrients to be used several times by plants and animals in a short period, contrary to what happens in plants not defoliated by grazing. It has been reported that excreta contain roughly the desired proportions of nutrients needed by plants (Herrero *et al.*, 2013).

Greater methane (CH_4) emissions exist in animals that consume more DM. However, enteric CH_4 emissions in ruminants can be modified by the presence of secondary metabolites, which include condensed tannins (Tiemann *et al.*, 2008), which occur at significant levels in *L. leucocephala* (Barahona *et al.*, 2003). In general, the presence of condensed tannins has been associated with reductions of 13-16% in CH_4 emissions (Patra and Saxena, 2010).

In this study, enteric methane emissions for the production of one kilogram of live weight were estimated to be 340 and 793 g for steers in the ISS and *M. maximus* pastures, respectively. Opio *et al.* (2013) estimated that in the production of one kg of beef carcass weight, there is an emission of 67.8 kg of CO_2 equivalent, of which 28.9 kg correspond to enteric methane emissions. In our experiment, with a dressing percentage of 52% (Montoya, 2015), enteric methane emissions for the production of one kg of beef carcass weight would be 13.7 and 34.2 kg of CO_2 equivalent in ISS and *M. maximus* pastures, respectively.

CONCLUSIONS

In intensive silvopastoral systems, DMI of grazing animals is increased, leading to nutrient intakes in excess of maintenance requirements, thus leading to increased animal productivity. Animals in ISS have greater fecal excretion, leading to greater pasture nutrient availability. Despite similar methane emissions (kg per animal per year) in ISS and in *M. maximus* pastures, given differences in animal productivity which are directly related to changes in DMI, methane emissions per kg of weight gain would be around 40% in ISS compared to the *M. maximus* pasture. Further research should be carried out to determine if this benefit occurs in other (not *L. leucocephala* based) ISS systems.

REFERENCES

Ali, H.A., Mayes, R.W., Hector, B.L., Orskov, E.R. 2005. Assessment of n-alkanes, long chain fatty alcohols and long-chain fatty acids as diet composition markers: The concentrations of these compounds in rangeland species from Sudan. Animal Feed Science and Technology.121:257-271.

Barahona, R., Sanchez, S. 2005. Limitaciones físicas y químicas de la digestibilidad de pastos tropicales y estrategias para aumentarla. Revista CORPOICA. 6:69-82.

Barahona, R.R., Lascano, C.E., Narvaez, N., Owen, E., Morris, P., Theodorou, M.K. 2003. *In vitro* degradability of mature and immature leaves of tropical forage legumes differing in condensed tannin and non-starch polysaccharide content and composition. Journal of the Science of Food and Agriculture. 83:1256-1266.

Boval, M., Dixon, R.M. 2012. The importance of grasslands for animal production and other functions: a review on management and methodological progress in the tropics. Animal. 6:748-762.

Cuartas, C.A., Naranjo, J.F., Tarazona, A.M., Barahona, R.R. 2013. Energy use in cattle in intensive silvopastoral systems with *Leucaena leucocephala* and its relationship to animal performance. Revista CES Medicina Veterinaria y Zootecnia. 8:70-81.

Cuartas, C.A., Naranjo, J.F., Tarazona, A.M., Murgueitio, E., Chara, J.D., Ku, J., Solorio, F.J., X. Flores, M.X., Solorio, B., Barahona, R. 2014. Contribution of intensive silvopastoral systems to animal performance and to adaptation and mitigation of climate change. Revista Colombiana de Ciencias Pecuarias. 27:76-94.

Delgado, D.C., Olivera, M.C., Navarro, A. 2000. Composición de n-alcanos cuticulares en plantas tropicales. Su potencialidad como marcadores para estimar consumo y selección de rumiantes en pastizales. Revista Cubana de Ciencias Agrícolas. 34:151.

Dove, H., Freer, M., Foot, J.Z. 2000. The nutrition of grazing ewes during pregnancy and lactation: a comparison of alkane-based and chromium/in vitro-based estimates of herbage intake. Australian Journal of Agricultural Research. 51:765-777.

Dove, H., Mayes, R.W. 2006. Protocol for the analysis of n-alkanes and other plant wax compounds and for their use as markers for quantifying the nutrient supply of large mammalian herbivore. Nature Protocols. 1:1680-1697.

Estermann, B.L., Wettstein, H.R., Sutter, F., Erdin, D., Kreuzer, M. 2003. Effect of calving period on herbage intake and nutrient turnover of Simmental and Angus suckler cows with Angus sired calves grazing subalpine and alpine pastures. Livestock Production Science. 79:169-182.

Faria, V.P., Mattos, W.R. 1995. Nutrição de bovinos tendo em vista performances econômicas máximas. In: Peixoto, A.M., Moura, J.C., Faria, V.P., editors. Nutrição de bovinos: conceitos básicos e aplicados Piracicaba; FEALQ; p.199-222.

Ferreira, L.M., Olivan, M., Garcia, U., Rodrigues, M., Osoro, K. 2005. Validation of the alkane technique to estimate diet selection of goats grazing heather–gorse vegetation communities. Journal of the Science of Food and Agriculture. 85:1636-1646.

Gaviria-Uribe, X., Naranjo-Ramírez, J.F., Bolívar-Vergara, D.M., Barahon- Rosales, R. 2015. Consumo y digestibilidad en novillos cebuínos en un sistema silvopastoril intensivo. Archivos de Zootecnia. 245:21-27.

Gordon, I.J., Prins, H.H. 1975. Introduction: Grazers and Browsers in a Changing World. In: Gordon, I.J., Prins, H.H.T., editors. The Ecology of Browsing and Grazing Ecological

Studies. Heidelberg, Germany; Springer–Verlag; 2008. p.1-20.

Haydock, K.P., Shaw, N.H. 1975. The comparative yield method for estimating dry matter yield of pasture. Australian Journal of Experimental Agriculture and Animal Husbandry. 15:169-171.

Hendricksen, R.E., Reich, M.M., Roberton, R.F., Reid, D.J., Gazzola, C., Rideout, J.A., Hill, R.A. 2002. Estimating the voluntary intake and digestibility of buffel-grass and Lucerne hays offered to Brahman-cross cattle using n-alkanes. Animal Science. 74:56.

Herrero, M., Grace, D., Njuki, J., Johnson, N., Enahoro, D., Silvestri, S., Rufino, M.C. 2013. The roles of livestock in developing countries. Animal. 7 (s1):3–18.

IPCC. 2006. IPCC Guidelines for National Greenhouse Gas Inventories. Vol 4. Agriculture, Forestry and Other Land Use. Chap 10. Emissions from Livestock and Manure Management 2006; [access date: August 21, 2013] URL: http://www.ipccnggip.iges.or.jp/public/2006gl/pdf/4_Volume4/V4_10_Ch10_Livestock.pdf

Laredo, M.A., Simpson, G.D., Minson, D.J., Orpin, C.G. 1991. The potential for using n-alkanes in tropical forages as a marker for the determination of dry matter intake by grazing ruminants. Journal of Agricultural Science. 117:355-361.

Mahecha, L., Angulo, J. 2012. Nutrient management in silvopastoral systems for economically and environmentally sustainable cattle production: A Case Study from Colombia. In: Whalen J editor. Soil fertility improvement and integrated nutrient management - A global perspective. Rijeka, Croatia; InTech; p.201–216.

Mayes, R.W., Lamb, C.S., Colgrove, P.M. 1986. The use of dosed and herbage n-alkanes as markers for the determination of herbage intake. Journal of Agricultural Science. 107:161-170.

Molina, I.C., Donney´s Lemos, G., Montoya Uribe, S., Angarita Amaya, E., Rivera Herrera, J.E., Villegas Sánchez, G., Cantet, J.M., Correa Londoño, G., Mayorga Mogollón, O.L., Chará, J., Barahona Rosales, R. 2015a. Emisiones de metano en sistemas de producción con y sin inclusión de Leucaena leucocephala. En: Peri, P.L., 3° Congreso Nacional de Sistemas Silvopastoriles – VIII Congreso Internacional de Sistemas Agroforestales, Misiones, Argentina, pp. 611-615.

Molina, I.C., Donney`s, G., Montoya, S., Rivera, J.E., Villegas, G., Chara, J., Barahona, R. 2015b. La inclusión de *Leucaena leucocephala* reduce la producción de metano de terneras Lucerna alimentadas con *Cynodon plectostachyus* y *Megathyrsus maximus*. Livestock Research for Rural Development. 27(5): #96.

Montoya, C. 2015. Caracterización de algunas variables de calidad de carne en bovinos manejados bajo diferentes condiciones de producción en el trópico colombiano. MSc Thesis, Universidad Nacional de Colombia, Medellín, 115 pp.

Murtagh, G.J., Halligan, E.A., Greer, D.H. 1987. Components and growth and dark respiration of kikuyu (*Pennisetum clandestinum* Chiov.) at various temperatures. Annals of Botany. 59:140-143.

Niderkorn, V., Baumon, R. 2009. Associative effects between forages on feed intake and digestion in ruminants. Animal. 3:951–960.

Opio, C., Gerber, P., Mottet, A., Falcucci, A., Tempio, G., MacLeod, M., Vellinga, T., Henderson, B., Steinfeld, H. 2013. Greenhouse gas emissions from ruminant supply chains – A global life cycle assessment. Rome; FAO.

Patra, A.K., Saxena, J. 2010. A new perspective on the use of plant secondary metabolites to inhibit methanogenesis in the rumen. Phytochemistry. 71:1198-1222.

Poppi, D.P., France, J., McLennan, S.R. 2000. Intake, Passage and Digestibility. In: Theodorou, M.K., France, J., editors. Feed Evaluation for Animal Production. Wallingford (UK); CABI Publishing; p.35-52.

Rivera, J.E., Molina, I.C., Donney`s, G., Villegas, G., Chará, J., Barahona, R. 2015. Dinámica de fermentación y producción de metano en dietas de sistemas silvopastoriles intensivos con *L. leucocephala* y sistemas convencionales orientados a la producción de leche. Livestock Research for Rural Development. 27(4):#76.

Rodriguez, N.M., Saliba, E.O.S., Guimaraes-Junior, R. 2007. Uso de Indicadores para estimar consumo y digestibilidad de pasto. LIPE, lignina purificada y enriquecida. Revista Colombiana de Ciencias Pecuarias. 20:518-525.

Rodriguez, R.M., Mota, C., Castrillo, M. 2010. *In vitro* rumen fermentation of the tropical grass *Pennisetum purpureum* and mixtures with browse legumes: effects of tannin contents. Journal of Animal Physiology and Animal Nutrition. 94:696-705.

Sanchez, T., Pedraza, R., Mayes, R.W., Oskorv, B. 2009. Determinación de compuestos hidrocarbonados en la pared celular de *M. maximus* y *L. leucocephala* en silvopastoreo. Pastos y Forrajes. 32:1-6.

Smith, H.J., Taweel, H.Z., Tas, B.M., Tamminga, S., Elgersma, A. 2005. Comparison of techniques for estimating herbage intake of grazing dairy cows. Journal of Dairy Science. 88:1827-1836.

Tarazona, A.M., Ceballos, M.C., Cuartas, C.A., Naranjo, J.F., Murgueitio, E., Barahona, R. 2013. The relationship between nutritional status and bovine welfare associated to adoption of intensive silvopastoral systems in tropical conditions. In: Makkar, H., editor. Enhancing animal welfare and farmer income through strategic animal feeding - Some case studies. Rome; FAO; p.69-78.

Tiemann, T.T., Lascano, C.E., Wettstein, H.R., Mayer, A.C., Kreuzer, M., Hess, H.D. 2008. Effect of the tropical tannin-rich shrub legumes *Calliandra calothyrsus* and *Flemingia macrophylla* on methane emission and nitrogen and energy balance in growing lambs. Animal. 2:790–799.

Unal, Y., Garnsworthy, P.P. 1999. Estimation of intake and digestibility of forage based diets in group-fed dairy cows using alkane as markers. Journal of Agricultural Science. 133:419-425.

Permissions

All chapters in this book were first published in TSA, by Universidad Autónoma de Yucatá¡n; hereby published with permission under the Creative Commons Attribution License or equivalent. Every chapter published in this book has been scrutinized by our experts. Their significance has been extensively debated. The topics covered herein carry significant findings which will fuel the growth of the discipline. They may even be implemented as practical applications or may be referred to as a beginning point for another development.

The contributors of this book come from diverse backgrounds, making this book a truly international effort. This book will bring forth new frontiers with its revolutionizing research information and detailed analysis of the nascent developments around the world.

We would like to thank all the contributing authors for lending their expertise to make the book truly unique. They have played a crucial role in the development of this book. Without their invaluable contributions this book wouldn't have been possible. They have made vital efforts to compile up to date information on the varied aspects of this subject to make this book a valuable addition to the collection of many professionals and students.

This book was conceptualized with the vision of imparting up-to-date information and advanced data in this field. To ensure the same, a matchless editorial board was set up. Every individual on the board went through rigorous rounds of assessment to prove their worth. After which they invested a large part of their time researching and compiling the most relevant data for our readers.

The editorial board has been involved in producing this book since its inception. They have spent rigorous hours researching and exploring the diverse topics which have resulted in the successful publishing of this book. They have passed on their knowledge of decades through this book. To expedite this challenging task, the publisher supported the team at every step. A small team of assistant editors was also appointed to further simplify the editing procedure and attain best results for the readers.

Apart from the editorial board, the designing team has also invested a significant amount of their time in understanding the subject and creating the most relevant covers. They scrutinized every image to scout for the most suitable representation of the subject and create an appropriate cover for the book.

The publishing team has been an ardent support to the editorial, designing and production team. Their endless efforts to recruit the best for this project, has resulted in the accomplishment of this book. They are a veteran in the field of academics and their pool of knowledge is as vast as their experience in printing. Their expertise and guidance has proved useful at every step. Their uncompromising quality standards have made this book an exceptional effort. Their encouragement from time to time has been an inspiration for everyone.

The publisher and the editorial board hope that this book will prove to be a valuable piece of knowledge for researchers, students, practitioners and scholars across the globe.

List of Contributors

Ana Veruska Cruz da Silva and José Guedes de Sena-Filho
Embrapa Coastal Tablelands

Allívia Rouse Carregosa Rabbani and Camila Santos Almeida
Universidade Federal de Sergipe (UFS)

Rosana Barroso Feitosa
Universidade Federal do Vale do São Francisco

Melvin González Arcia
Universidad Autónoma de Zacatecas, Unidad Académica de Medicina Veterinaria y Zootecnia. Km 31.5 Carretera Zacatecas-Fresnillo, El Cordovel, Enrique Estrada, Zacatecas, México

Braulio Valles de la Mora, Miguel Ángel Alonso Díaz, Epigmenio Castillo Gallegos, Eliazar Ocaña Zavaleta and Jesús Jarillo Rodríguez
Universidad Nacional Autónoma de México, Facultad de Medicina Veterinaria y Zootecnia, Centro de Enseñanza, Investigación y Extensión en Ganadería Tropical. Km 7.5 Carretera Federal Martínez de la Torre-Tlapacoyan, CP 93600, Martínez de la Torre, Veracruz, México

Rodrigo Grazinoli Garrido
Institute of Forensic Genetic Research, DGPTC, PCERJ, Brazil
Scholl of Health Science an Environment, UCB

Fabiola de S. R. G. Garrido
Environmental and Administrative Science Department, ITR, UFRRJ, Brazil

Sonia Regina de Souza
Chemistry Department, ICE, UFRRJ, Brazil

Manlio Silvestre Fernandes
Soils Department, IA, UFRRJ, Brazil

Artemio Carrillo-Parra, Humberto González and Fortunato Garza
Universidad Autónoma de Nuevo León, Facultad de Ciencias Forestales

Martha Rosales
Centro Interdisciplinario de Investigación para el Desarrollo Integral Regional Unidad Durango. Instituto Politécnico Nacional

Christian Wehenkel
Instituto de Silvicultura e Industria de la Madera. Universidad Juárez del Estado de Durango

Rahim Foroughbakhch
Universidad Autónoma de Nuevo León, Facultad de Ciencias Biológicas

Jacob Banuelos, Dora Trejo and Liliana Lara
Universidad Veracruzana, Facultad de Ciencias Agrícolas, Laboratorio de Organismos Benéficos. Circuito Gonzalo Aguirre Beltrán s.n., Zona Universitaria Xalapa, Veracruz. CP 91000, México

Mayra Gavito
CIEco, UNAM, Antigua Carretera a Pátzcuaro No 8701, Col. Exhacienda de San José de La Huerta C.P. 58190, MoreliaMichoacán, México

Yazmín Carreón
Universidad Michoacana de San Nicolás Hidalgo, Av. Francisco J. Mujica S/N Ciudad Universitaria, C.P. 58030, Morelia, Michoacán, México

Daniel Alejandro Vazquez-Cahuich, Judith Espinosa Moreno, Dora Centurion Hidalgo and Jose Rodolfo Velazquez Martinez
División Académica de Ciencias Agropecuarias de la Universidad Juárez Autónoma de Tabasco Km. 25+2 de la Carretera Villahermosa-Teapa, R/a La Huasteca 2a Sección. Villahermosa, Tabasco, México

Rocio Borges-Argaez and Mirbella Caceres Farfan
Unidad de Biotecnología,Centro de Investigación Científica de Yucatán, Calle 43 No. 130 Chuburná de Hidalgo, C.P. 97200, Mérida, Yucatán, Mexico

Martha Elena Nava-Tablada and Enrique Martínez-Camarillo
El Colegio de Veracruz. Carrillo Puerto # 26, Zona Centro. CP. 91000, Xalapa, Veracruz, México

Dora Trejo, Cesar Moreira, Liliana Lara and Guillermo Alafita
Universidad Veracruzana, Facultad de Ciencias Agrícolas, Laboratorio de Organismos Benéficos. Circuito Gonzalo Aguirre Beltrán s.n., Zona Universitaria Xalapa, Veracruz. CP 91000, México

Isaac Bañuelos
Unidad de Servicios de Apoyo en Resolución Analítica (SARA), Dr. Luis Castelazo Ayala s/n, Col. Industrial-Animas Xalapa, Veracruz. CP 91190, México

Araceli Reyes
Ciencias Básicas Dr. Luis Castelazo Ayala s/n, Col. Industrial-Animas Xalapa, Veracruz. CP 91190, México

Roberto Gregorio Chiquito-Contreras, Ramón Zulueta-Rodríguez and Doris Guadalupe Castillo-Rocha
Universidad Veracruzana, Facultad de Ciencias Agrícolas, Campus Xalapa. Circuito Gonzalo Aguirre Beltran s/n Zona Universitaria, CP 91091, Xalapa, Veracruz, México

Francisco Osorio-Acosta, Eliseo García-Pérez and Juan Antonio Villanueva-Jiménez
Colegio de Postgraduados, Campus Veracruz, Km 88.5 Carretera Federal Xalapa-Veracruz, CP 91700, Tepetates, Veracruz, México

Ana Veruska Cruz da Silva, Ana da Silva Lédo and Maria Salete Alves Rangel
The Brazilian Agricultural Research Corporation's (Embrapa)

Allívia Rouse Ferreira dos Santos
Universidade de Santiago de Compostela (USC)

Rosana Barroso Feitosa, Camila Santos Almeida and Gilvânia Melo da Silva
Universidade Federal de Sergipe (UFS)

M. M. El-Adawy
Department of Animal production, Faculty of Agriculture (El-Shatby), Alexandria University, Egypt

M. K. Mohsen, G. S. El-Santiel and M. Z. Dakron
Department of Animal production, Faculty of Agriculture, Kafr El-Shikh University, Egypt

A. Z. M. Salema
Department of Animal production, Faculty of Agriculture (El-Shatby), Alexandria University, Egypt
Facultad de Medicina Veterinaria y Zootecnia, Universidad Autónoma del Estado de México, México

B. M. A. Mariezcurrena
Facultad de Medicina Veterinaria y Zootecnia, Universidad Autónoma del Estado de México, México

Servando Rueda-López
Programa de Licenciatura en Oceanología, Facultad de Ciencias Marinas, Universidad Autónoma de Baja California (UABC), Ensenada BC, México

María Teresa Viana and Gabriel Correa-Reyes
Instituto de Investigaciones Oceanológicas, UABC, Ensenada BC 22860, México

Daniel Badillo-Zapata
Programa de Maestría en Oceanografía Costera, Facultad de Ciencias Marinas e Instituto de Investigaciones Oceanológicas, UABC, Ensenada BC, México

Carlos A. Martinez-Palacios
Instituto de Investigaciones Agropecuarias y Forestales UMSNH, Morelia, Michoacán

Tania Escalante and Enrique Martínez-Meyer
Museo de Zoología "Alfonso L. Herrera", Dpto. Biología Evolutiva, Facultad de Ciencias, UNAM. Apdo. Postal 70-399, 04510 México, D. F., México
Laboratorio de Análisis Espaciales, Dpto. Zoología, Instituto de Biología, Universidad Nacional Autónoma de México, Apdo. Postal 70-153, 04510 México, D. F., México

Jesús Mao Estanislao Aguilar-Luna and Pedro Antonio Macario-Mendoza
El Colegio de la Frontera Sur - Unidad Chetumal. C.P. 77014. Chetumal, Quintana Roo, México

Eduardo García-Villanueva
Colegio de Postgraduados Campus Montecillo. C.P. 56230. Edo. de México, México

Esperanza Huerta-Lwanga
El Colegio de la Frontera Sur - Unidad Villahermosa. C.P. 86280. Villahermosa, Tabasco, México

Ramiro de Alba-Becerra
Fumigaciones Integrales de Quintana Roo. C.P. 77000. Chetumal, Quintana Roo, México

José Roberto Villagómez Ibarra, Javier Castro Rosas and César Abelardo González Ramirez
Academic Area of Chemistry, Institute of Basic Science and Engineering, Hidalgo State University. City of Knowledge, Pachuca - Tulancingo highway Km 4.5. C. P. 42184. Mineral de la Reforma, Hidalgo, Mexico

Otilio Arturo Acevedo Sandoval
Engineering Academic Area of Agribusiness and Food Engineering, Institute of Agricultural Sciences, Hidalgo State University. Km 1 University Avenue, Ex-Hacienda of Aquetzalpa, Tulancingo, Hidalgo

Margarita Islas Pelcastre
Academic Area of Chemistry, Institute of Basic Science and Engineering, Hidalgo State University. City of Knowledge, Pachuca - Tulancingo highway Km 4.5. C. P. 42184. Mineral de la Reforma, Hidalgo, Mexico
Engineering Academic Area of Agribusiness and Food Engineering, Institute of Agricultural Sciences, Hidalgo State University. Km 1 University Avenue, Ex-Hacienda of Aquetzalpa, Tulancingo, Hidalgo

Alfredo Madariaga Navarrete
Division of Professional studies. Monterrey Tech Institute, Hidalgo Campus. Felipe Angeles Boulevard No. 2003, Pachuca, Hidalgo

Gabriela González-Mariscal
Centro de Investigación en Reproducción Animal, CINVESTAV-Universidad Autónoma de Tlaxcala, Apdo. Postal 62, Tlaxcala Tlax 90000, México

Ifeoluwa Akin Babalola and Kabir Kayode Salman
Department of Agricultural Economics, University of Ibadan, Ibadan, Nigeria

K. Sri Rama Murthy
Department of Biotechnology, Montessori Mahila Kalasala, Vijayawada, Andhra Pradesh - 520 010, India

P. V. Bramhachari
Department of Biotechnology, Krishna University, Machilipatnam -521 001, Andhra Pradesh, India

M. Chandrasekhara Reddy
Department of Biotechnology, Montessori Mahila Kalasala, Vijayawada, Andhra Pradesh - 520 010, India
Department of Biotechnology, Krishna University, Machilipatnam -521 001, Andhra Pradesh, India

Luis Ramírez y Avilés, Francisco J Solorio Sánchez and Marcos Barros-Rodriguez
Department of Animal Nutrition, Faculty of Veterinary Medicine and animal science, University of Yucatan (UADY), Mérida, Yucatán, México, KM 15.5 carretera Mérida, Xmatkuil Apdo. Itzimna Merida

Ahmed M. M. Ahmed
Department of Animal Nutrition, Faculty of Veterinary Medicine and animal science, University of Yucatan (UADY), Mérida, Yucatán, México, KM 15.5 carretera Mérida, Xmatkuil Apdo. Itzimna Merida
Department of Plant Protection, Faculty of Agriculture, Assiut University, Assiut Egypt

F. A. Al-Zyoud
Department of Plant Protection and Integrated Pest Management, Faculty of Agriculture, Mu'tah University, Karak Jordan

Alejandro Ley de Coss, Juan F. Aguirre Medina, Consepción Arce Espino and Saúl Posada Cruz
Faculty of Agronomic Sciences, Campus IV, Autonomous University of Chiapas (UNACH), Coast Highway and Station Huehuetán Railway, Huehuetán, Chiapas, Mexico. CP. 36670

Willians de León de León
Master in Tropical Agropecuary Production Student at UNACH

René Pinto Ruíz and Francisco Guevara Hernandez
Faculty of Agronomic Sciences, Campus V, UNACH. Villaflores, Chiapas, Mexico

Susana G. Gayón Amaro
Faculty of Veterinary Medicine and Zootechny, Autonomous University of the State of Mexico, Toluca, Estado de Mexico

Ana Veruska Cruz da Silva, Evandro Neves Muniz and Ana da Silva Ledo
Researcher at Embrapa Coastal Tablelands, 49025-040, Aracaju-SE, Brazil

Camila Santos Almeida and Marina Ferreira da Vitória
Federal University of Sergipe, 49100-000, São Cristóvão-SE, Brazil

Marília Freitas de Vasconcelos Melo
Faculty of Agricultural Sciences, State University of Sao Paulo, 18610-307, Botucatu-SP, Brazil

Allivia Rouse Carregosa Rabbani
Federal Institute of Education, Science and Technology of Bahia, 45810000, Porto Seguro-BA, Brazil

G. Ramírez-Zúñiga, R. F. García-Castillo, Juan D. Hernández-Bustamante, A. Valdéz-Oyervides and J. M. Fuentes-Rodríguez
Universidad Autónoma Agraria Antonio Narro, Buenavista, Saltillo, Coahuila, México. CP. 25315

J. Salinas-Chavira
Universidad Autónoma de Tamaulipas. Facultad de Medicina Veterinaria y Zootecnia, Cd. Victoria, Tamaulipas, México

A. Vega
Universidad Autónoma de Chiriquí. Facultad de Recursos Naturales

M. H. Ruiloba
Universidad de Panamá. Facultad de Ciencias Agropecuarias. David, Chiriquí, República de Panamá

Cesar A Cuartas C.
Doctorado en Ciencias Animales, Universidad de Antioquia

Juan F Naranjo R.
Grupo INCA-CAES, Facultad de Medicina Veterinaria y Zootecnia, Universidad CES, Medellín. Colombia

Ariel M Tarazona M. and Rolando Barahona Rosales
Departamento de Producción Animal, Facultad de Ciencias Agrarias, Universidad Nacional de Colombia, Facultad de Ciencias Agrarias, Departamento de Producción Animal, AA 1779, Sede Medellín, Colombia

Guillermo A Correa L.
Departamento de Ingeniería Agronómica, Facultad de Ciencias Agrarias, Universidad Nacional de Colombia, Sede Medellín, Colombia

Index

R

S

T

V

CPSIA information can be obtained
at www.ICGtesting.com
Printed in the USA
LVHW060850030122
707669LV00002B/56

9 781641 165150